Analysis and Design of the Power-Split Device for Hybrid Systems

Xiaohua Zeng · Jixin Wang

Analysis and Design of the Power-Split Device for Hybrid Systems

北京理工大学出版社
BEIJING INSTITUTE OF TECHNOLOGY PRESS

Springer

Xiaohua Zeng
Jilin University
Changchun
China

Jixin Wang
Jilin University
Changchun
China

ISBN 978-981-10-4270-6 ISBN 978-981-10-4272-0 (eBook)
https://doi.org/10.1007/978-981-10-4272-0

Jointly published with Beijing Institute of Technology Press, Beijing, China

The print edition is not for sale in China Mainland. Customers from China Mainland please order the print book from: Beijing Institute of Technology Press

Library of Congress Control Number: 2017952524

Printed on acid-free paper

This Springer imprint is published by Springer Nature
The registered company is Springer Nature Singapore Pte Ltd.
The registered company address is: 152 Beach Road, #21-01/04 Gateway East, Singapore 189721, Singapore

Preface

Energy-saving technology is one of the most important research directions nowadays in auto industry. It improves product competitiveness and strives for market initiative for vehicle companies. At present, the hybrid technology is hot in the field of energy-saving technology. Its cores comprise the configurations of the power-split hybrid system and the design methods of the power-split device (PSD).

In this book, the actual energy consumption of the vehicle is analyzed from the vehicle power demand through visual chart statistics in driving cycles. Energy consumption and power demand are the original input of the power-spilt vehicle. Research in this aspect is one of the important topics of energy-saving technology. In addition, how to calculate the power demand of the power-split vehicle is one of the most important problems to be solved. This book presents a practical method to calculate the total power of the vehicle directly through the acceleration capability, and it is verified by examples.

Nowadays, a large number of hybrid vehicles appear in the market and series–parallel hybrid vehicle has become the mainstream. Most series–parallel hybrid vehicles use planetary gears as the PSD, which can achieve the function of electric continuously variable transmission (EVT) and have obvious advantages on dynamic performance and fuel economy performance. Toyota Prius, the most successful hybrid vehicle, is equipped with the PSD. Detailed introduction to the typical PSD configuration adopted by Toyota and GM corporation is given in this book. The former is mainly about input power-split configuration and Toyota Hybrid System (THS) is the representative. The latter is mainly about the combined power-split configuration, which is generally called Advanced Hybrid System (AHS). Then the working principle and analysis method for the two configurations are discussed in detail. A new PSD configuration, Differential-based Hybrid System (DHS), is also proposed and analyzed in this book.

As an important technology in the research and development of hybrid vehicle, simulation technology can not only adjust design scheme and optimize design parameters flexibly, but also reduce research cost and shorten development cycle. At present, the widely used softwares in the hybrid vehicle simulation include ADVISOR, CRUISE, and AMESim. Combined with simulation examples, the typical applications of hybrid vehicle modeling on the above simulation platforms are introduced in detail.

The power-split hybrid system is currently the promising hybrid system. This book proposes a multi-factor integrated parametric design method for a power-split Hybrid Electric Bus (HEB). This method takes the power source efficiency, transmission efficiency, driving cycle, and primary control strategy into account. Then it is applied to a specified city bus. Followed by that, system efficiency is analyzed, and the parametric design method is validated on the co-simulation platform of MATLAB/Simulink and CRUISE. The above research will contribute to the design method for the power-split HEB.

Since multi-power sources are highly coupled in PSD during mode shifting, the research on Dynamic Coordination Control Strategy (DCCS) to reduce system shock is rather important. This book proposes a predictive-model-based DCCS and builds the dynamic model of the objective power-split vehicle. Based on the model, the mode shifting process is analyzed to determine the reason for the system shock. According to the principle of the nonlinear observer, the book designs the engine torque estimation algorithm and establishes the prediction model of the degree of shock based on the model predictive control. Then, the adaptive DCCS for a complex driving cycle is realized by combining the feedback control and the prediction model.

In this book, a design method of PSD is mainly illustrated in the condition of multi-field coupling. A redesigned model of a DHS (as an example of the PSD) is developed and its bench test is conducted. The thermal analysis method and contact analysis method of the DHS are introduced. Finally, the multi-objective optimization design method based on surrogate model for the DHS is illustrated in detail.

We believe that readers could master the configuration, performance, analysis, and design method of PSD in general according to the methods and figures described in this book.

Part of study in this book is funded by National Natural Science Foundation of China (No 51575221, 51675214).

Gratitude to State Key Laboratory of Automotive Simulation and Control, Jilin University for supporting the publication of this book. Also thanks to the publisher for their support in the preparation of this book. And owing to the postgraduates DAI Yutong, SHI Xiangyun, SUN Kehua, DONG Bingbing, LI Lixin, LI Hongcheng, Wang Xinming, WANG Wei, ZHANG Yichao et al. in the research group, and pay special attitude to GAO Bingzhao giving a constructive suggestion for the book.

Due to the limitation of editors, errors and inaccuracies are inevitable. We expect readers and colleagues feel free to criticize and correct.

For the convenience of feedback, two e-mails are provided: zengxh@jlu.edu.cn and jxwang@jlu.edu.cn.

Changchun, China

Xiaohua Zeng
Jixin Wang

Contents

Chapter 1
Energy Consumption Analysis and Vehicle Power Demand

Energy saving technology is one of the most important research directions in the automotive industry [1–4]. For the major auto companies, it has significant influence to develop new products, enhance competitiveness and own their initiative in the market. In this chapter, on one hand, the actual vehicle energy consumption characteristics are analyzed with the visual graphics. On the other hand, the guidance about how the hybrid vehicle can save energy by the hybrid technology is given. On the basis of the above analysis, a practical solution based on the acceleration capability is proposed to calculate vehicle power demand directly.

1.1 Energy Consumption Analysis

The conventional vehicle has only one power source, i.e. the engine. The engine power must satisfy the requirement of the maximum velocity, the gradeability and the acceleration capability. To meet the dynamic requirements, the maximum power of engine could be 10 times as large as the power demand which the vehicle need when cruising on the horizontal road at speed of 100 km/h, or 3–4 times as at 6% grade at the speed of 100 km/h [5]. The inherent shortcomings of the single power source of the traditional vehicle create the energy consumption characteristics, which are of guiding significance to the research of the hybrid vehicle.

In this section, based on the fact that the conventional vehicle has poor fuel consumption performance, a visual and effective method on different driving cycles is proposed to analyze the fuel consumption characteristics. Then, an energy saving design method is given.

X. Zeng and J. Wang, *Analysis and Design of the Power-Split Device for Hybrid Systems*, https://doi.org/10.1007/978-981-10-4272-0_1

1.1.1 Energy Consumption

In this section, the representative driving cycles, including China Car Driving Cycle (CCDC), New European Driving Cycle (NEDC), the Japanese 10–15 Driving Cycle (10–15) and Urban Dynamometer Driving Schedule (UDDS), are chosen to analyze energy consumption.

1.1.1.1 Parameters of Conventional Vehicle

Table 1.1 shows the parameters of a conventional vehicle. This conventional vehicle has a 64 kW 4-cylinder engine. Its 0–100 km/h acceleration time is 16.6 s and the maximum velocity is 180 km/h.

1.1.1.2 Fuel Consumption of Conventional Vehicle

Simulation model for the conventional vehicle is established. After simulating under the driving cycles mentioned above, the fuel consumption results are shown in Table 1.2.

According to the data in Table 1.2, some differences in fuel consumption under different driving cycles can be seen. The fuel consumption under 10–15 driving cycle is the maximum while that under the NEDC is the minimum. Overall, this conventional vehicle consumes much fuel under all of the driving cycles and the average fuel consumption is 7.5 L/100 km.

The fuel consumption of traditional vehicle should be analyzed not only from the amount of fuel consumption, but also from the characteristics of fuel consumption, which can provide a solution for further improving the vehicle fuel economy.

Table 1.1 Parameters of a conventional vehicle

Vehicle parameters	Mass/kg	1500
	Transmission	5-Manual
Engine parameters	[Maximum power/kW]/[Speed/rpm]	64/6000
	[Peak torque/Nm]/[Speed/rpm]	120/4200

Table 1.2 Fuel consumption of the conventional vehicle

Driving cycle	CCDC	NEDC	10–15	UDDS
Fuel consumption/L	7.6	7.0	8.1	7.2

1.1.2 *Characteristics Analysis*

The analysis of the characteristics of vehicle fuel consumption is mainly to study the distribution of engine operating points during vehicle driving, i.e. according to the simulation results of fuel consumption, the output speed, torque and fuel rate of the engine are investigated.

The early research on the distribution of engine operating points is mainly done according to the engine 2-dimensional distribution points [6–8], which can only get the distribution characteristics qualitatively and is difficult to analyze the proportion of the operating time and the proportion of fuel consumption on the engine MAP.

In this section, the distribution characteristics on the engine MAP is studied quantificationally based on the statistical analysis method.

(1) Study the time proportion of different engine operating points (area) and search for the time distribution of engine load.
(2) Study the accumulated fuel consumption of different engine operating points (area) and search for the main fuel consumption points (area).

The above quantitative analysis can provide more practical guidance for engine MAP design and vehicle energy savings by hybrid technology.

1.1.3 *Statistical Analysis*

The vehicle model established in ADVISOR is simulated under the different driving cycles mentioned above. Then, the distribution characteristics of the engine MAP is studied quantificationally based on the statistical analysis.

1.1.3.1 Statistical Analysis Under CCDC

Based on the driving cycle shown in Fig. 1.1, the engine MAP within its operating range and the operating points are shown in Fig. 1.2.

1. Analysis of the time proportion

According to the simulation results, the time proportion of engine operating points under CCDC is calculated and the operating time distribution is shown in Table 1.3 and Fig. 1.3 with the histogram.

2. Analysis of fuel proportion

The proportion of the engine accumulated fuel consumption in different areas under CCDC is shown in Table 1.4 and Fig. 1.4 with the histogram.

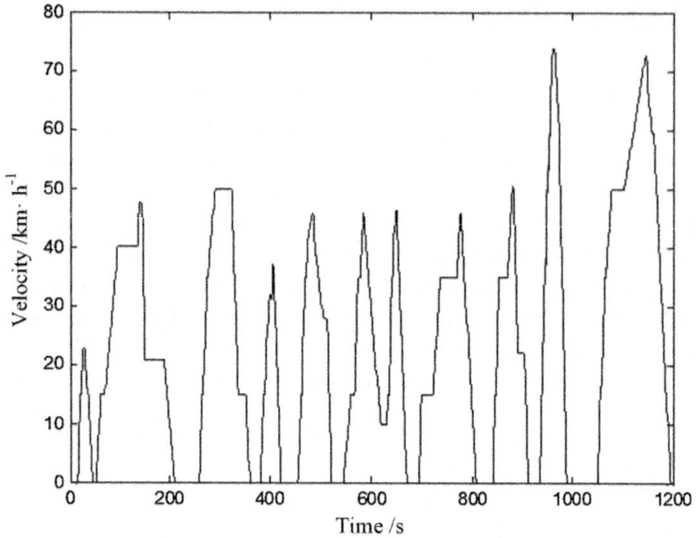

Fig. 1.1 Velocity-time curve of CCDC

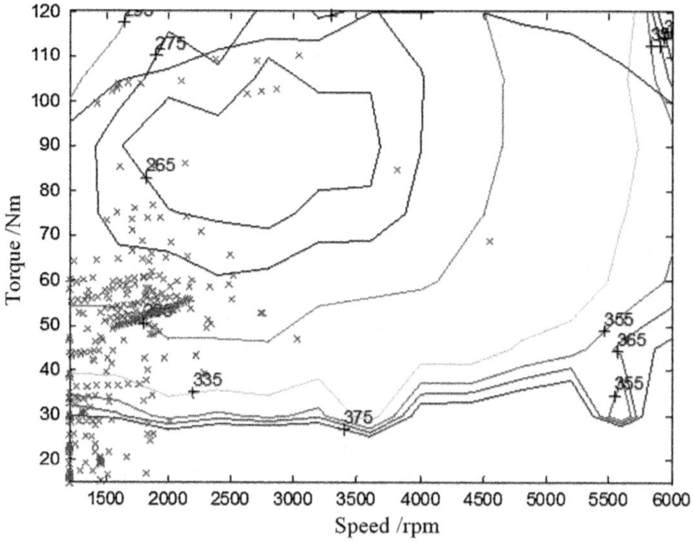

Fig. 1.2 Engine operating points under CCDC

Table 1.3 Operating time proportion under CCDC (%)

Speed/rpm	Torque/Nm						
	0	20	40	60	80	100	120
1200	22.1	36.8	28.5	6.8	0.7	2.0	0
2400	0	0	1.0	0.5	0	1.0	0
3600	0	0	0	0.2	0.2	0	0
4800	0	0	0	0	0	0	0
6000	0	0	0	0	0	0	0

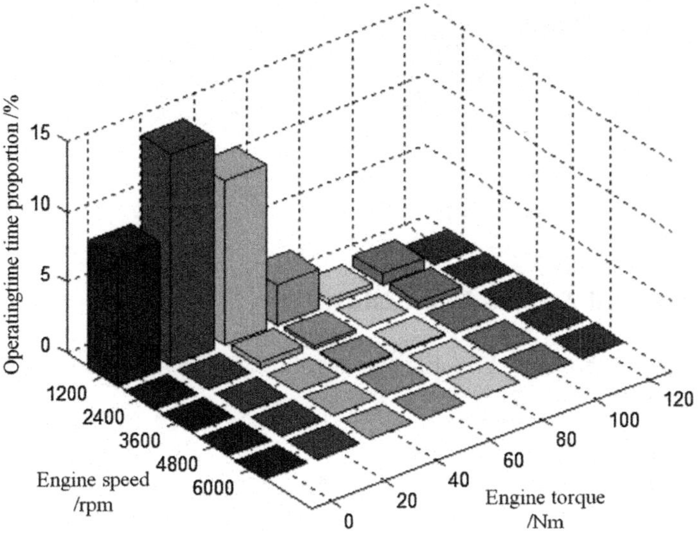

Fig. 1.3 Operating time proportion under CCDC

Table 1.4 Fuel proportion under CCDC (%)

Speed/rpm	Torque/Nm						
	0	20	40	60	80	100	120
1200	11.5	26.5	36.2	10.8	1.7	4.6	0
2400	0	0	2.1	1.0	0	3.8	0
3600	0	0	0	1.0	0.8	0	0
4800	0	0	0	0	0	0	0
6000	0	0	0	0	0	0	0

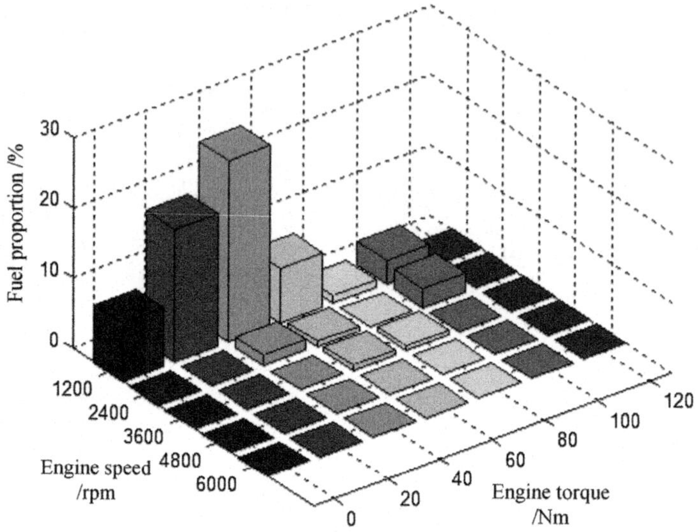

Fig. 1.4 Fuel proportion under CCDC

1.1.3.2 Statistical Analysis Under NEDC

Based on NEDC shown in Fig. 1.5, the distribution of engine operating points is shown in Fig. 1.6.

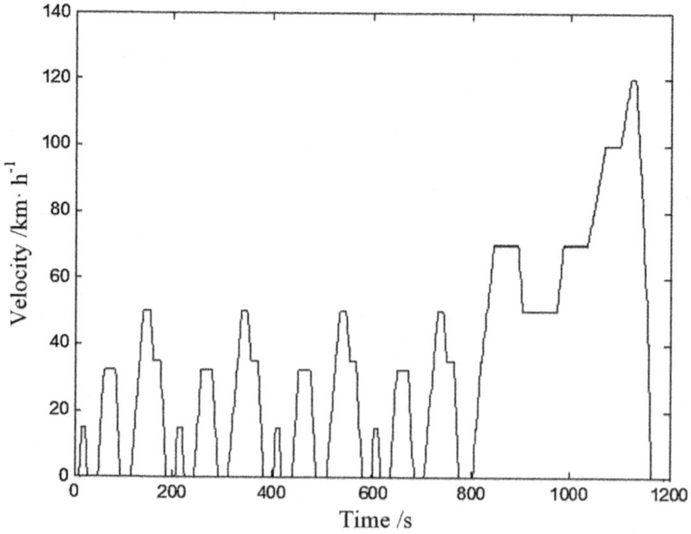

Fig. 1.5 Velocity-time curve of NEDC

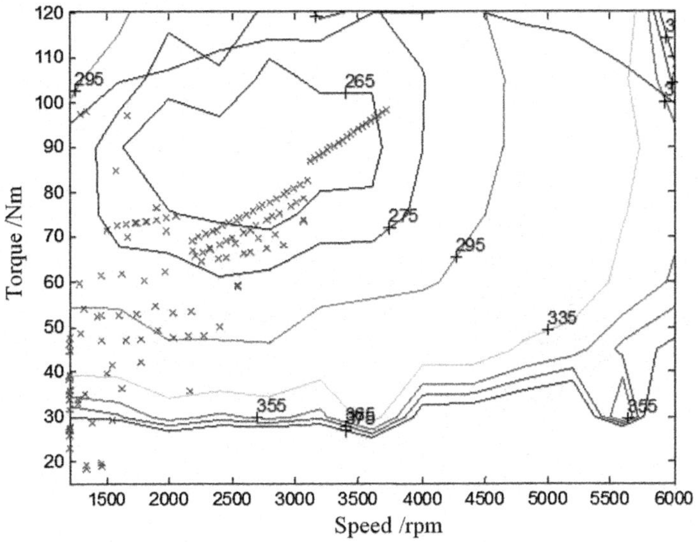

Fig. 1.6 Engine operating points under NEDC

Table 1.5 Operating time proportion under NEDC (%)

Speed/rpm	Torque/Nm						
	0	20	40	60	80	100	120
1200	17.9	28.5	25.9	10.1	0.6	0	0
2400	0	0	5.5	7.8	3.2	0	0
3600	0	0	0	0	0.6	0	0
4800	0	0	0	0	0	0	0
6000	0	0	0	0	0	0	0

1. Analysis of the time proportion

According to the simulation results, the time proportion of engine operating points under NEDC is calculated and the operating time distribution is shown in Table 1.5 and Fig. 1.7 with the histogram.

2. Analysis of fuel proportion

The proportion of the engine accumulated fuel consumption in different areas under NEDC is shown in Table 1.6 and Fig. 1.8 with the histogram.

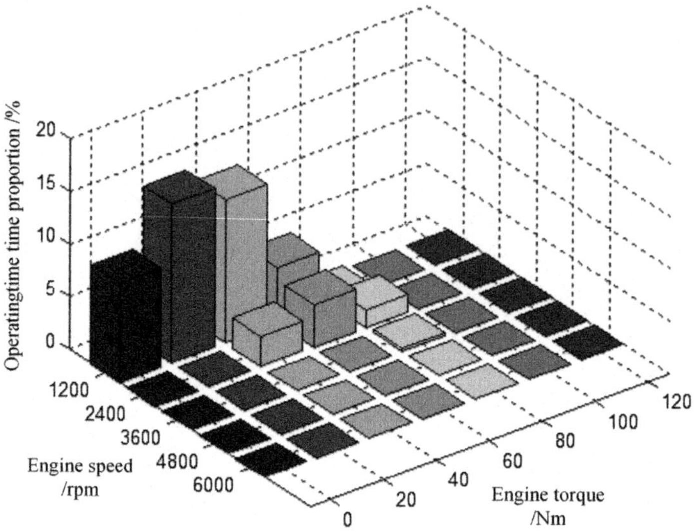

Fig. 1.7 Operating time proportion under NEDC

Table 1.6 Fuel proportion under NEDC (%)

Speed/rpm	Torque/Nm						
	0	20	40	60	80	100	120
1200	7.9	17.0	24.0	14.5	1.0	0	0
2400	0	0	8.5	15.6	9.3	3.8	0
3600	0	0	0	0	2.3	0	0
4800	0	0	0	0	0	0	0
6000	0	0	0	0	0	0	0

1.1.3.3 Statistical Analysis Under 10–15 Driving Cycle

Based on 10–15 driving cycle shown in Fig. 1.9, the distribution of engine operating points is shown in Fig. 1.10.

1. Analysis of the time proportion

According to the simulation results, the time proportion of engine operating points under 10–15 driving cycle is calculated and the operating time distribution is shown in Table 1.7 and Fig. 1.11 with the histogram.

2. Analysis of fuel proportion

The proportion of the engine accumulated fuel consumption in different areas under 10–15 driving cycle is shown in Table 1.8 and Fig. 1.12 with the histogram.

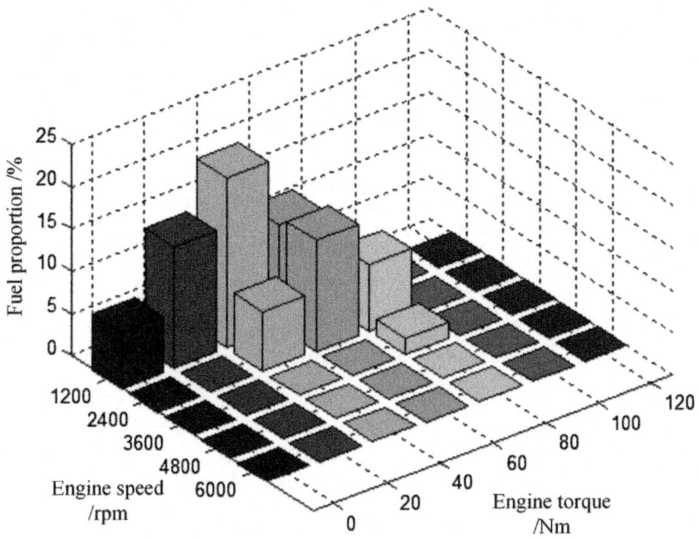

Fig. 1.8 Fuel proportion under NEDC

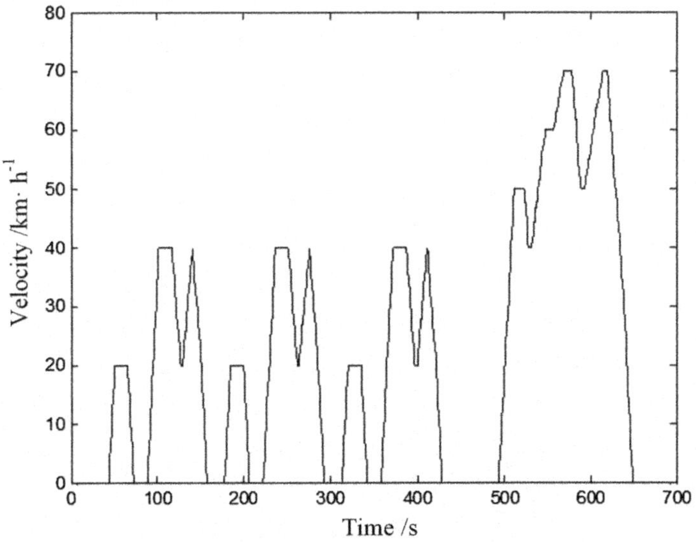

Fig. 1.9 Velocity-time curve of 10–15 driving cycle

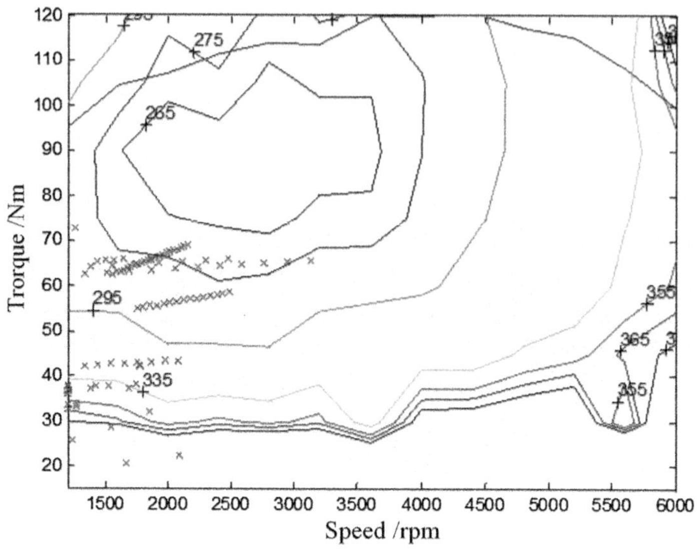

Fig. 1.10 Engine operating points under 10–15 driving cycle

Table 1.7 Operating time proportion under 10–15 driving cycle (%)

Speed/rpm	Torque/Nm						
	0	20	40	60	80	100	120
1200	17.6	38.0	17.6	23.6	0	0	0
2400	0	0	0.7	2.8	0	0	0
3600	0	0	0	0	0	0	0
4800	0	0	0	0	0	0	0
6000	0	0	0	0	0	0	0

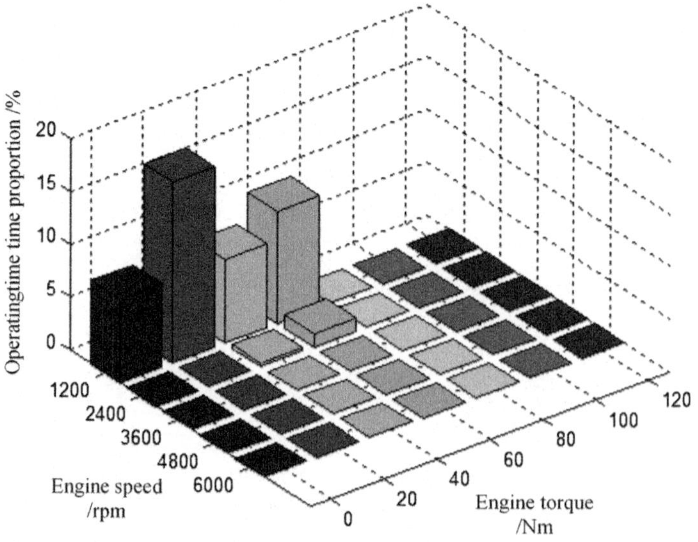

Fig. 1.11 Operating time proportion under 10–15 driving cycle

Table 1.8 Fuel proportion under 10–15 driving cycle (%)

Speed/rpm	Torque/Nm						
	0	20	40	60	80	100	120
1200	7.5	27.4	21.1	36.6	0	0	0
2400	0	0	1.3	6.1	0	0	0
3600	0	0	0	0	0	0	0
4800	0	0	0	0	0	0	0
6000	0	0	0	0	0	0	0

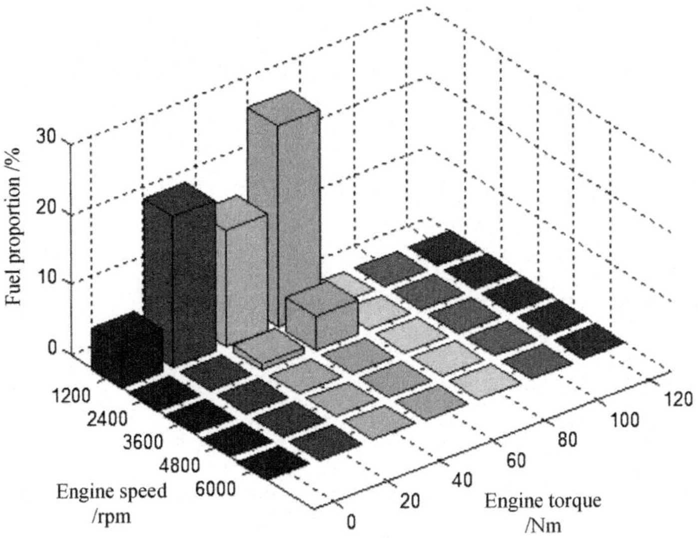

Fig. 1.12 Fuel proportion under 10–15 driving cycle

1.1.3.4 Statistical Analysis Under UDDS

Based on UDDS shown in Fig. 1.13, the distribution of engine operating points is shown in Fig. 1.14.

1. Analysis of the time proportion

According to the simulation results, the time proportion of engine operating points under UDDS is calculated and the operating time distribution is shown in Table 1.9 and Fig. 1.15 with the histogram.

2. Analysis of fuel proportion

The proportion of the engine accumulated fuel consumption in different areas under UDDS is shown in Table 1.10 and Fig. 1.16 with the histogram.

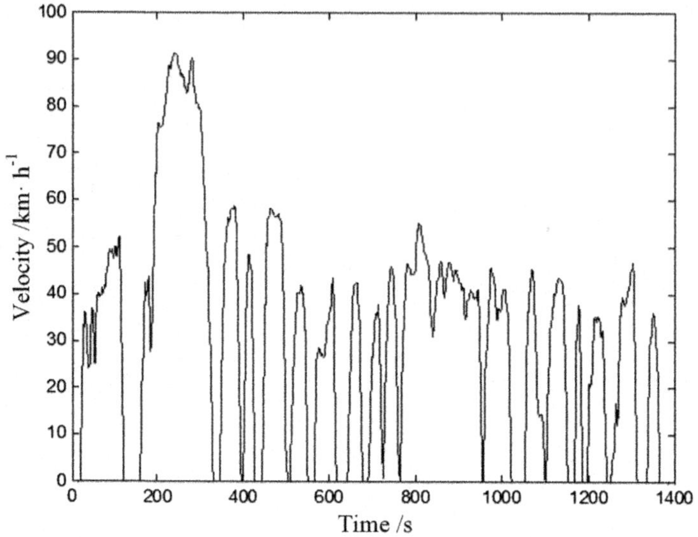

Fig. 1.13 Velocity-time curve of UDDS

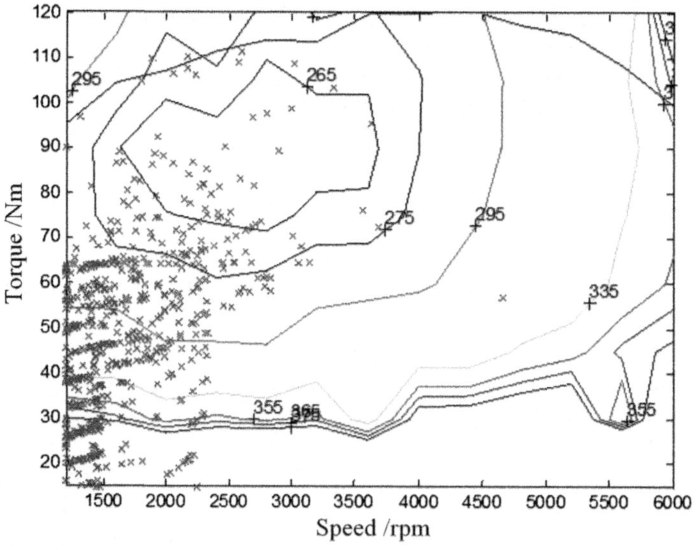

Fig. 1.14 Engine operating points under UDDS

1.1.3.5 Statistical Analysis Under Different Driving Cycles

The results based on the statistical analysis of fuel consumption characteristics under the above driving cycles show that the engine of conventional vehicle often works in low speed and low load area, which causes much fuel consumption.

Table 1.9 Operating time proportion under UDDS (%)

Speed/rpm	Torque/Nm						
	0	20	40	60	80	100	120
1200	15.7	28.2	29.2	16.2	2.8	0.9	0
2400	0	0	1.0	3.9	1.0	0.5	0
3600	0	0	0.2	0.2	0.2	0	0
4800	0	0	0	0	0	0	0
6000	0	0	0	0	0	0	0

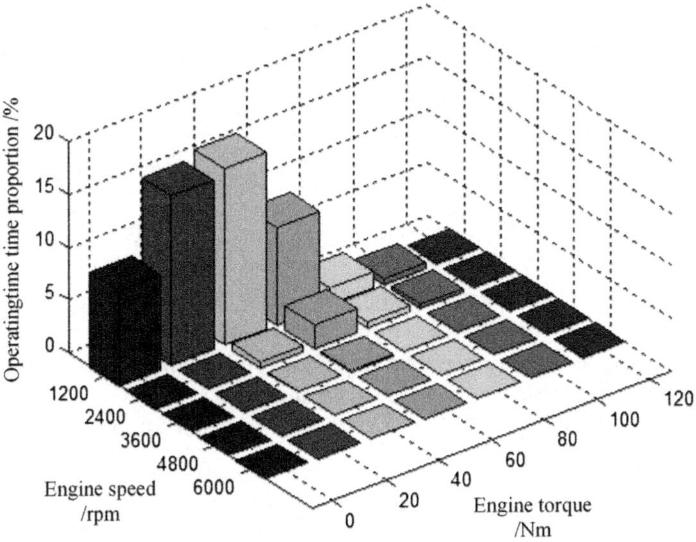

Fig. 1.15 Operating time proportion under UDDS

Table 1.10 Fuel proportion under UDDS (%)

Speed/rpm	Torque/Nm						
	0	20	40	60	80	100	120
1200	5.6	18.5	30.3	21.7	5.0	2.2	0
2400	0	0	1.9	8.4	2.9	1.9	0
3600	0	0	0.6	0.5	0.6	0	0
4800	0	0	0	0	0	0	0
6000	0	0	0	0	0	0	0

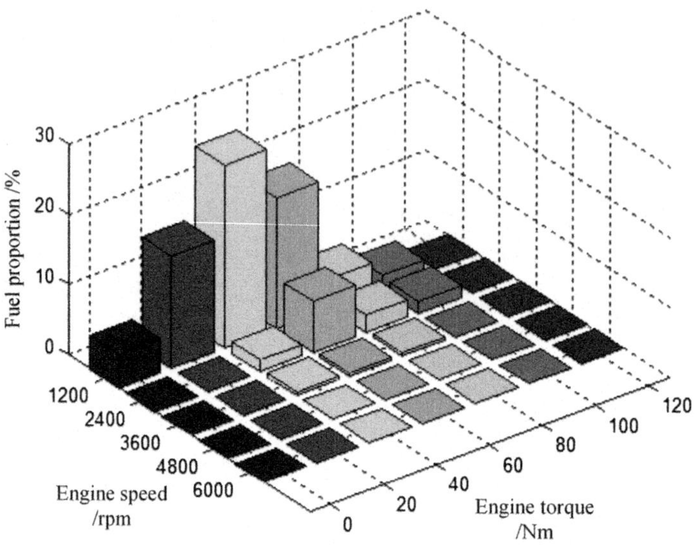

Fig. 1.16 Fuel proportion under UDDS

In the area which is enclosed by engine torque below 40 Nm (load factor is about 30%) and speed below 1200 rpm (20% of the maximum speed), the time proportion of engine operating points and fuel proportion under different driving cycles are shown in Tables 1.11, 1.12, 1.13 and 1.14.

Time proportion under CCDC is 87.4% and fuel consumption in this area is 74.2%.

Table 1.11 Analysis of engine operating points under low load (CCDC)

Speed/rpm	Torque/Nm			
	0–20		20–40	
	Time proportion/%	Fuel consumption proportion/%	Time proportion/%	Fuel consumption proportion/%
0–1200	58.9	38.0	28.5	36.2

Table 1.12 Analysis of engine operating points under low load (NEDC)

Speed/rpm	Torque/Nm			
	0–20		20–40	
	Time proportion/%	Fuel consumption proportion/%	Time proportion/%	Fuel consumption proportion/%
0–1200	46.4	24.9	25.9	24.0

Table 1.13 Analysis of engine operating points under low load (10–15 driving cycle)

Speed/rpm	Torque/Nm			
	0–20		20–40	
	Time proportion/%	Fuel consumption proportion/%	Time proportion/%	Fuel consumption proportion/%
0–1200	55.6	34.9	17.6	21.1

Table 1.14 Analysis of engine operating points under low load (UDDS)

Speed/rpm	Torque/Nm			
	0–20		20–40	
	Time proportion/%	Fuel consumption proportion/%	Time proportion/%	Fuel consumption proportion/%
0–1200	43.9	24.1	29.2	30.3

Time proportion under NEDC is 72.3% and fuel consumption in this area is 48.9%.

Time proportion under 10–15 driving cycle is 73.2% and fuel consumption in this area is 56.0%.

Time proportion under UDDS is 73.1% and fuel consumption in this area is 54.5%.

It can be seen that the time proportion of the engine under low load is very large. Although the fuel rate is not high in this area, the final calculated fuel consumption under low load is still large. Therefore, by choosing a smaller power engine, eliminating the idle or driving in pure electric mode in this area, the overall efficiency of the vehicle can be improved effectively, for example, by hybrid technology. This analysis method and results have practical value to improve vehicle fuel consumption and optimize the design of engine MAP.

1.2 Vehicle Power Demand

One of the cores for hybrid vehicle control is the appropriate power distribution between the two power sources, whose foundation is to find the solution for vehicle power demand. In this section, the traditional power calculation method is introduced. Then, in order to overcome its disadvantages, a reasonable and practical method to calculate the vehicle power demand directly based on the acceleration index is proposed.

1.2.1 Traditional Power Calculation Method

A traditional method to calculate vehicle power demand is usually an iterative, step-by-step method by simulation.

According to the maximum vehicle speed, the maximum vehicle power demand P_1 is calculated using the following equation [9]

$$P_1 = P_f + P_w = \frac{1}{3600 \cdot \eta_t}\left(m \cdot g \cdot f \cdot v + \frac{C_D \cdot A}{21.15} v^3\right) \tag{1.1}$$

where P_f is the power of rolling resistance, P_w is the power of air resistance, C_D is the air resistance coefficient, A is frontal area, f is the rolling resistance coefficient, η_t is mechanical efficiency of the powertrain, m is total vehicle mass, g is gravitational acceleration and v is vehicle speed.

To ensure that the vehicle can meet the requirement of gradeability, the vehicle power demand P_2 at a particular speed at particular grade is calculated by

$$P_2 = P_f + P_w + P_i = \frac{1}{3600 \cdot \eta_t}\left(m \cdot \cos\theta \cdot g \cdot f \cdot v + \frac{C_D \cdot A}{21.15} v^3 + mg \cdot \sin\theta \cdot v\right) \tag{1.2}$$

where P_i is the power of grade resistance and θ is the road grade.

Finally, the larger value in P_1 and P_2 is used as the power demand, which cannot meet the demand of the extreme acceleration because it only considers the maximum velocity and climbing performance. So, generally, the following steps will be needed to ensure the vehicle power demand correctly:

(1) Set a maximum power and then match parameters of power sources of the hybrid vehicle.
(2) Establish models for the hybrid vehicle and simulate the acceleration time.
(3) If the acceleration time cannot meet the design demand, the power demand should be modified. Then, repeat step 1 and step 2 until meeting the design demand and the vehicle power demand is finally determined.

The traditional method to calculate vehicle power combines theory and simulation and is an iterative, step-by-step method. However, it is time consuming and not intuitive [7, 9, 10].

1.2.2 Practical Power Calculation Method and Application

To address the issue that the traditional method is not intuitive and also costs a long time, in this section, a preliminary method for calculating the power demand based on acceleration index is discussed. The theoretical equations of this method are deduced in detail, the impact coefficients in the equations are discussed and compared with the simulation curves of the vehicle acceleration process to verify the rationality and practicability of the method.

1.2.2.1 Practical Power Calculation Method

For most of the vehicles, the maximum power depends on the acceleration capability, which means once acceleration index is satisfied, other indices will be satisfied as well [10]. Simulation comparison between power demand of acceleration of a vehicle weighted 1000 kg and driving at a constant velocity (100 km/h) is shown in Fig. 1.17 and it can be seen that the power of acceleration is larger than power at constant velocity.

1.2.2.2 Theoretical Equations for Vehicle Acceleration Capability

When calculating the total power of power source, there are following assumptions:

(1) Characteristics curve for power source should be power contour (conventional vehicle with multi-gear transmission and electric vehicle (EV) both have external characteristics approximate to power contour).
(2) The rolling resistance and air resistance during acceleration are ignored (these road resistances are very small at low velocity).

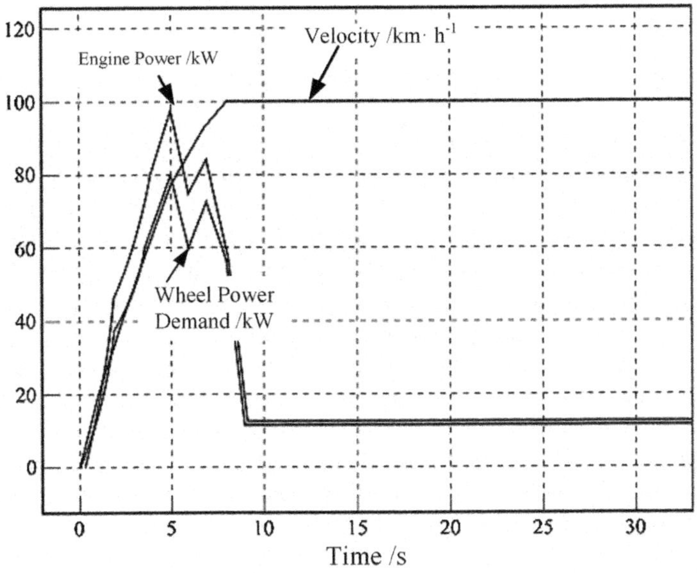

Fig. 1.17 Comparison of power demand between acceleration and constant velocity

Based on the above assumptions, vehicle kinetic equilibrium equation is shown as:

$$F \cdot v = P \to F = \frac{P}{v} \tag{1.3}$$

The power value P is set to a constant A and equation is derived as below:

$$a = \frac{F}{m} = \frac{dv}{dt} \xrightarrow{F(v)=A/v} \frac{A}{m \cdot v} = \frac{dv}{dt} \xrightarrow{B=A/m} B \cdot dt = v \cdot dv \tag{1.4}$$

$$\int_0^{t_m} Bdt = \int_0^{v_m} vdv \tag{1.5}$$

$$2B = \frac{v_m^2}{t_m} = \frac{v^2}{t} \to v = v_m \left(\frac{t}{t_m}\right)^{0.5} \tag{1.6}$$

So, vehicle speed could be approximated as [11]:

$$v = v_m \left(\frac{t}{t_m}\right)^{x} \tag{1.7}$$

where x is fitting coefficient, which is 0.5 generally; t_m, v_m are acceleration time and final velocity respectively; t is time; v is current velocity at time t; F, v are driving force of power source and velocity respectively; m is vehicle mass.

1.2.2.3 Impact Factors of Vehicle Acceleration Capacity

From last section, it can be seen that it's easy to estimate vehicle dynamic using Eq. (1.7) to represent velocity curves of acceleration. So what is the relationship between the real acceleration process and the factor x in Eq. (1.7)? It is the effect of x on the vehicle acceleration capacity that is the key to the question. Reference [12] makes a preliminary discussion and proposes that x should be about 0.47–0.53. Since the external characteristics of vehicle power source is approximate power contour, if vehicle has a better acceleration capacity, i.e. higher than power contour, which means the vehicle has better acceleration capacity and the factor x is tend to be a smaller one. But for this question, Ref. [12] has a contrary statement. While x equals to different values, from Eq. (1.7) it can be concluded that it is a curve of exponential function a^x. When a is smaller than 1, the smaller x is, the larger the value of exponential will be, which means the velocity is higher. The conclusion is verified by comparing simulation curves shown in Fig. 1.18.

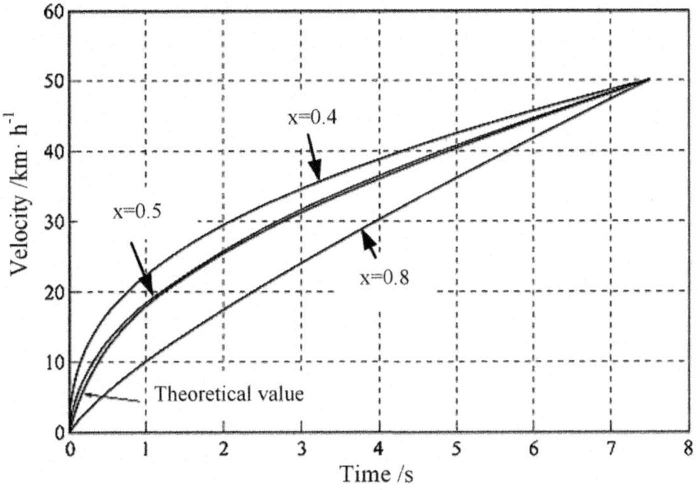

Fig. 1.18 Comparison of acceleration fitting curves

Through the above derivation and comparison, the following conclusions can be drawn:

The fitting factor x is 0.5, which is assumed that the power source is ideal and ignores the rolling resistance and air resistance.

For the actual conditions:

(1) When taking resistance into account, velocity increases slowly. So fitting factor $x > 0.5$, the bigger x is, the bigger vehicle resistance will be.
(2) While not considering about resistance, or for the dynamic stronger vehicle, its external characteristics is higher than power contour and fitting factor $x < 0.5$. So the smaller x is, the stronger vehicle dynamic will be. Reference [12] has a wrong statement about this.

1.2.2.4 Total Power Demand

According to acceleration theoretic equations above, a method to solve total power of power source based on acceleration index can be got. Assuming that vehicle accelerates on a flat road, based on vehicle acceleration kinetic equation, transient power is:

$$P_{total} = P_j + P_f + P_w = \frac{1}{3600 \cdot \eta_t}\left(\delta \cdot m \cdot v\frac{dv}{dt} + m \cdot g \cdot f \cdot v + \frac{C_D \cdot A}{21.15}v^3\right) \quad (1.8)$$

where P_{total} is total power of acceleration process, P_j is power of acceleration and δ is rotational mass conversion factor.

Because the maximum power is assumed to equal power, the average power output during the acceleration process is equal to the maximum power of the power source.

$$
\begin{aligned}
P_{total} &= \frac{W}{t_m} = \frac{\int_0^{t_m} (P_j + P_f + P_w)dt}{t_m} \\
&= \frac{1}{3600 \cdot t_m \cdot \eta_t}\left(\delta \cdot m \cdot \int_0^{v_m} vdv + m \cdot g \cdot f \cdot \int_0^{t_m} vdt + \frac{C_D \cdot A}{21.15}\int_0^{t_m} v^3 dt\right)
\end{aligned}
\tag{1.9}
$$

Make $v = v_m(t/t_m)^{0.5}$ and substitute it into equation above:

$$
P_{total} = \frac{1}{3600 \cdot t_m \cdot \eta_t}\left(\delta \cdot m \cdot \frac{v_m^2}{2} + m \cdot g \cdot f \cdot \int_0^{t_m} v_m\left(\frac{t^{0.5}}{t_m^{0.5}}\right)dt + \frac{C_D \cdot A}{21.15}\int_0^{t_m} v_m^3\left(\frac{t^{1.5}}{t_m^{1.5}}\right)dt\right)
\tag{1.10}
$$

Simplify it as

$$
P_{total} = \frac{1}{3600 \cdot t_m \cdot \eta_t}\left(\delta \cdot m \cdot \frac{v_m^2}{2} + m \cdot g \cdot f \cdot \frac{v_m}{1.5} \cdot t_m + \frac{C_D \cdot A \cdot v_m^3}{21.15 \times 2.5} \cdot t_m\right)
\tag{1.11}
$$

The first term is the main part of vehicle power, which is power of acceleration. Generally, the power of acceleration is much larger than the other two terms.

1.2.3 Application of Power Calculation Method

1.2.3.1 Assumption of External Characteristics

To explain whether the theoretical equation can be used in the design of the total power of vehicle power source, it is assumed that there is a driving motor, whose external characteristics is close to power contour. But for the real power source, it's impossible to meet the power contour. For example, an EV has a driving motor whose external characteristics is close to power contour as its power source, after matching a multi-shift transmission, power source will be much closer to power contour and its external characteristics curve is shown in Fig. 1.19.

According to the external characteristics of this power source, the theoretical calculation and simulation are carried out respectively. Respectively, as for considering and not considering other resistances of acceleration process, different calculation methods to calculate the vehicle total power are compared.

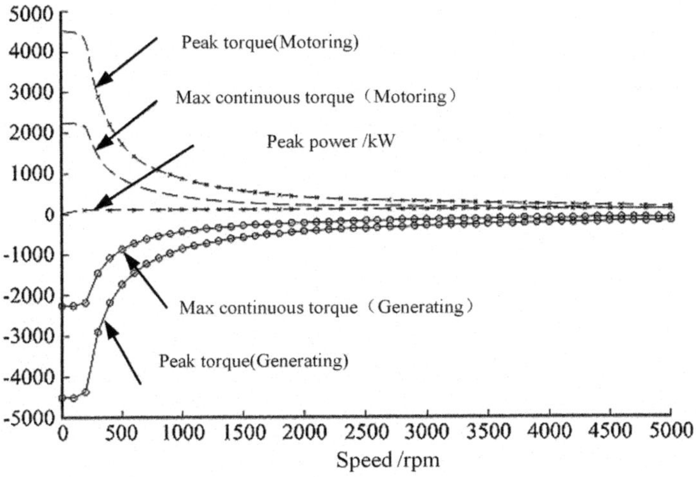

Fig. 1.19 External characteristics curve

1.2.3.2 Comparison of Simulation and Theoretical Results

The vehicle uses the power source close to the above power contour. Comparison of simulation curve, theoretical integration acceleration curve and fitting curve are shown in Fig. 1.20.

It can be seen that the above acceleration curves fit well, especially the power contour fitting curve fits the theoretical integration acceleration curve perfectly and they fit well with simulation curve at later period of the acceleration process.

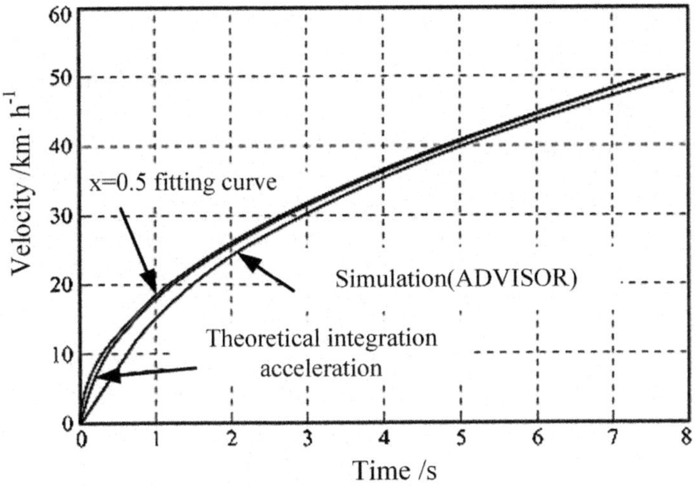

Fig. 1.20 Velocity curves during acceleration

Table 1.15 Acceleration time and deviation

Calculation method	Acceleration time of 0–50 $km \cdot h^{-1}$/s	Deviation with simulation result/s	Error rate/%
Simulation	7.9	\	\
Fitting curve $x = 0.5$	7.51	0.39	4.9
Theoretical integration method without resistance	7.51	0.39	4.9
Theoretical integration method with resistance	8.17	0.27	3.4

After adding actual resistance factor into this three curves, which means considering vehicle air resistance and rolling resistance, the acceleration curves and acceleration time are calculated with different methods. Comparison of results are shown in Table 1.15.

These results show that curves from vehicle theory equation fit well with power contour fitting curves. Because both the methods are in a presumed ideal state (i.e., no resistance term is considered), they are slightly different in the initial phase of acceleration and then substantially completely coincide. This is due to the calculation of the vehicle theory method is not fully calculated by power contour, velocity will be smaller when accelerating at low speed.

Theoretical integration calculation has deviation with actual simulation curves. One reason for this deviation is that the simulation process does not ignore the rolling resistance and air resistance. And the other reason is that there is power loss on the ADVISOR platform, which makes acceleration time 7.5 s longer than the theoretical value, which is 7.9 s in simulation. But from comparison of the simulation results, it is indicated that theoretical integration calculation power source can be well fit with the simulation and the deviation is small in the extent permitted.

Now, further study on the influence of considering resistance during the acceleration or not is taken. As shown in Fig. 1.21, the theoretical calculation curve considering resistance fits well with that not considering resistance at low speed. Comparison of driving torque with or without resistance is shown in Fig. 1.22.

It shows that, for the small resistance at low velocity, the two driving torques are close and the curves fit well. Resistance rises as the velocity increases and the curves do not fit very well with each other at high velocity.

After considering the resistance, acceleration curves from theoretical integrate calculation and ADVISOR simulation are very close. Acceleration time of theoretical calculation is 8.17 s, the deviation with ADVISOR simulation result 7.9 s is smaller, just is 3%.

Through the discussion on vehicle acceleration process, the conclusions are as follows: (1) The total power of vehicle is mainly determined by the acceleration process. (2) The acceleration process of vehicle can be simplified as an equal power

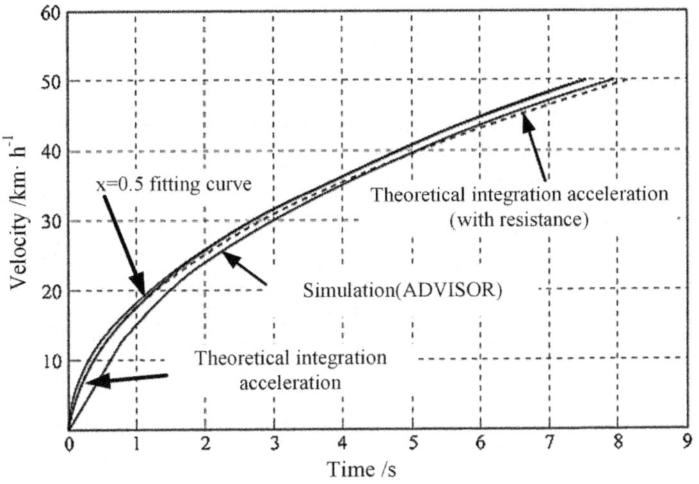

Fig. 1.21 Velocity curves during acceleration

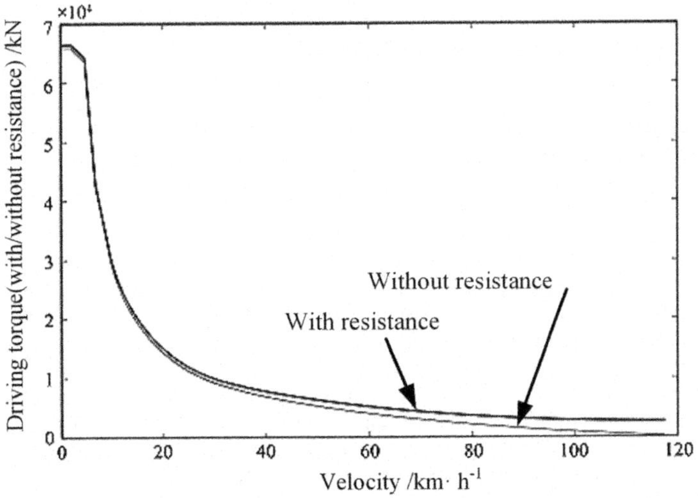

Fig. 1.22 Comparison of driving torque with/without resistance

process. Besides, through the comparison of calculation methods and actual simulation results, it can be seen that the direct curve-fitting method is the intuitive and effective way to calculate the total power, which makes it possible to estimate the total source power with acceleration performance of vehicle. This method can provide practical guidance for the enterprises to design their power source.

1.3 Summary

In this chapter, the actual energy consumption of conventional vehicle is analyzed with visual graphics. It can be concluded that choosing an engine with smaller power, eliminating engine idling and driving in pure electric mode in hybrid vehicle can improve the vehicle efficiency effectively. Then, based on the traditional power calculation method, a solution of directly designing the vehicle power through the acceleration performance index is proposed. After computing the actual simulation calculation, it is verified that the proposed power calculation method is intuitive, effective and accurate when solving the total power demand of the power source.

References

1. Chen Q, Sun F. Hybrid vehicle foundation. Beijing Institute of Technology Press; 2001.
2. Jun Li. The automobile powertrain core technology route and FAW's strategy for environmental protection and energy conservation. Eng Sci. 2009;11(08):64–71.
3. Raymond A, Sutula RA, Heitner KL, Rogers SA, et al. Advanced automotive technologies energy storage R&D programs at the US department of energy-recent achievements and current status. SAE Technical Paper, 2000-01-1604.
4. Hongyan Fang, Jin Wang, Keqiang Liu. Automotive industry energy analysis and energy-saving technology research. Automobile & Parts. 2009;35:42–5.
5. Gao Y, Rahman KM, Ehsani M. Parametric Design of the Drive Train of an Electrically Peaking Hybrid (ELPH) Vehicle. SAE Technical paper; 1997. p. 145–150.
6. Zeng X. Study on mechanism of energy saving and method of parameter design for hybrid electric bus. Jilin University; 2006.
7. Chu L. Control strategy and matching method for hybrid powertrain. Jilin University; 2006.
8. Qingnian Wang, Xiaohua Zeng. The application of hybrid power technology in military vehicles. J Jilin University (Engineering and Technology Edition). 2003;1:38–42.
9. Yu Z. Automotive Theory. China Machine Press; 1982.
10. Liu M. Study on vehicle control strategy and assembly parameter matching for hybrid electric bus. Jilin University; 2005.
11. Andersson T, Groot J. Alternative energy storage system for hybrid electric vehicles. Department of Electric Power Engineering Chalmers University of Technology; 2003.
12. Ng HK, Anderson JL, Santini DJ, Vyas AD. The prospects for electric and hybrid electric vehicles: second-stage results of a two-stage Delphi study; 1996.

Chapter 2
Configuration of Planetary Hybrid Power-Split System

In Chap. 1, the energy consumption characteristics of traditional vehicle have been analyzed by intuitive graphical method. Accordingly, the basic way of traditional vehicle energy saving is proposed. It's a good choice to apply the hybrid technology to traditional vehicle in the short term. In recent years, the series-parallel hybrid vehicle has become the stream in the market. Most kinds of the series-parallel hybrid vehicles use planetary gears as the PSD. In this chapter, detailed introduction to the typical PSD configuration is given firstly, including THS of Toyota corporation and AHS of GM corporation. Then the working principle and analysis method for the two kinds of configurations are discussed in detail. Finally, a new kind of PSD configuration, DHS which is developed independently, is proposed and analyzed.

2.1 Typical Configuration of Planetary Hybrid System

2.1.1 Toyota Planetary Hybrid System

2.1.1.1 Toyota Single Planetary Configuration

Toyota corporation has granted a patent for invention in October 17th, 2000 [1], and the THS configuration was released which the first and second generation Prius adopted. The structure of THS is as shown in Fig. 2.1. Where 120 is the shell, 128 is sprocket wheel of ring output, 129 is the chain, 112 is a duplicate gear, 111 is the drive gear of final drive, 114 is differential, 139, 149 are torsional dampers, 150 is the engine, 156 is the power input shaft and MG1, MG2 are electric motors. The lever model of THS is shown in Fig. 2.2.

X. Zeng and J. Wang, *Analysis and Design of the Power-Split Device for Hybrid Systems*, https://doi.org/10.1007/978-981-10-4272-0_2

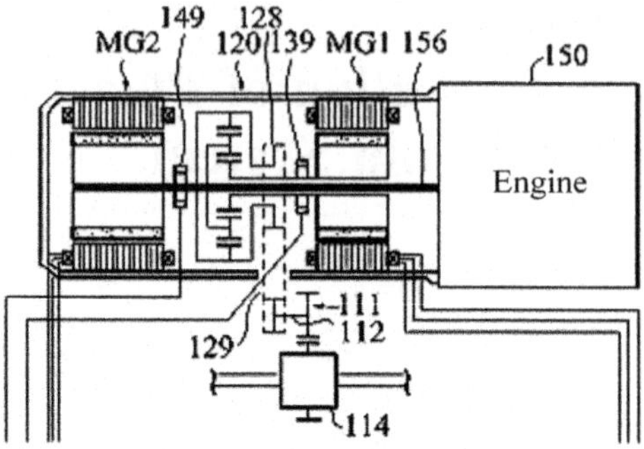

Fig. 2.1 First generation of THS

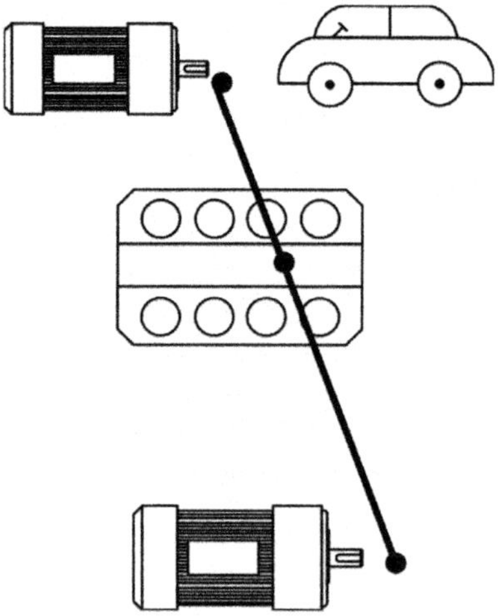

Fig. 2.2 Lever model of THS

As can be seen from Fig. 2.1, in the THS configuration, the engine is connected to the planet carrier, and it transfers the power to the ring gear and the sun gear via planetary gears, the left end of the ring gear connects with the motor through the torsional damper, the right side of the ring gear connects with the drive shaft through the drive chain, the sun gear connects with the generator. According to the lever analysis method [2], the speed and torque relationships of the configuration can be got, which is shown as:

$$\mathrm{n}_{out} = \frac{1+k}{k} n_e - \frac{1}{k} n_{m1} \quad (2.1)$$

$$\mathrm{T}_{out} = T_e \frac{k}{1+k} + T_{m2} \quad (2.2)$$

where T_{out} is the output torque of the system, n_{out} is the speed of output shaft; n_e and n_{m1} are engine speed and motor speed of MG1; T_e and T_{m2} are engine torque and torque of motor MG2; k is the characteristic parameter of the planetary system, and it is equal to the ratio of the number of ring gear teeth and the number of sun gear teeth.

This system transmits most of the engine torque to the drive shaft directly through the ring gear, and few of the torque is transient to the generator by sun gear. Electric energy generated by the generator is used to charge the battery or drive the motor to enhance the driving force according to orders. This structure can decouple the speed and torque of the engine, and it can work as a continuously variable transmission by controlling the speed of engine and the torque of driving motor, by which it can allow the engine to operate continuously in the high-efficiency zone or low-emission zone. However, this configuration only has the input-split mode. Its specific operation modes are shown in Fig. 2.3.

This configuration is mainly used in the first and second generation of Prius vehicle. In December, 1997, Toyota corporation started to sale their first generation Prius, and released the second generation Prius in 2003 and the third in 2006. The main modification focused on the engine and electric drive system, and the power-split device was still the THS. In this system, Prius adopted the gear ratio 2.6 (gear ratio = number of ring gear teeth/number of sun gear teeth). The smart and precise planetary mechanism can redistribute the output power of the engine to achieve a reasonable balance of engine load.

Toyota raised another configuration in their patent in 1999 [3], and the configuration diagram is shown in Fig. 2.4. Where, 10 is the whole hybrid power system, 12 is the engine, 14 is the torsional damper, 16 is generator, 22 is driving motor, 22r is the rotor of electric motor, 18 is the power output component, 26 is the power output gear, 30 is the transmission gear, 28 is the gear shaft, 32 is transmission pinion, 34 is differential, 20 is planetary mechanism, 20r is the ring gear, 20 s is the sun gear, 24 is shaft of generator, 36 is generator controller, 38 is the drive motor controller, 40 is high voltage power supply.

This configuration is the third generation of Prius, the difference from the previous two generations is the connection way between the ring gear and drive axle. The first and second generations are four-axle structure, the engine, torsional damper, power coupling device, generator MG1 and motor MG2 are arranged in the

Fig. 2.3 Operation pattern of Prius

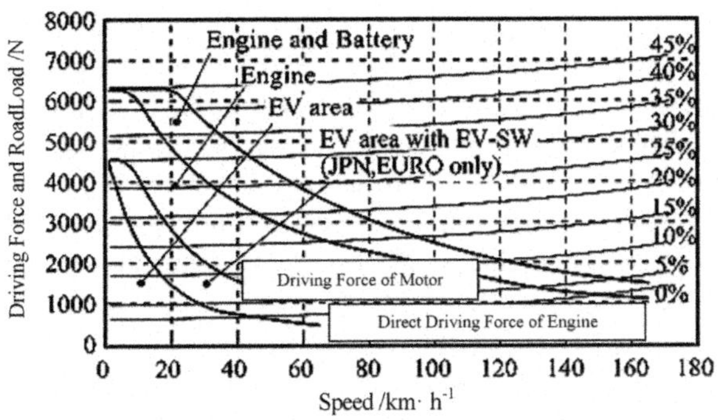

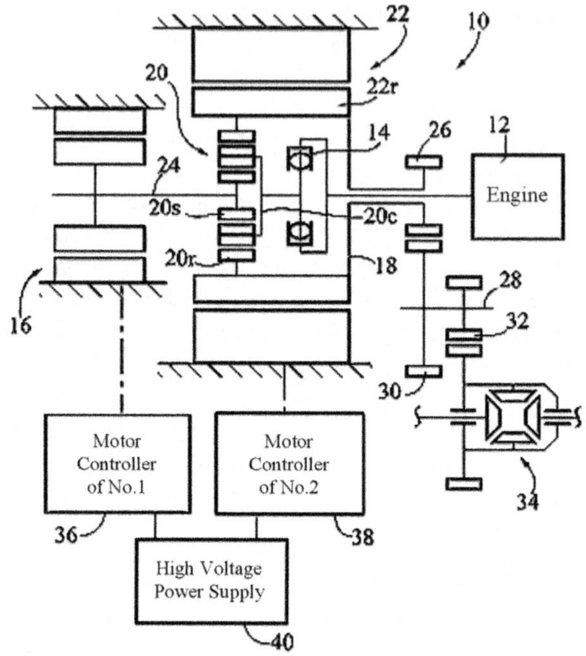

Fig. 2.4 New configuration diagram of THS

first axle, the transmission chain connects the first axle with the second one. Between the 2nd axle and 3rd axle is the intermediate gear, between the 2nd and 3rd shafts are the main drive gear. In contrast, the third generation Prius cancels the transmission chain and intermediate gear, which makes the weight lighter and the structure compact.

Toyota corporation got another patent license in 2000 [4], which introduced another single planetary power output device, and the configuration diagram is shown in Fig. 2.5. Where, 150 is engine, 130 and 140 are electric motors, 120 is the planetary gear, 121 is the sun gear, 124 is the planet gear, 122 is the ring gear, 123 is the planet carrier, 156 is the engine crank, 133 and 143 are stators of the two motors, 132 and 142 are rotors of the two motors, 160 is a clutch, 162 is a brake, 117,118 and 119 are rotational speed sensors, 165 is accelerator pedal position sensor, 166 is gear shift sensor, 114 is differential, 116 is the vehicle axle.

The main differences between this configuration and the third generation of THS are that it has a clutch (160) between planet gear (120) and electric motor (140) to realize the separation and connection of the two components, and that it sets a brake (162) to fix ring (122) while clutch (160) is separated in order to realize the series hybrid configuration, and come to a power-split hybrid configuration while clutch engaged. Changing modes according different running states of the vehicle can play the advantages of each mode, thereby improve the power and economy of vehicle. Detailed modes analysis are as follows:

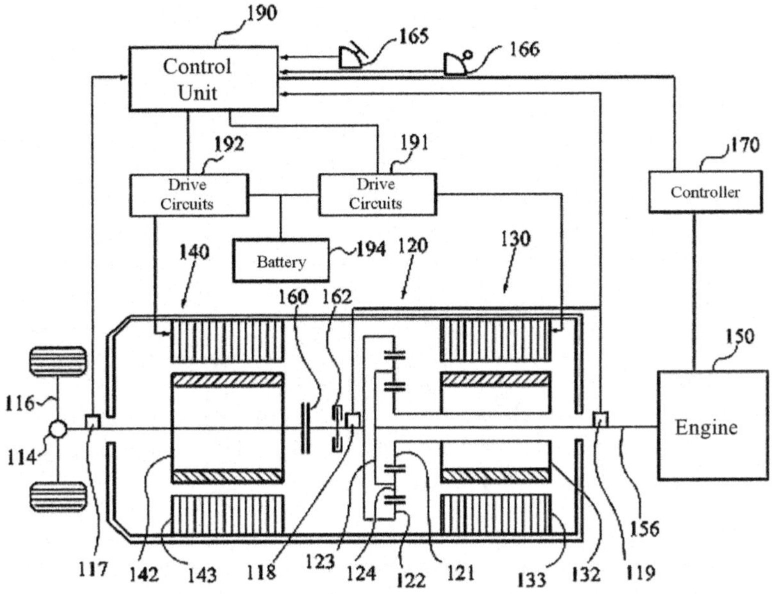

Fig. 2.5 Configuration diagram of the THS with clutch

Mode1: clutch is separated and brake is engaged. The relationships of torque and speed are shown as Eqs. (2.3) and (2.4).

$$T_{out} = T_{m2} \tag{2.3}$$

$$n_{out} = n_{m2} \tag{2.4}$$

Mode 2: clutch is engaged and brake is separated. Under this mode, connecting style of the configuration is same to the first and second generation of Prius THS system. The relationships of torque and speed are shown as Eqs. (2.1) and (2.2).

From the above, it's easy to see that the separation of clutch, in this configuration, can realize the separation between drive motor and engine inertial system. Through it, the driving ability of electric motor can be played out easily, at the same time pure electric driving ability will be promoted. The engine efficiency can be increased by series mode at low speed.

2.1.1.2 Toyota THS-C System

In 2001, Toyota launched their THS-C system [5]. It is a hybrid system consisting of THS and continuously variable transmission (CVT). Its configuration diagram is shown in Fig. 2.6.

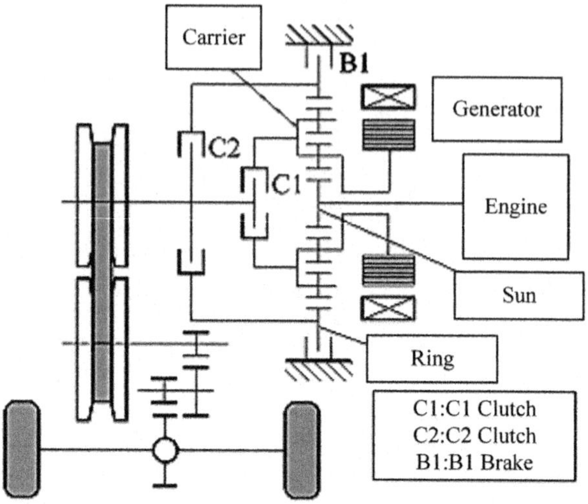

Fig. 2.6 Simplified diagram of THS-C configuration

It has some differences between the THS configuration shown below with THS. In this system the engine is connected with the sun gear and generator is connected with planet carrier. There is a brake B1 on the ring gear, the ring gear and the planet carrier are connected with CVT by clutch C1, C2 respectively, and CVT outputs the power to the drive axle. Furthermore, there are two rows of planets here, but they share the same planetary carrier. Its main view is shown in Fig. 2.7.

The speed equation is

$$(k - 1)\mathrm{n}_c = k \cdot n_r - n_s \tag{2.5}$$

where n_c, n_r and n_s are the speeds of the planet carrier, ring gear and sun gear, respectively.

THS-C is mainly used in Estima and Alphard. Estima has been launched in June, 2001. In the autumn of 2003, Toyota released Alpard based on Estima. These two types have the similar powertrain, the main improvement is about the control and security system. In Estima and Alphard, THS-C system is used in the front drive unit, the rear drive unit is powered by a separate rear motor, its configuration diagram is shown as Fig. 2.8.

Estima and Alphard are both MPVs (multi-purpose vehicle). Compared with sedans, MPV has a big and heavy body and consumes more fuel. Applying hybrid system in this kind of vehicle will be more effective in terms of CO_2 emission reduction. The original THS system hugely dependent upon the drive motor, if the size of components of the THS system are enlarged and then applied the system on the MPV, the size of motor, inverter, power battery and so on also need to be enlarged, which will lead to a lot of problems in terms of cost, quality, loading and so on. So in Toyota corporation, the belt type CVT is used to supply the drive torque with a smaller motor. At the same time, the bulks of inverter and drive

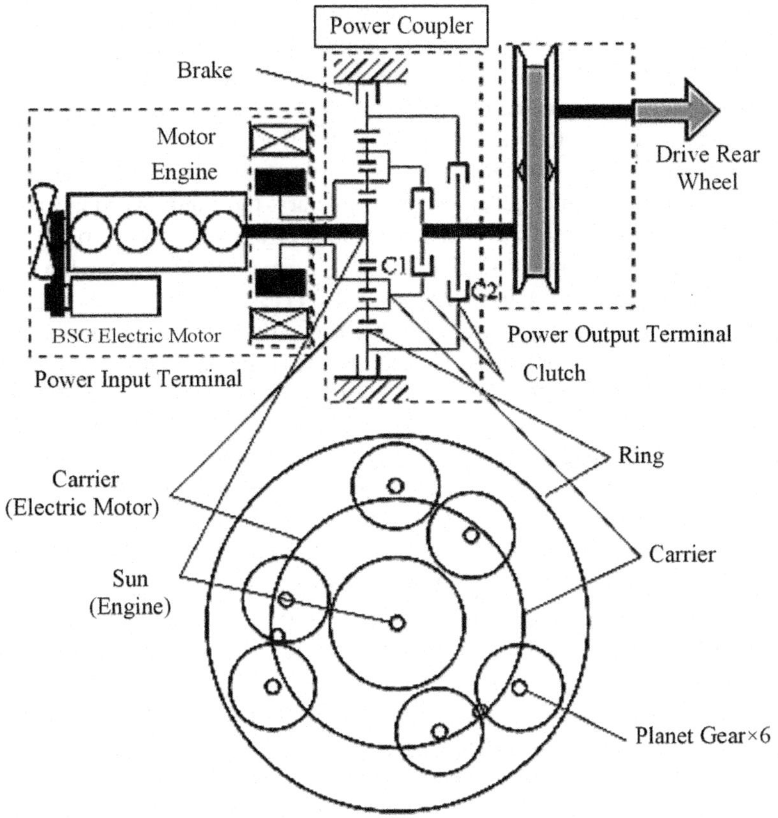

Fig. 2.7 Main view of THS-C configuration

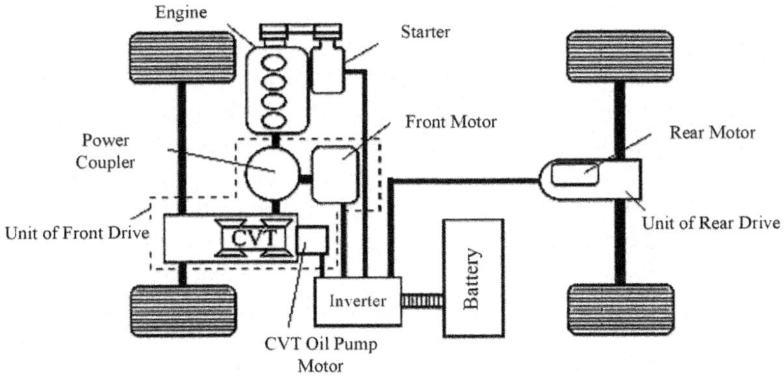

Fig. 2.8 Configuration diagram of THS-C dynamic system

battery are reduced. Besides, for a MPV which is four-wheel drive (4WD), the increased mass of drive shaft and other components will make it consume much more fuel compared with a 2WD vehicle, but in the electric 4WD, the designer could abolish some components like the drive shafts. And it is in 4WD mode only if necessary, and absorb braking energy through the front and rear motor to minimize the bad fuel consumption caused by 4WD. Furthermore, it changes the link relationship between the engine, generator and planetary, the engine is connected to the sun gear, generator is connected to the planetary, power is output through the ring gear. In this way, the torque of engine could be increased through the bigger ratio.

However, there are also many weaknesses in THS-C system. Most of the time, the system is a parallel configuration, for the reason that THS-C system in the front shaft and drive motor in rear shaft drive the vehicle at the same time, so it is difficult to ensure the balance of SOC, there is a risk that the battery cannot be used when the battery is insufficiently charged. Another problem is THS-C could realize both the function of THS system and CVT, and its mechanism is complex and costs more, which could be a key factor that limits its application. And the addition of CVT causes the system to require a motor to drive the hydraulic pump in it, which further exacerbates the complexity of the system structure. So the overall effect of THS-C is not so ideal. In January, 2006, the hybrid Estima was re-modified, not using the THS-C system any longer, but using the THS configuration directly. Because motor and generator are separated in THS II, the motor can be used to drive in any time during driving, and if the battery is undercharging, the generator can recharge it immediately, which improves the fuel economy. Experiments show that fuel consumption of Estima that adopt THS-II under 10–15 driving cycle is 5 L/100 km, while the fuel consumption of Estima that adopt THS-C is 5.5 L/100 km, that's to say the new version has decreased 9% fuel consumption.

2.1.1.3 Toyota Dual-Planetary Configuration

Toyota corporation applied for an invention patent in 2005 [6], this patent described a kind of hybrid vehicle power output device, which has a dual-planetary configuration and is shown as Fig. 2.9. Where, 20 is the whole hybrid system, 22 is the engine, 24 is the engine controller, 28 is the vibration damper, 26 is the crank, 30 is the planetary gear, 31 is the sun gear, 32 is the ring gear, 33 is the planet, 34 is the planet carrier, 32a is the ring gear shaft, 35 is the reduction gear, 60 is gear mechanism, 62 is the differential, 63a, 63b are drive wheels, 41, 42 are inventers, 50 is the battery, 54 is the power line, 40 is the motor controller, 43, 44 are rotor position sensors, 46 is the motor speed sensor, 47 is the current sensor, 70 is the vehicle control unit, 52 is the battery controller, 51 is the temperature sensor, 80 is the ignite signal, 81 is the transmission control lever, 82 is the position sensor of transmission control lever, 83 is the acceleration pedal, 84 is the position sensor of acceleration pedal, 85 is the brake pedal, 86 is the position sensor of brake pedal, 88 is vehicle speed sensor.

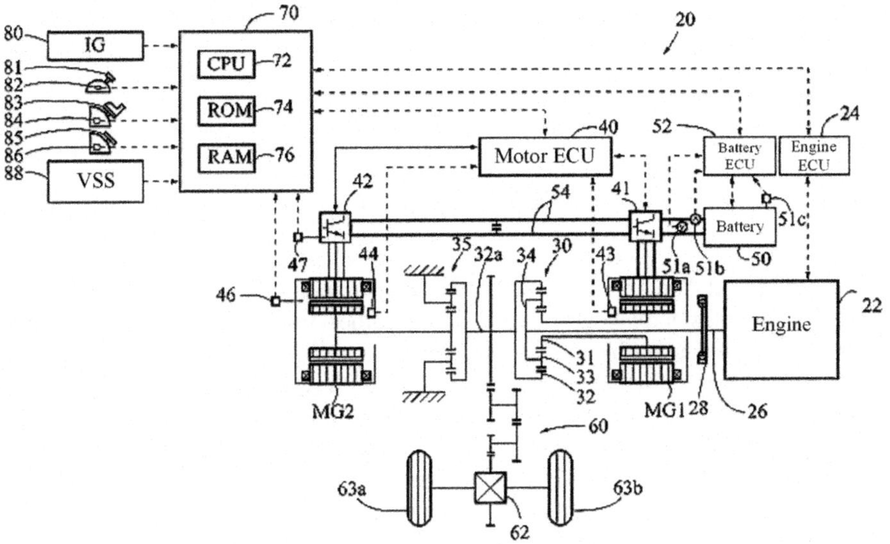

Fig. 2.9 Dual-row planetary configuration diagram of THS

As can be seen from the above figure, engine connects with the front planetary carrier through the torsional damper, the front sun gear connects with motor MG2, the rear planetary carrier is fixed, so the motor MG2 can transfer power to the rear ring gear in fixed ratio. At the same time, the front ring gear connects with the rear ring gear, power is coupled here and then be transferred to the drive axle (Fig. 2.10).

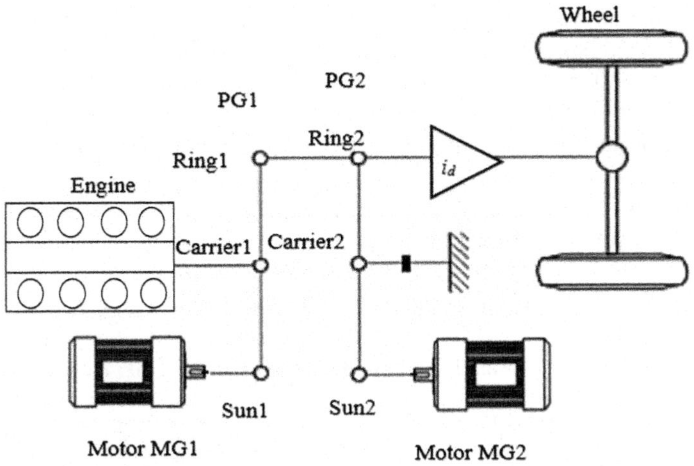

Fig. 2.10 Dual-row planetary lever model of THS

From the configuration diagram, the relationships of torque and speed are shown as below:

$$T_{out} = T_e \frac{k_1}{1+k_1} + T_{m2} \cdot k_2 \tag{2.6}$$

$$n_{out} = \frac{1+k_1}{k_1} n_e - \frac{1}{k_1} n_{m1} \tag{2.7}$$

Hermance and Abe have also made some research on this configuration [7]. This configuration adds a planetary gear, and becomes triple-axle from the original four-axle, which causes the configuration more compact. And it realizes the torque increase and speed reduction of the motor through fixing the rear planetary carrier, which improves the torque output capacity of the drive motor. But it has only one operating mode due to the rear planetary is a simple gear mechanism [8].

The configuration described in this patent is applied in three types of HEV of Toyota corporation, Lexus RX400h, Camry and Lexus Highlander [9]. After comparing Eqs. (2.2) and (2.6), several conclusions can be drew: this configuration is modified on the basis of the Prius configuration so that the drive motor is not directly connected to the front planetary ring gear but is connected to the front ring gear by the rear transmission. For the rear planetary, $k_2 = 2.478$. As a result, the system can output more torque, so Lexus RX400h, Camry and Lexus Highlander can achieve better power and off-road than the Prius without increasing the size of the drive motor.

2.1.1.4 Ravigneaux Configuration of Toyota Corporation

Toyota corporation launched Lexus GS450h vehicle in 2007, which adopts a special dual-planetary configuration. This configuration belongs to the second generation hybrid system of Toyota—THS-II, which is also called input-split Multiplication, and the configuration diagram is shown in Fig. 2.11 [10].

As can be seen from the above figure, the front planetary is just a normal one, but the rear is a Ravigneaux planetary mechanism, which is also called a compound planetary gear. It combines two common planetary gears into a power coupling device, includes two sun gears, the front sun gear 2 and the rear sun gear 3. And it also contains a compound planetary carrier, which is a combination of two planetary carriers with different radius, the small planet carrier mesh with the front sun gear and the inside of the larger one mesh with the rear sun gear, its outside gears engage with the ring gear. It is obviously that this kind of compound planetary configuration could be seen as dual-planetary which sharing the same planet carrier and ring gear.

In 22nd, March, 2005, Toyota launched the other two high power full hybrid SUVs (sport utility vehicle), Harrier and Kluger, which both adopt the THS-II configuration shown above. Harrier and Kluger hybrid SUV have good fuel economy. America Environmental Protection Agency (EPA) published that fuel

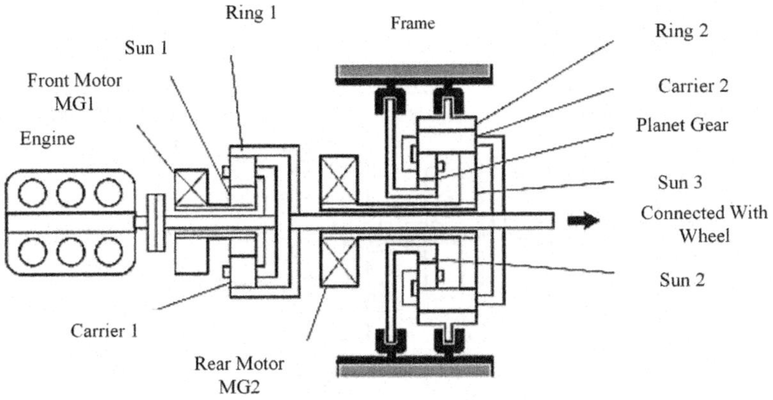

Fig. 2.11 Simplified configuration diagram of Lexus GS450h

economy of Harrier reached 7.8 L/100 km, 30% lower than conventional vehicles. Later, this kind of configuration became the main configuration of Toyota corporation. In 2007, it was applied to the Lexus LS600hL.

This configuration includes an engine, a dual-planetary gear, a generator, a drive motor and two brakes. The rear planetary mechanism can provide two ratios for 1.9 and 3.9 by different states of the brakes. So that it can achieve two different operation modes to fulfil the high speed cycles and the low speed cycles. The curves of mode switch are shown in Fig. 2.12.

(1) Low speed mode (high load)

In low speed mode, the rear ring gear R2 of Lexus GS450h is braked, the rear small sun gear is in idle state, the configuration diagram and lever model are shown as Figs. 2.13 and 2.14.

The torque relationship can be derived from the above configuration diagram:

$$T_{out} = T_{\mathrm{e}} \frac{k_1}{1 + k_1} + T_{m2}(1 + k_2) \tag{2.8}$$

Fig. 2.12 Mode switch curves of Lexus GS450h

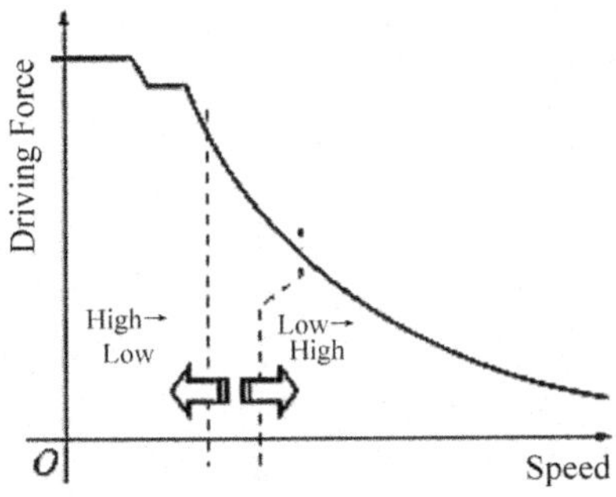

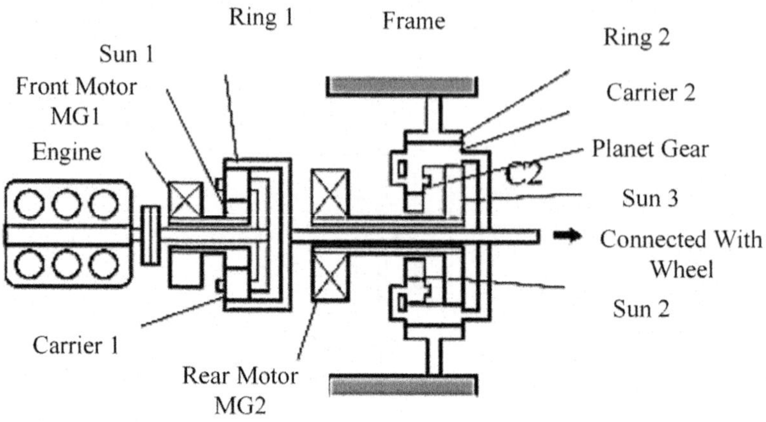

Fig. 2.13 Simplified configuration diagram of Lexus GS450h at low speed

Fig. 2.14 Lever model of Lexus GS450h at low speed

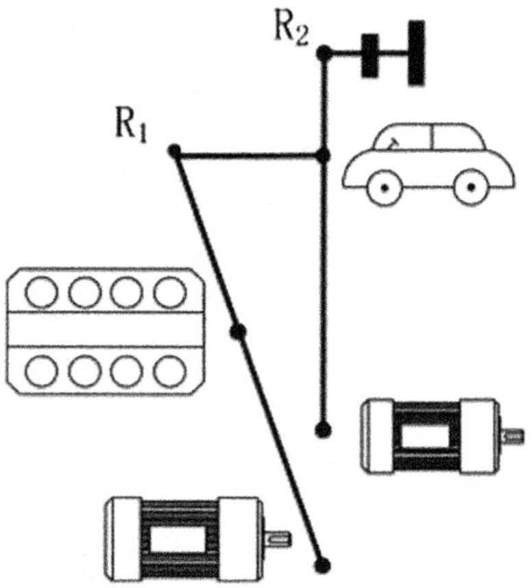

(2) High speed mode (low load)

In high speed mode, the rear smaller sun gear is braked, the rear ring gear is in idle state. Configuration diagram and lever model are shown in Figs. 2.15 and 2.16.

The relationships of torque and speed are shown as below:

$$T_{out} = T_{\mathrm{e}} \frac{k_1}{1 + k_1} + T_{m2} \frac{k_2 + k_3}{k_3} \tag{2.9}$$

$$n_{out} = n_{\mathrm{e}} \frac{1 + k_1}{k_1} - \frac{1}{k_1} n_g \tag{2.10}$$

where $k_1 = 2.6$, $k_2 = 2.9$, $k_3 = 3.2$.

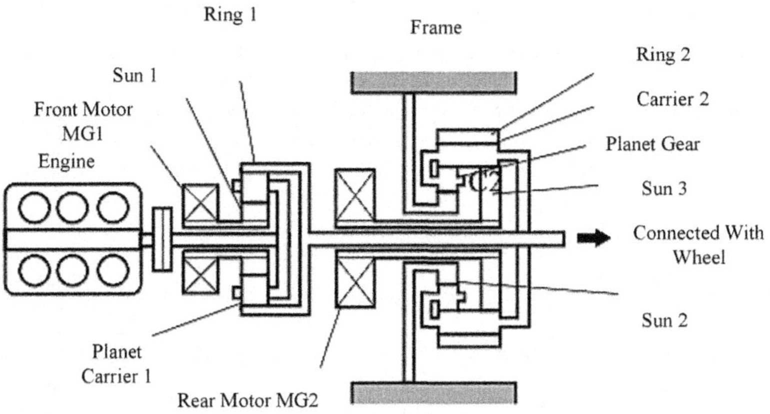

Fig. 2.15 Simplified configuration diagram of Lexus GS450h at high speed

Fig. 2.16 Lever model of Lexus GS450h at high speed

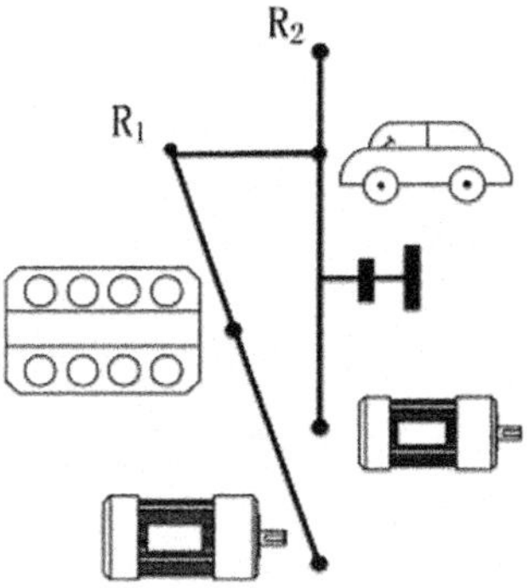

After putting the specific values into the formula, it can be calculated out that: Lexus GS450h can output a larger torque in the low-speed mode, and can also get better power performance in high-speed mode. So at high speed, it can achieve better acceleration performance than the average hybrid constructions.

2.1.1.5 Summary of EVT Configurations of Toyota Corporation

Looking at the development route and technical achievements of the Toyota corporation in the EVT hybrid coupling system, the technical route of the dynamic coupling configuration can be concluded as shown in Table 2.1 and Fig. 2.17. From single-row to the dual-row, from Prius to Lexus GS450h, the development of EVT configurations of Toyota corporation is extremely clear: always with the initial THS configuration as the core, and on this basis for the development, diversification, complexity, and thus produce a variety of other superior performance configurations.

Table 2.1 EVT configurations of Toyota

Single-row	Invention time	Source of literature	Vehicle	Configuration	Feature
	2000	US 6131680	The first/second Prius	THS	Former Prius —EVT classic structure, realize power-split
	1999	US 5934395	The third Prius	THS	Change the chain link of the original Prius to the gear transmission
	2001	SAE2002-01-0931 paper	Estima Alphard	THS—C System	The characteristics of planetary and the CVT are integrated, which is suitable for MPV
	2000	CN1336879A		THS + Clutch	For the improvement of the original Prius, more than one clutch and a model
Dual-row	2005	CN1819934A	Lexus RX400h Camry Lexus highlander	Single + single	Prius increase a deceleration of the speed ratio
	2007	SAE2007-01-0296	Lexus GS450h Harrier Kluger Lexus LS600hL	Single + Ravigneaux	Prius increase two deceleration of the speed ratio

2.1.2 *GM Planetary Hybrid System*

2.1.2.1 Dual-Planetary Configuration [11]

American GM corporation has published their patent early in 1996 [12], describing an input-split hybrid dual-planetary gear mechanism. It is an early planetary configuration, the engine and electric motor are not connected with the planetary gear

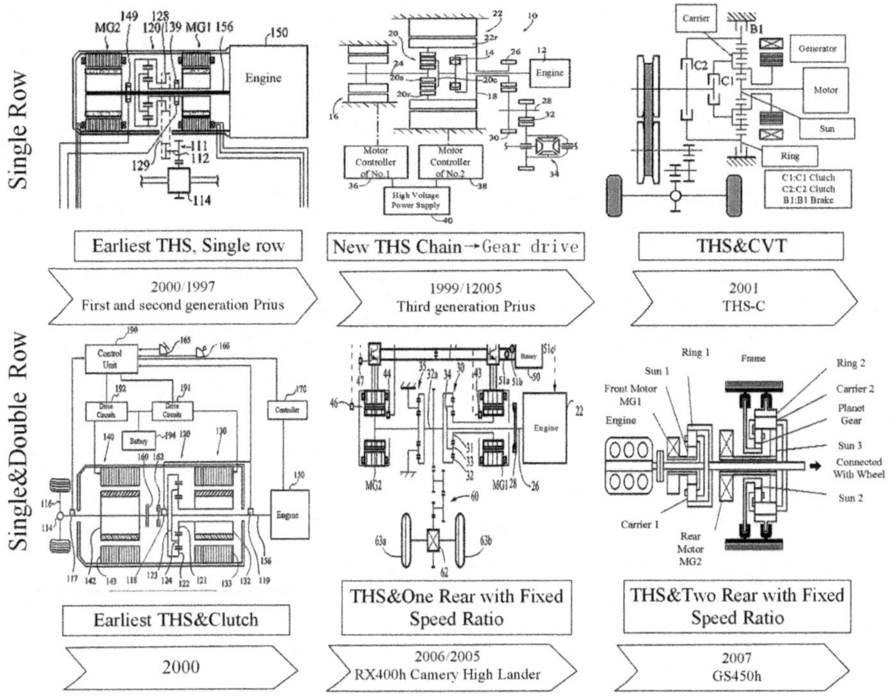

Fig. 2.17 Configurations evolution roadmap of Toyota EVT

directly. Therefore, its structure is complex and has not been used for a long time. The configuration diagram is shown in Fig. 2.18. Where, 110 stands for the whole hybrid system, 124 is engine, 112 is engine crank, 120 is generator, 122 is electric motor, 168, 142 and 154 are clutches, 114 is planetary gear, 132 is ring gear, 136 is planetary gear, 134 is sun gear, 138 is planetary carrier, 139 is output shaft of ring gear, 140 is output shaft of sun gear, 166 and 163 are transmission gears, 160A is output gear, 118 is power output shaft, 174 is differential, 170 and 172 are half axles, 176 is driving wheel, 126 is battery, 128 is electric control unit, 130A, 130B, 130C, 130D, 130E and 130F are power lines.

In 2002, GM corporation published a patent which introduces a new dual-planetary hybrid power-split device [13], called the Timken system, its diagram is shown in Fig. 2.19.

There are one engine, two motors, two clutches and two brakes in this configuration. The engine connects to the front ring gear, the front planetary carrier is connected to the rear planetary carrier, the front sun gear connects to the generator and also connects with the rear ring gear through the clutch, the rear sun gear is connected to the drive motor, the rear ring gear is connected to the brake, both the front and rear carriers connect to the output shaft. By controlling the clutches and brakes, it can realize different operating modes. Analysis about this is shown as below:

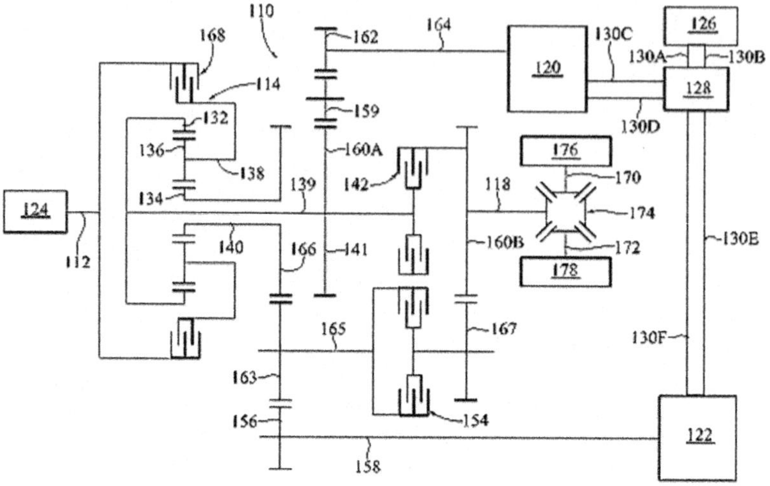

Fig. 2.18 Triple-row configuration diagram of GM in early time

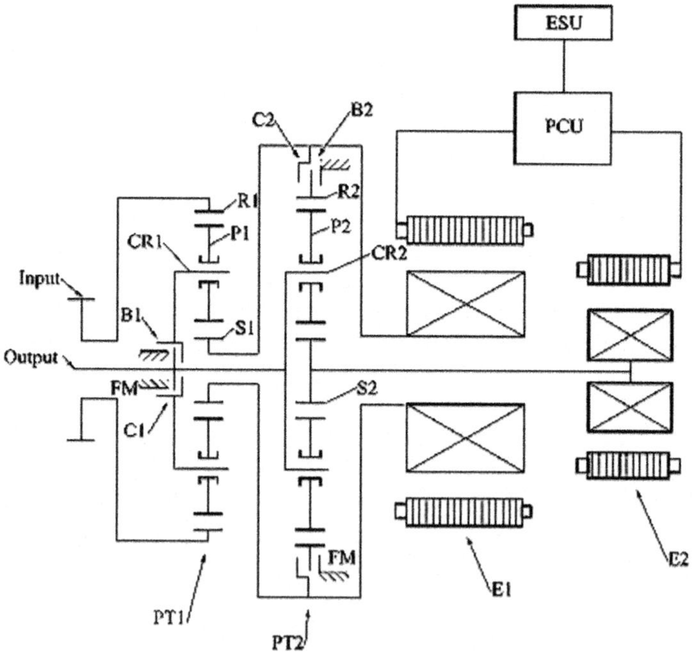

Fig. 2.19 Timken system configuration diagram

(1) Low speed mode (high load)

C1 is engaged, C2 is separated, B2 is engaged, B1 is separated. The lever model is shown in Fig. 2.20.

The relationships between torque and speed are as follows:

$$T_{out} = T_e \frac{k_1 + 1}{k_1} + T_{m2}(\mathrm{k}_2 + 1) \tag{2.11}$$

$$n_{out} = \frac{1}{1 + k_2} n_{m2} \tag{2.12}$$

(2) High speed mode (low load)

C1 and C2 are engaged, B1 and B2 are separated. The lever model is shown in Fig. 2.21.

The relationship between these speeds is:

$$\mathrm{n}_{out} = \frac{k_2}{1 + k_2} n_{m1} + \frac{1}{1 + k_2} n_{m2} \tag{2.13}$$

In 2005, *Xiaolan Ai* from the Timken corporation and *Scott Anderson* [14] from Ricardo Inc. simulated and analyzed this configuration, its torque curves are shown in Fig. 2.22 and its mode classifications are shown in Table 2.2.

The results show that this configuration improves the acceleration performance and reduces fuel emissions compared to conventional 4WD vehicles. When calculated in unit of MPG, the fuel economy increased by 158% under FTP driving cycle, and increased by 92% under HWFET driving cycle. This configuration demands a less motor capability than the Prius does: the maximum motor torque request is only one third of the Prius, and maximum power is two thirds. Through comparing this configuration with Prius in economy efficiency and dynamic, the

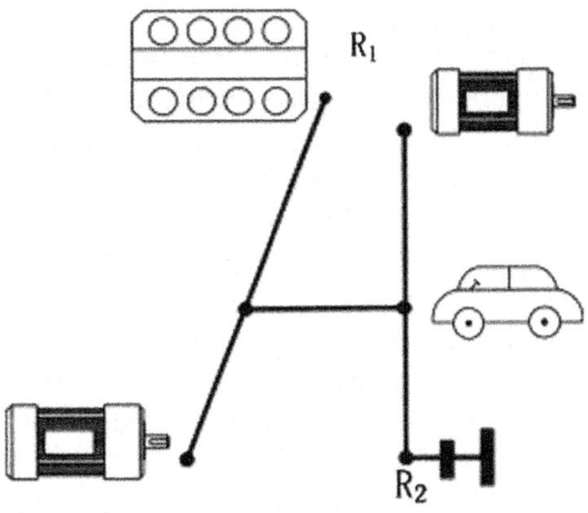

Fig. 2.20 Lever model at low speed

Fig. 2.21 Lever model at high speed

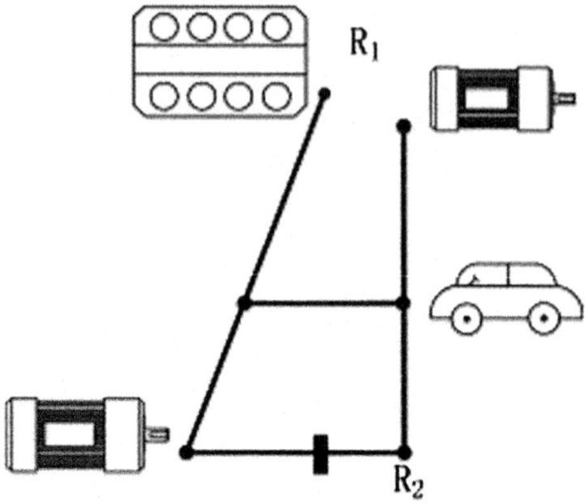

Fig. 2.22 Torque curves of the Timken system

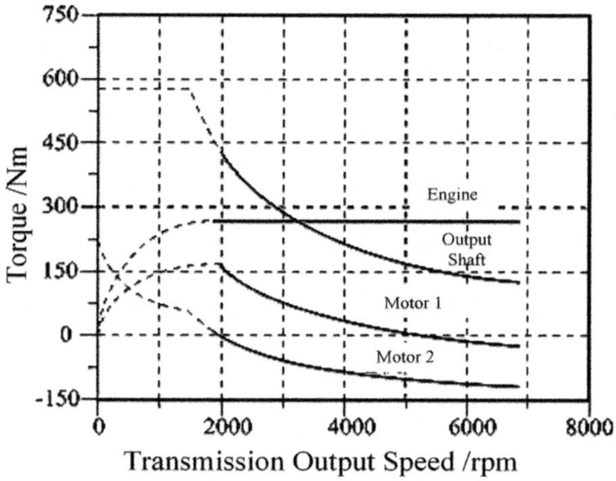

conclusion is that this configuration is better, some of the contrast curves are shown as Figs. 2.23, 2.24, 2.25 and 2.26:

In 2003, GM corporation issued another patent [15], which developed a dual-row dual-mode planetary power-split device, referred to as AHS (2PG), and its configuration diagram is shown in Fig. 2.27.

It can be seen from the configuration diagram: engine 12 connects with front planetary carrier 26 by clutch 16, the front sun gear 24 is connected to the generator 56 and is connected to the rear ring gear 32 by a clutch 64. The rear ring gear connects to the vehicle frame 54 by clutch 62, the front ring gear 22 connects to rear sun gear 36 and the drive motor 68. This configuration can realize two modes which are suitable for the high speed low load condition and the low speed high load condition respectively by controlling the connection and separation condition of clutch 64 and 62, in order to optimize the engine operation points and improve the

Table 2.2 Mode classifications of the Timken system

Mode/state	Brake combination		Clutch combination		Motor state	
	B1	B2	C1	C2	E1	E2
Parking/engine off	Yes	Yes	Yes	Yes	Off	Off
Parking/engine on	Yes	Yes	Yes	Yes	Generate	Off
Parking/engine on	Yes	No	Yes	No	Generate	Generate
Neutral	No	No	No	No	\	\
Start engine/stop	Yes	Yes	Yes	No	Off	Electric
Start engine/neutral	Yes	No	No	No	Electric	\
Forward/low speed	No	Yes	Yes	No	Generate	Electric
Forward/high speed	No	No	Yes	Yes	Electric/generate	Generate/electric
Reverse 1/series mode	Yes	Yes	No	No	Generate	Electric
Reverse/output power-split	No	Yes	Yes	No	Generate	Electric
Zero emission 1/single motor drive	No	Yes	No	No	\	Electric
Zero emission 2/dual motor drive	No	No	No	Yes	Electric	Electric

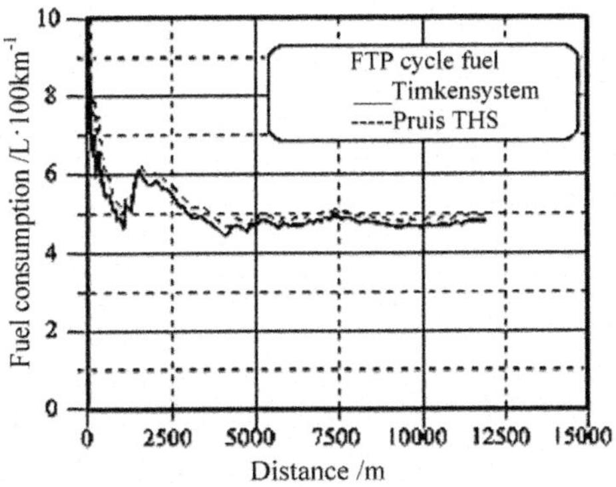

Fig. 2.23 Fuel economy curve(FTP)

fuel economy. This configuration is mainly used in Yukon and Tahoe of GM corporation, and is suitable for the SUV with high power requirement.

(1) Low speed mode (high load)

Clutch 64 of the AHS (2PG) is separated, clutch 62 is engaged, which means the rear ring gear be locked. The diagram for the configuration in low speed mode and its lever model are shown in Figs. 2.28 and 2.29 respectively.

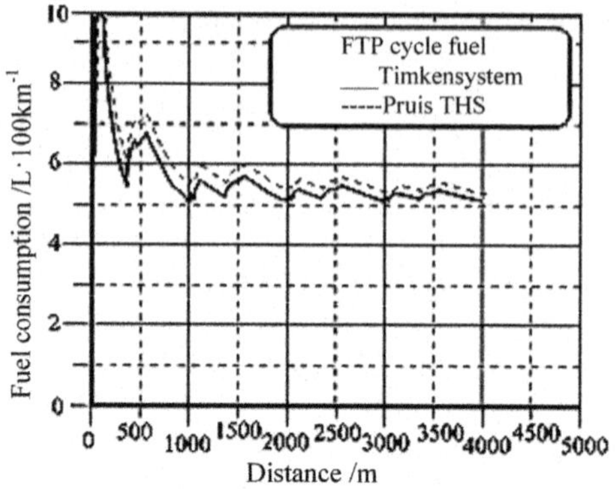

Fig. 2.24 Fuel economy curve(ECE)

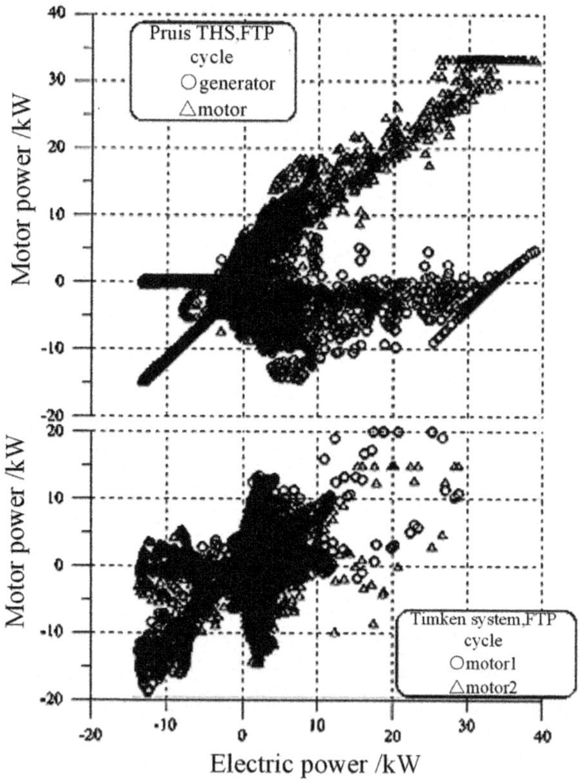

Fig. 2.25 Power demand comparison

The torque relationship between the components is as:

$$T_{out} = (1 + \mathrm{k}_2)\left(T_e \frac{k_1}{1 + k_1} + T_{m2}\right) \tag{2.14}$$

Fig. 2.26 Power demand comparison

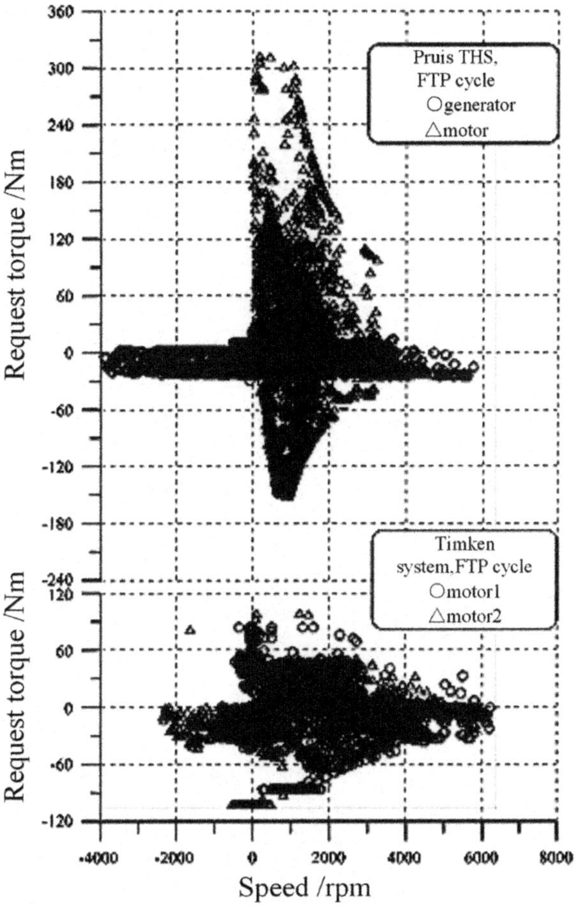

(2) High speed mode (low load)

Clutch 62 of the AHS (2PG) is separated and clutch 64 is engaged. The diagram in high speed mode and its lever model are shown in Figs. 2.30 and 2.31 respectively.

The torque and speed relationships between the components are as follows:

$$T_{out} = (1 + \mathrm{k}_2)\left(T_e \frac{k_1}{1 + k_1} + T_{m2}\right) \tag{2.15}$$

$$n_{out} = \left(\frac{1 + k_1}{k_1} \cdot \frac{1}{1 + k_2}\right) n_e - \frac{1 - k_1 k_2}{k_1 (1 + k_2)} n_{m1} \tag{2.16}$$

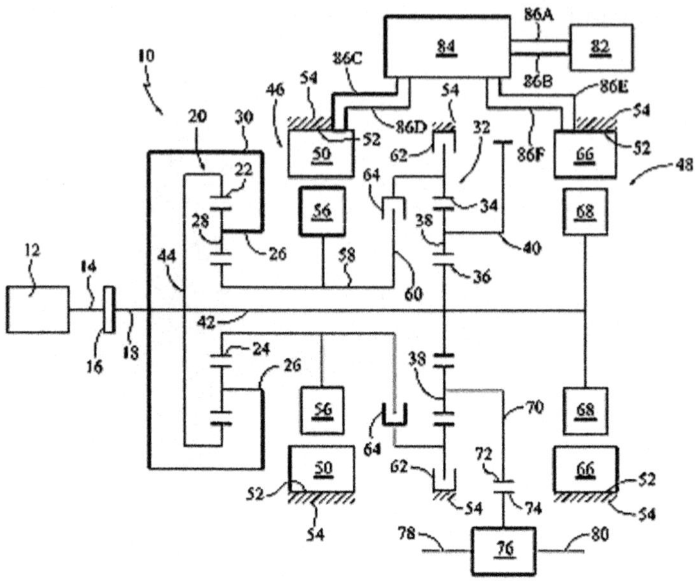

Fig. 2.27 AHS (2PG) configuration diagram

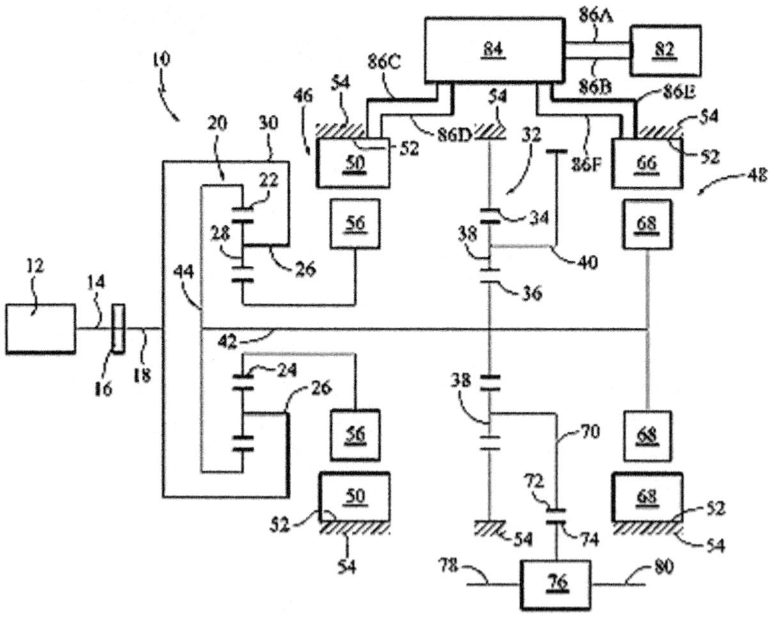

Fig. 2.28 Low speed mode configuration diagram of Yukon

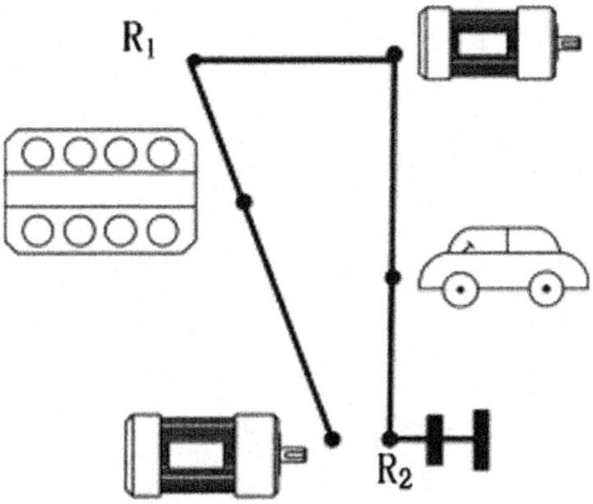

Fig. 2.29 Low speed lever model of Yukon

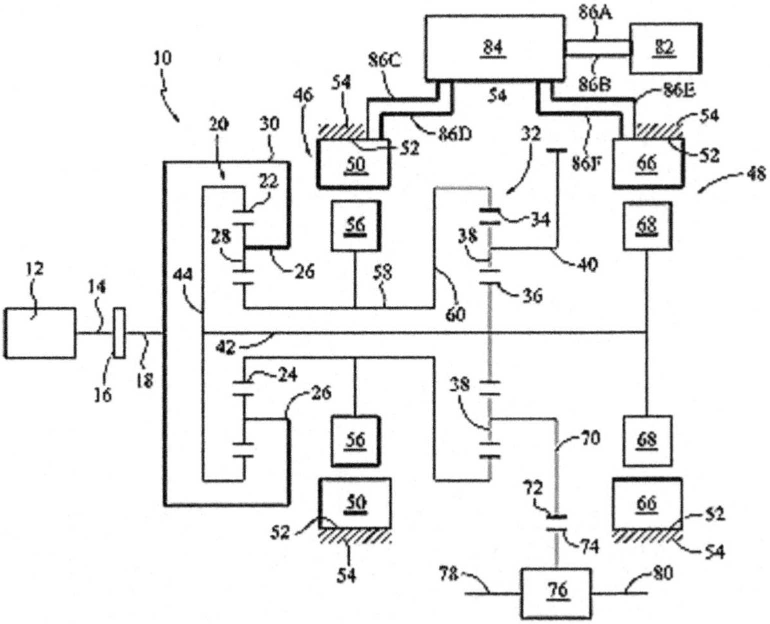

Fig. 2.30 Diagram in high speed mode of Yukon

The following conclusions can be drawn by taking the specific value into the formula: In high speed mode, the effect of acceleration for Yukon is less obvious, furthermore it needs MG2 operating at high speed. In the low speed mode, Yukon has an output torque larger than the usual hybrid vehicles and thus has more powerful actuation, so Yukon is a large-scale SUV.

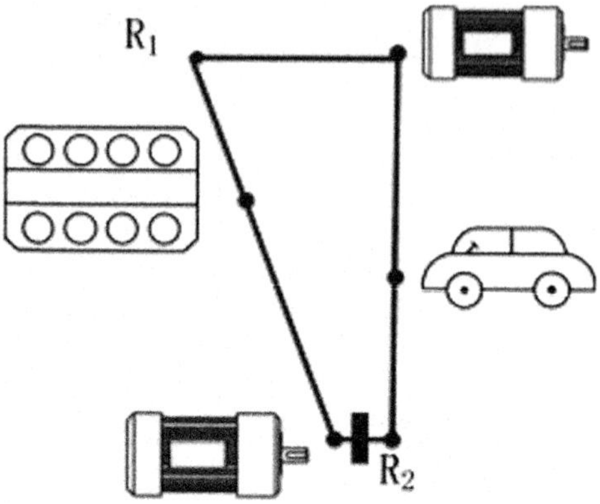

Fig. 2.31 Lever model in high speed mode of Yukon

2.1.2.2 Triple-Planetary Gear Configuration [16]

1. AHS (3PG) configuration

In 1999, Schmidt published a patent [17], which introduced a kind of triple-planetary gear power-split configuration, referred to as AHS (3PG). Its configuration diagram and lever model are shown in Figs. 2.32 and 2.33 respectively.

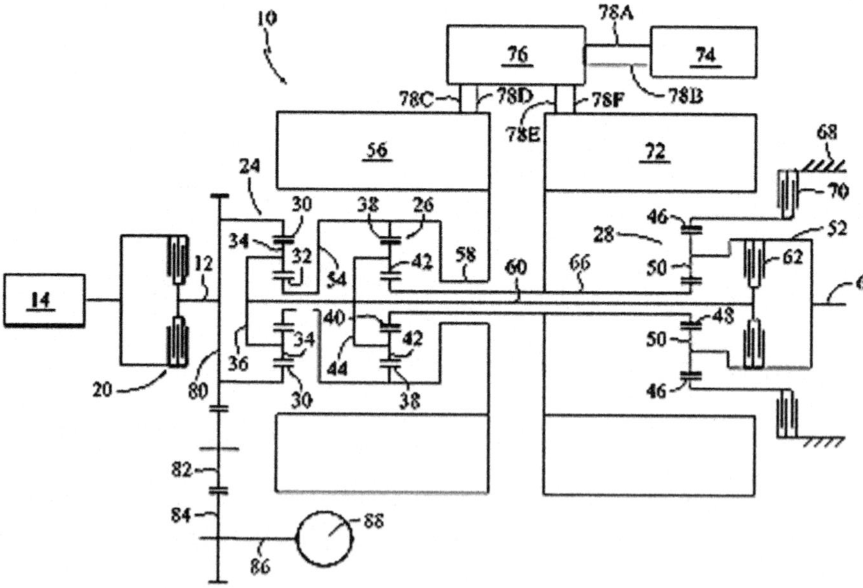

Fig. 2.32 Diagram of AHS (3PG)

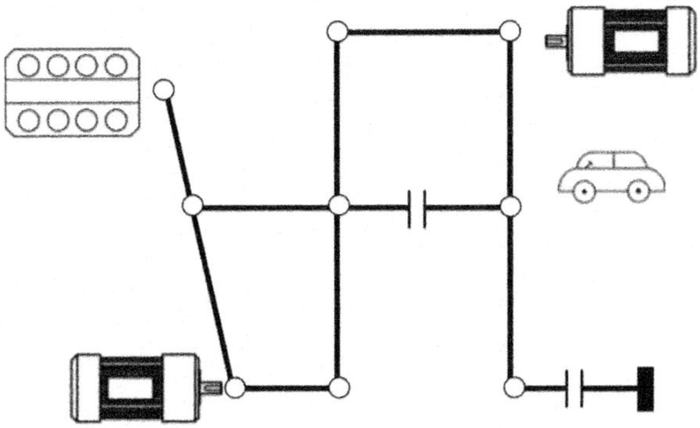

Fig. 2.33 Lever model of AHS (3PG)

The engine 14 connects to the ring gear of the first planetary, generator 56 connects to sun gear 32 of the first planetary and the ring gear of the second planetary, the planetary carrier 36 of the first planetary connects to the planetary carrier 44 of the second planetary and it also connects to planetary carrier 52 of the third planetary by clutch 62, the planetary carrier of the third planetary connects to output shaft 64, the sun gear 40 of the second planetary connects to the sun gear 48 of the third planetary and drive motor 72, the ring gear 46 of the third planetary connects to frame 68 by clutch 70. This system realizes two modes of high speed and low speed by controlling the separation and connection of the clutches.

2. Configuration of AHS (3PG) with 4FG ratio

In 2007, *Tim M. Grewe*, *Brendan M. Conlon* and *Alan G. Holmes* from GM r introduced a new triple-row planetary gear configuration, which was produced on the basis of AHS by adding two clutches C3 and C4. Its configuration diagram and lever model are shown in Figs. 2.34 and 2.35 respectively.

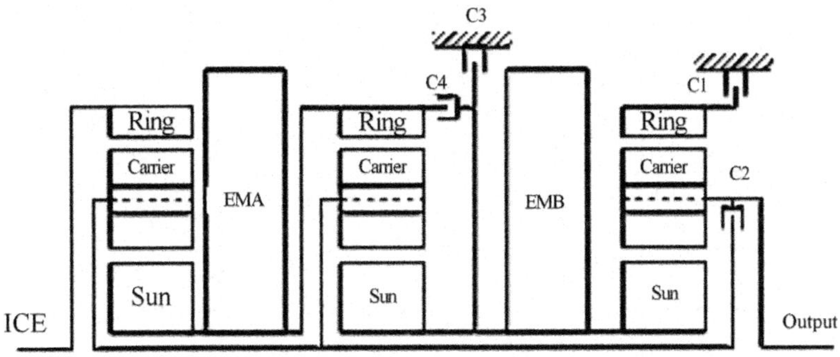

Fig. 2.34 Diagram of AHS (3PG) with constant ratio

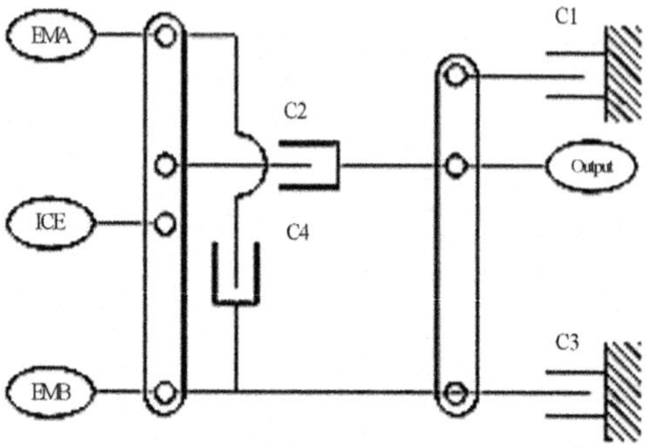

Mode/Gear	C1	C2	C3	C4
EVT1	X			
EVT2		X		
FG1	X			X
FG2	X	X		
FG3		X		X
FG4		X	X	

Fig. 2.35 Modes classification of AHS (3PG) with constant ratio

After increasing the two clutches C3 and C4, four operating modes are also added: the dual-mode (high speed and low speed) and the four fixed speed ratio (FG) modes. This configuration is applied to the new version Yukon and Tahoe vehicles. Kaehler from GM also analyzes the new configuration [18], results show that this configuration is suitable to the heavy vehicle and SUV with large power, for it optimizes the fuel economy and meets the off-road performance of SUV. Its various models are analyzed as follows

(1) EVT1 mode (for low speed)

The EVT1 mode is amount to an input-split configuration, in this mode, only clutch C1 is engaged and all the other clutches are separated. The simplified model is shown as Fig. 2.36.

The torque and speed relationships between the components are as follows:

$$T_{out} = \left(T_e \frac{k_1+1}{k_1}\frac{1}{k_2+1} + T_{mB}\right)(k_3+1) \tag{2.17}$$

$$(1+k_3)n_{out} = \frac{1-k_1k_2}{1+k_1}n_{mA} + \frac{k_1+k_1k_2}{1+k_1}n_e \tag{2.18}$$

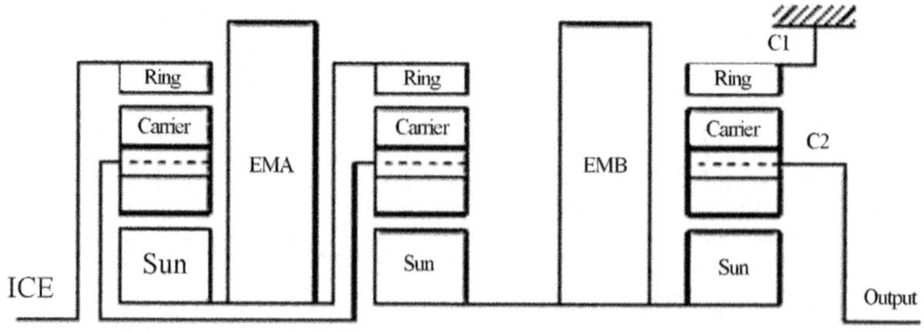

Fig. 2.36 Diagram in EVT1 mode

From the equations above, EVT1 has the function of continuous variable transmission, which could adjust ratios by controlling the speed of the electric motor.

(2) EVT2 mode (for high speed)

EVT2 mode is amount to a compound power-split configuration, only the clutch C2 is engaged, and the other clutches are separated, the simplified model is shown in Fig. 2.37.

The torque and speed relationships between the components are as follows:

$$T_{out} = T_e \frac{k_1 + 1}{k_1} + T_{mB}(k_2 + 1) \tag{2.19}$$

$$(1 + k_1)n_{out} = n_{mA} + k_1 n_e \tag{2.20}$$

Similar to the EVT1, EVT2 also has the function of continuous variable transmission, which could adjust ratios by controlling the speed of the electric motor.

(3) FG1: clutch C1 and C4 are engaged (Fig. 2.38)

The torque and speed relationships between the components are as follows:

$$n_{out} = \frac{1}{k_3 + 1} n_e \tag{2.21}$$

$$T_{out} = (T_e + T_{mB})(k_3 + 1) \tag{2.22}$$

Both ratios of the two planetary are 1 at this time, and the PG1, PG2 are self-locking.

(4) FG2: clutch C1, C2 are engaged (Fig. 2.39)

The torque and speed relationships between the components are as follows:

$$n_{out} = \frac{k_1 k_2}{k_1 k_2 + k_3} n_e \tag{2.23}$$

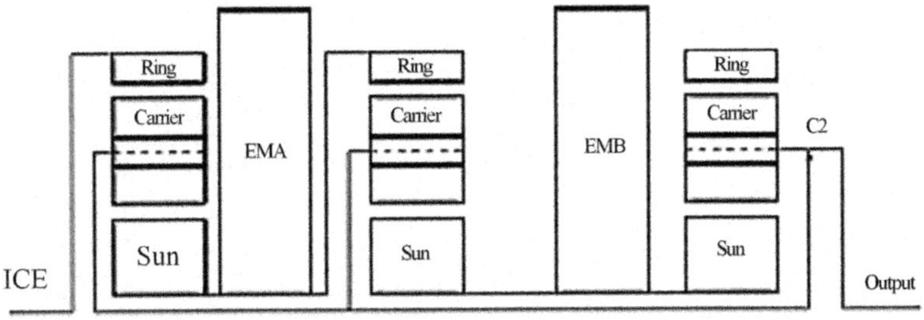

Fig. 2.37 Diagram of EVT2 mode

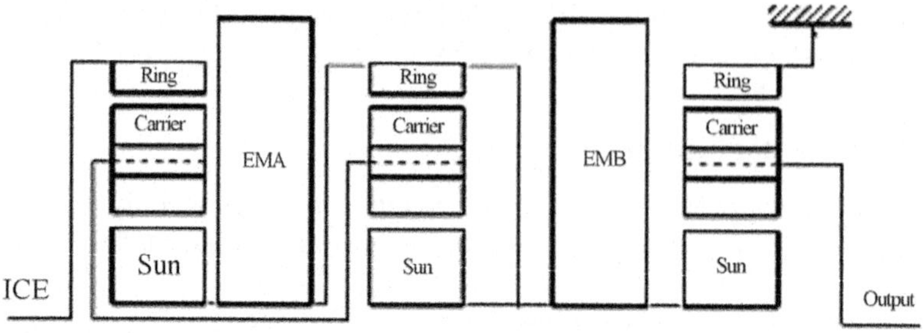

Fig. 2.38 Diagram of FG1 mode

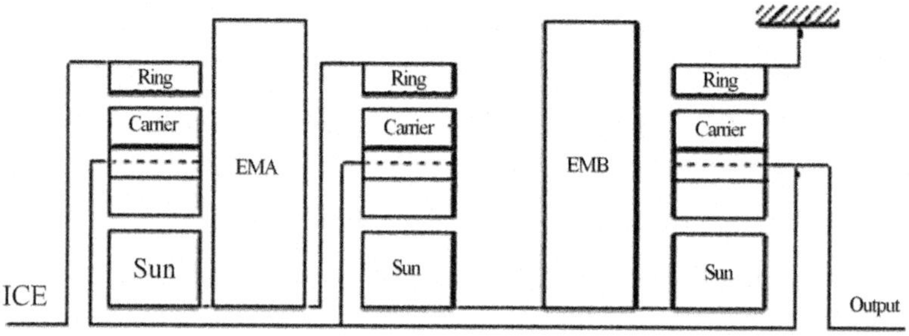

Fig. 2.39 Diagram of FG2 mode

$$T_{out} = T_e\left(\frac{k_3}{k_1 k_2} + 1\right) T_{mB}(k_3 + 1) \tag{2.24}$$

(5) FG3: clutch C2, C4 are engaged (Fig. 2.40)

The torque and speed relationships between the components are as follows:

$$n_{out} = n_e \tag{2.25}$$

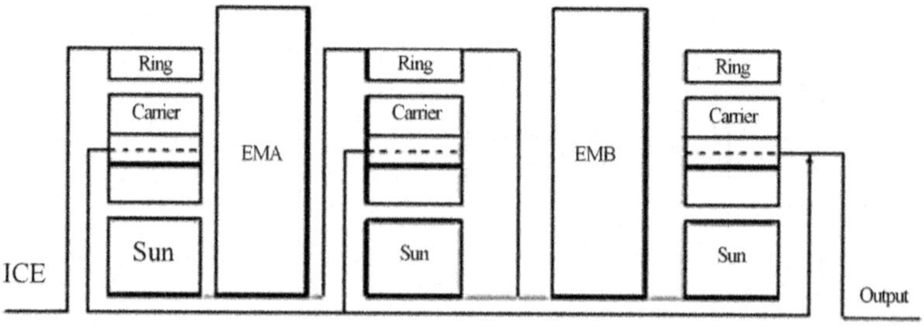

Fig. 2.40 Diagram of FG3 mode

$$T_{out} = T_{\mathrm{e}} + T_{mB} \tag{2.26}$$

At this time, PG1, PG2, PG3 are all self-locking, and the gear ratio is 1.

(6) FG4: clutch C2, C4 are engaged (Fig. 2.41)

The torque and speed relationships between the components are as follows:

$$n_{out} = \frac{k_1 k_2}{k_1 k_2 - 1} n_e \tag{2.27}$$

$$T_{out} = T_e \left(1 - \frac{1}{k_1 k_2} \right) \tag{2.28}$$

The comparisons of EVT1 mode and EVT2 mode are shown as below:

Comparison of torque: make subtraction of the torque equation in EVT1 and EVT2, the difference is:

$$\Delta T_{out} = \left[T_e \frac{k_1 + 1}{k_1 (k_2 + 1)} + T_{mB} \right] (k_3 - k_2) \tag{2.29}$$

Generally, $k_3 > k_2$, so $\Delta T_{out} > 0$, which means EVT1 has a larger torque thus suitable for the low speed and starting mode. Substituting parameters of Lexus GS450h, $k_1 = 2.6$, $k_2 = 2.9$, $k_3 = 3.2$, into the equation above, the following result can be drawn:

$$\Delta T_{out} = 0.11 T_e + 0.3 T_{mB} \tag{2.30}$$

It is obvious that EVT1 has a larger torque output than EVT2, and is more suitable for the low speed mode.

Comparison of rotary speed: substituting parameters into the equation, the following relationships can be drawn:

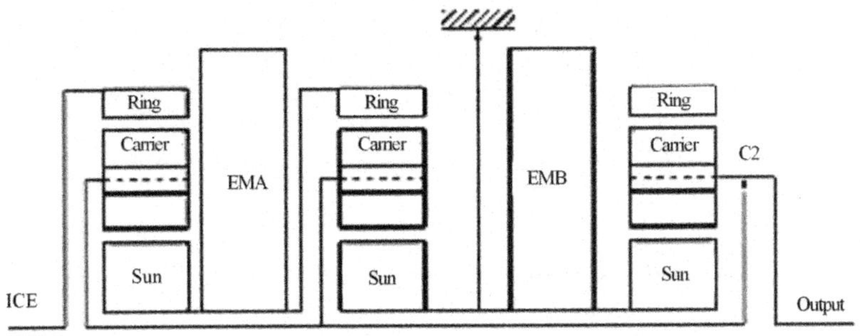

Fig. 2.41 Diagram of FG4 mode

$$\begin{aligned} \text{EVT1}: n_{out} &= -0.43n_{mA} + 0.67n_e \\ \text{EVT2}: n_{out} &= 0.28n_{mA} + 0.72n_e \end{aligned}$$

Above all, EVT1 is more suitable for low speed mode and EVT2 for high, the tipping point for switching EVT1 and EVT2 is FG2. The operating areas in various modes are shown in Figs. 2.42, 2.43 and 2.44, where the 4 black points stand for the 4 fixed gear ratio, and 'inverse transmission ratio' is the ratio of the output speed to the input speed.

3. 2MT with 4FG configuration

In 2009, Hendrickson et al. [19] and Meisel [20] introduced a new 2MT + 4FG configuration. The configuration is used in front drive vehicle and is also the first type of 2MT + 4FG for the front drive. It is used in the *Saturn Vue Green Line* vehicle of GM corporation and is named 2MT70. It is a configuration with tree planetary gears, but is usually considered as a dual-planetary configuration when analyzing its characteristics, Fig. 2.45 shows its configuration diagram.

Mode classification of the 2MT70 is shown in Table 2.3.

Each of the 4 fixed ratios has their own characteristics and actions:

(1) FG1 is equivalent to locking the PG1 and does not change the speed. So, it could provide the largest traction force in these 4 ratios. As a parallel mode, both the motor A and motor B can provide extra torque;
(2) FG2 is equivalent to locking motor A to make its rotary speed to be 0, it is the first mechanical point (that means all of the engine power is transferred through

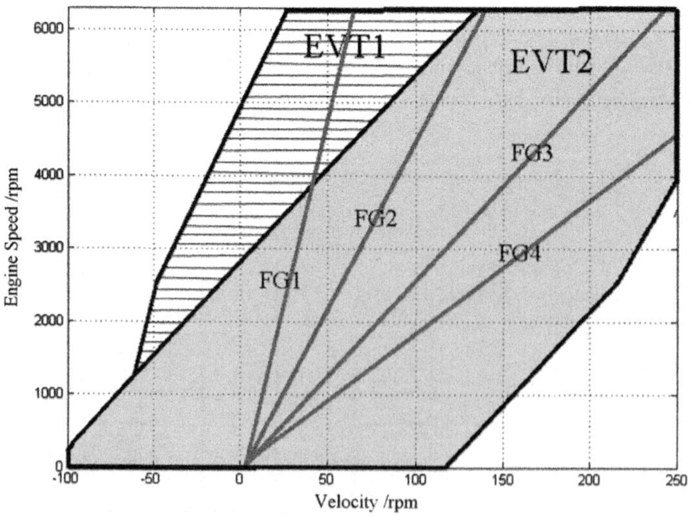

Fig. 2.42 Operating areas of AHS (3PG) with constant ratio

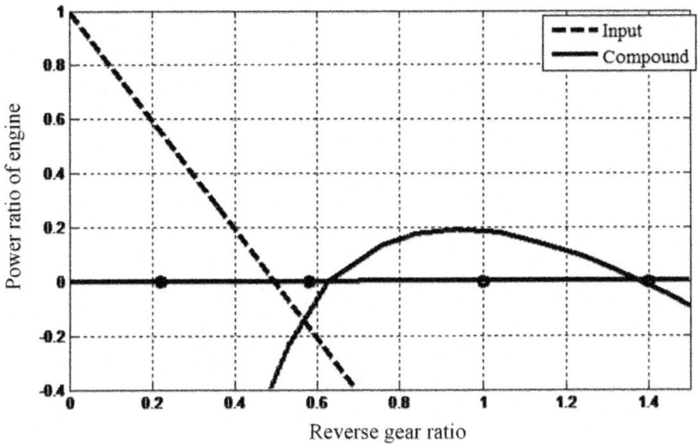

Fig. 2.43 Operating areas of AHS (3PG) with constant ratio

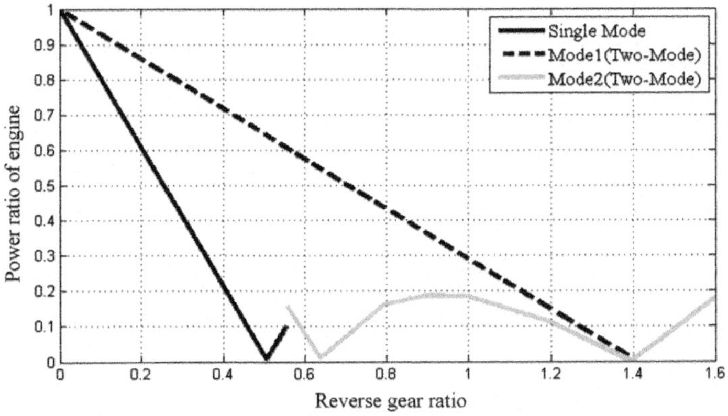

Fig. 2.44 Operating areas of AHS (3PG) with constant ratio

mechanical path). It can realize the switches between two EVT modes. As a parallel mode, motor B can provide the extra torque;

(3) FG3 is equivalent to locking PG1 and PG2, so that the overall ratio of the system is 1, it could provide an ideal grade ability and traction ability. As a parallel mode, both the motor A and motor B can provide extra torque;

(4) FG4 is equivalent to locking motor B to make its rotary speed to be 0, it is the second mechanical point. It is the fixed ratio which suits cruise at high speed. As a parallel mode, motor A could provide extra torque.

2MT70 configuration has many advantages: (1) the full-featured dual-mode hybrid drive unit successfully applied in the general front drive structure; (2) maximize the fuel economy on the premise that doesn't sacrifice much of the

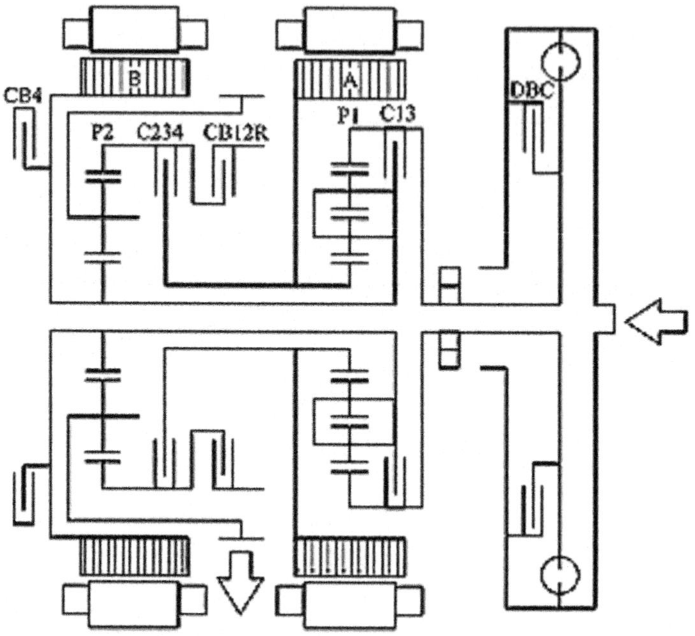

Fig. 2.45 Diagram of 2MT70

Table 2.3 Mode classification of 2MT70

Clutch work mode		CB12R	C234	C13	CB4	DBC
EV drive	EV	Engage	\	\	\	\
Engine on/off	ESS	Engage	\	\	\	Engage
Lows peed EVT	EVT1	Engage	\	\	\	\
Fixed ratio 1	FG1	Engage	\	Engage	\	\
Fixed ratio 2	FG2	Engage	Engage	\	\	\
High speed EVT	EVT2	\	Engage	\	\	\
Fixed ratio 3	FG3	\	Engage	Engage	\	\
Fixed ratio 4	FG4	\	Engage	\	Engage	\

power performance; (3) adding an innovative vibration damper bypath system, so that the engine starts and stops technology becomes the industry leader.

The performance improvements of 2MT70 are shown as Table 2.4.

The fuel economy improvements of 2MT70 are shown as Fig. 2.46.

Besides, GM corporation published a patent in 1999 [21], which introduced a triple-planetary power-split configuration. This configuration is similar to the configurations presented above and researchers have little research on it, so the utilization rate is not high. The engineer of GM published patents in 2007 [22], which introduced diagrams of 15 kinds of triple-planetary configurations, almost all the

Table 2.4 Performance improvement comparison of 2MT70

Project	Traditional models of Saturn Vue	Dual-mode hybrid models of Saturn Vue
0–60 $mi \cdot h^{-1}$ acceleration time	6.9 s	7.5 s
EPA fuel economy	Reference	Improve 50%

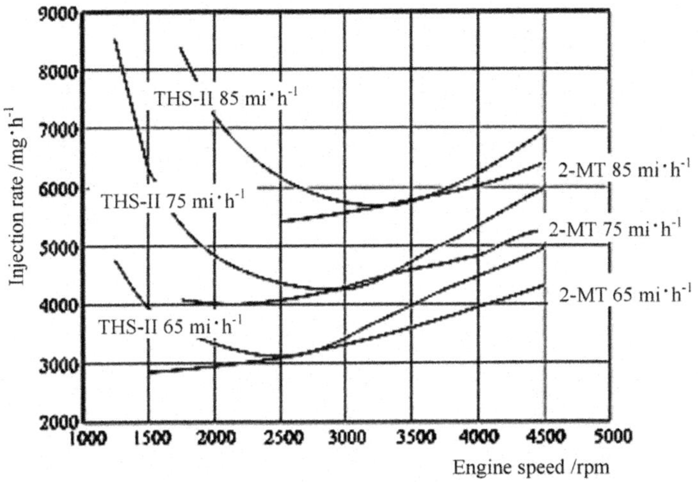

Fig. 2.46 Fuel economy of 2MT70

configurations. And samples for ratio designing were given, while the detail analysis was not conducted. One of the configurations in this patent is shown as Fig. 2.47.

2.1.2.3 Summary of EVT Configurations of GM Corporation

Looking at the development route and technical achievements of the GM EVT configurations, the summary of its scientific research achievements route is shown as Fig. 2.48. Configurations of GM corporation are more complex, mainly focus on the dual- and triple-planetary configurations, especially published some patents in the triple-planetary configurations respect, as shown in Table 2.5.

2.1.3 Summary

Through the comparison of the EVT configurations of Toyota and GM, it can be found that AHS system of GM is nimbler to control for the many power-split modes

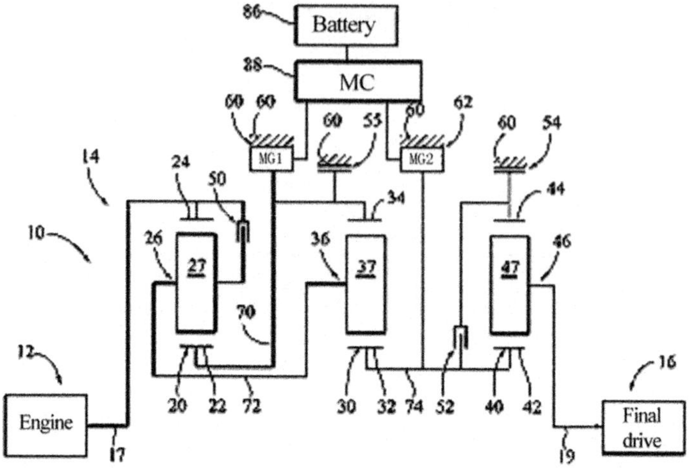

Fig. 2.47 Diagram of GM's early triple-planetary configuration

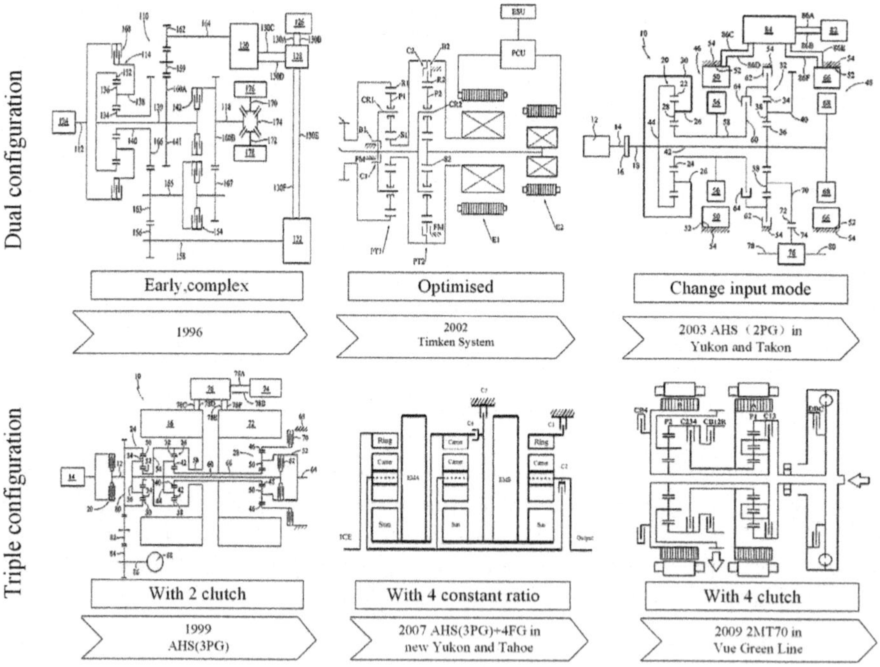

Fig. 2.48 Configurations evolution roadmap of GM corporation

Table 2.5 EVT configurations of GM corporation

Duel-planetary configuration	Invention time	Source of literature	Vehicle	Configuration	Feature
	1996	US 5558588	\	\	Original, complex
	2002	US 6478705 B1 SAE2005-01-0281	\	The Timken system	Achieve high, low two models, better than Prius
	2003	US 6527658 B2	Yukon Tahoe	AHS(2PG)	Dual-mode, suitable to SUV
Triple-planetary configuration	1999	US 5931757	\	AHS (3PG)	Two clutches, two modes
	2007	SAE 2007-01-0273	New version Yukon Tahoe	AHS (3PG) + 4FG	The same as the former, but also added a 2 clutch, two mode can be achieved 4 fixed speed
	2007	US 7238131B2	\	AHS (3PG)	Three 15 row configurations are listed
	2009	SAE 2009-01-0508, SAE 2009-01-1321	Vue Green Line	2MT70	For the first time in the configuration of the precursor, it is good to improve the economy

it could work in. So it is better than the Toyota THS system in both overall efficiency and power performance. But the AHS system always includes two or three planetary gears and several clutches and brakes, this makes it more complex, more difficult to manufacture, and its control strategy quite complicated.

2.2 Operating Principle of Planetary Hybrid System

Based on the typical EVT configurations, the general analysis model for the power-split device is built in this section, which is the core component of EVT system. Through the discussion of its speed and torque relations, the basic working principle of EVT system is obtained. Furthermore, the EVT system configuration is classified according to the characteristics of the planetary coupling device to explain the different types of features.

2.2.1 Basic Principles of Planetary Power-Split Device

Based on the lever mode for EVT parallel-series hybrid system, a static analysis mode is build, which is shown in Fig. 2.49.

where ω_A and T_A are the rotary speed and torque of MG1 respectively, ω_B, T_B stand for the rotary speed and torque of MG2, ω_i and T_i are rotary speed and torque of engine respectively, ω_o, T_o are the gear speed and torque of the planetary which connects to the drive axle.

According to the force balance of the system, the following conclusion can be drawn:

$$T_A + T_B + T_i = T_o \tag{2.31}$$

Seeking torque according to the point O, the following equation can be obtained:

$$T_i + \alpha T_A + \beta T_B = 0 \tag{2.32}$$

The relations of speed:

$$\frac{\omega_i - \omega_o}{1} = \frac{\omega_A - \omega_o}{\alpha} = \frac{\omega_B - \omega_o}{\beta} \tag{2.33}$$

Based on the Eqs. (2.31), (2.32), and (2.33), the following equation can be obtained:

$$\begin{aligned} &\omega_A = \alpha\omega_i + (1-\alpha)\omega_0 \\ &\omega_B = \beta\omega_i + (1-\beta)\omega_0 \\ &T_i = -\alpha T_A - \beta T_B \\ &T_o = (1-\alpha)T_A + (1-\beta)T_B \end{aligned} \tag{2.34}$$

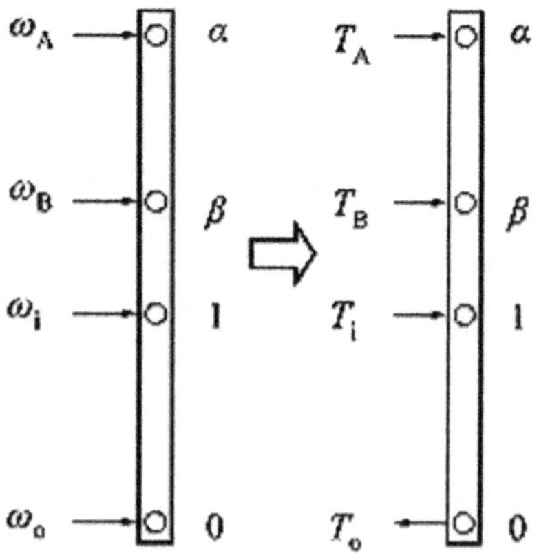

Fig. 2.49 Static characteristics analysis model of EVT hybrid system

For this type of EVT series-parallel hybrid system, the power of engine is transferred to the wheels in two paths to drive the vehicle. One is mechanical path in which the engine power is transferred to the wheels through the gear mechanism directly; the other is electric path (engine → motor MG1 → motor MG2), in this path, electric motor MG1 absorbs part of the engine power and translates them into electric power then drives motor MG2 to move the vehicle or charge the battery (SOC of the battery lower than the up limit). Obviously, the mechanical path has a higher efficiency than the second electric path for it has no second pass of energy. So, when the input/output speed ratio reach the mechanical point (that's to say one of the motors speed is 0), the efficiency of driveline is the largest.

For a input-split hybrid vehicle, Eq. (2.34) indicates that there is only one mechanical point, when $\omega_A = 0$, it can be expressed as Eq. (2.35):

$$\frac{\omega_i}{\omega_o} = \frac{\alpha - 1}{\alpha} \tag{2.35}$$

For compound-split hybrid vehicle, Eq. (2.34) indicates that there are two mechanical points, when $\omega_A = 0$ and $\omega_B = 0$, they can be expressed as Eq. (2.36):

$$\frac{\omega_i}{\omega_o} = \frac{\alpha - 1}{\alpha}, \frac{\beta - 1}{\beta} \tag{2.36}$$

2.2.2 *Classification of Planetary Device*

According to the connection location difference between the power source, vehicle and EVT, the parallel-series hybrid vehicles with EVT could be classified as Input-split Mode, Output-split Mode, Compound-split Mode and Combinations of Modes. The general model for parallel-series hybrid vehicle is shown in Fig. 2.50. Specifies the output node connected to the vehicle as the reference point (0 point), then the distance from the input to the output node is 1, the distance from the motor MG1 to the output node is α, and the distance from the motor MG2 to the output node is β. If α and β are negative, that's to say motor MG1 and MG2 are below the base point, if these values are positive they are above the base point.

(1) Input-split Mode

When $\alpha = 0$ or $\beta = 0$, it is the input-split hybrid system. Such as THS system of Prius and the AHS system in its low speed mode, as shown in Fig. 2.51.

(2) Output-split Mode

When $\alpha = 1$ or $\beta = 1$, it is the output-split hybrid system. In this mode, when the vehicle is driven the electric power always exist in the system, resulting in the lower efficiency, so this mode is not adopted so far.

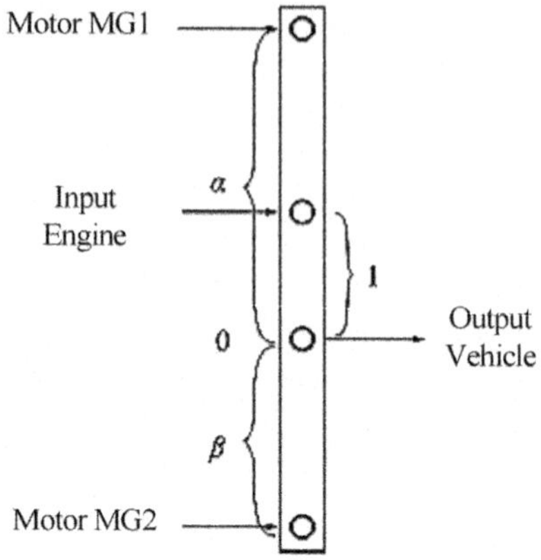

Fig. 2.50 General lever model of power-split system

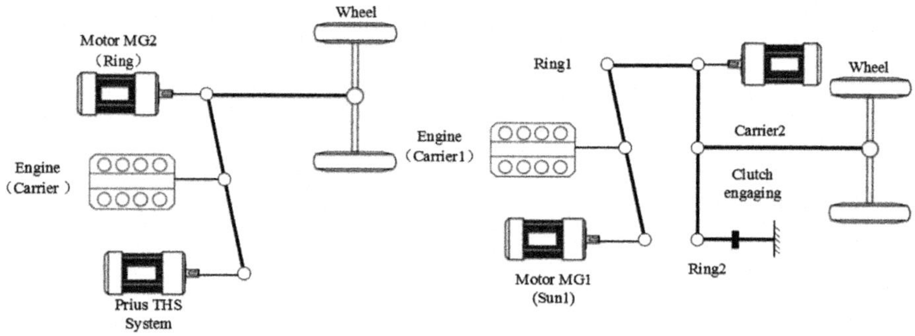

Fig. 2.51 Input-split hybrid system

(3) Compound-split Mode

When $\alpha \neq 0$ or $\beta \neq 0$, at the same time $\alpha \neq 1$ or $\beta \neq 1$, it is a four-node system like THS-II system of Toyota Lexus GS450h.

(4) Combinations of Modes

A combination of two or three of the above modes. Such as Allison EP-40/50 transmission (Fig. 2.52), when the front planetary row (Ravigneaux planetary gear set) is in the high load drive mode (when the vehicle starting-up) the clutch 1 is separated, the clutch 2 is engaged, and the power system is in input-split mode; when under the low load (cruise mode), clutch 1 is engaged, clutch 2 is separated, the system is in compound-split mode.

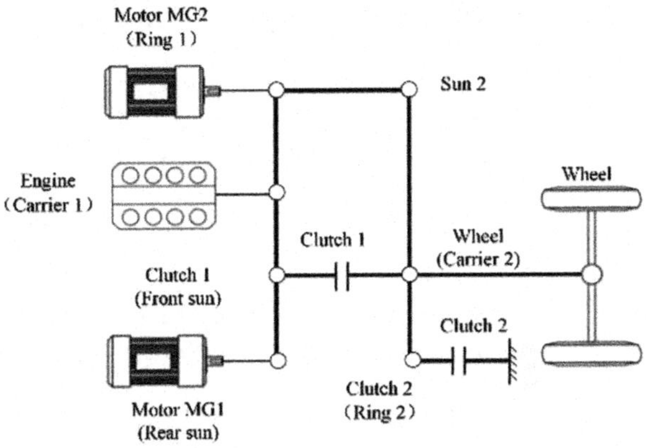

Fig. 2.52 EP-40/50 transmission lever model diagram of GM corporation

2.2.3 *Characteristics of Each Mode*

2.2.3.1 Characteristics Analysis for Input Power-Split Hybrid System

For an input power-split mode hybrid system, there should be $\beta = 0$, the configuration diagram is shown in Fig. 2.53.

Equation (2.34) can be simplified to Eq. (2.37).

$$\begin{aligned} \omega_A &= \alpha\omega_i + (1-\alpha)\omega_o \\ \omega_B &= \omega_o \\ T_i &= -\alpha T_A \\ T_o &= (1-\alpha)T_A + T_B \end{aligned} \tag{2.37}$$

In the definition $r_n = \frac{\omega_i}{\omega_o} / \frac{\alpha-1}{\alpha}$, r_n is obtained by dividing the ratio of EVT to the mechanical point. It's obvious that when $r_n = 1$, EVT reaches the mechanical point where the system has the highest efficiency. For electric motor MG1, the following equation can be obtained:

$$\begin{aligned} \frac{\omega_A}{\omega_i} &= \alpha\left(1 - \frac{1}{r_n}\right) \\ \frac{T_A}{T_i} &= \frac{-1}{\alpha} \\ P_{A\_elec/eng} &= \frac{\omega_A}{\omega_i}\frac{T_A}{T_i} = \frac{1-r_n}{r_n} \end{aligned} \tag{2.38}$$

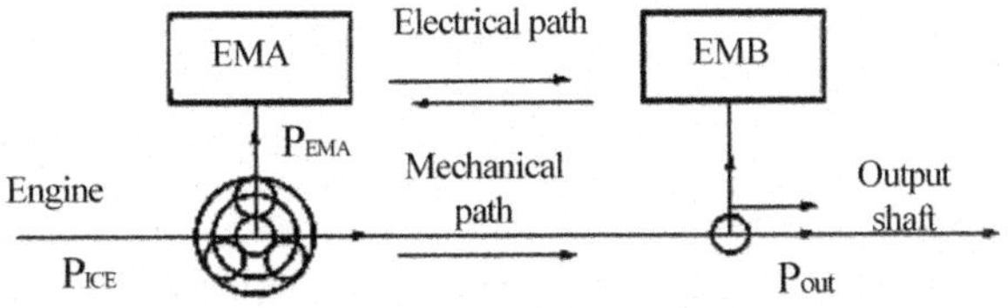

Fig. 2.53 Simplified diagram of the input power-split mode

Assuming that $T_A w_A = T_B w_B$ (the battery does not work), there is Eq. (2.39) for electric motor MG2.

$$\begin{aligned} \frac{\omega_B}{\omega_i} &= \frac{\alpha}{\alpha-1} \cdot \frac{1}{r_n} \\ \frac{T_B}{T_i} &= \frac{\alpha-1}{\alpha}(r_n - 1) \\ P_{B\_elec/eng} &= \frac{\omega_B}{\omega_i}\frac{T_B}{T_i} = -\frac{\omega_A}{\omega_i}\frac{T_A}{T_i} = \frac{r_n-1}{r_n} \end{aligned} \tag{2.39}$$

According to Eqs. (2.38) and (2.39), the static characteristics curves of an input power-split hybrid system are shown as Fig. 2.54. Where, horizontal axis stands for r_n, the vertical axis is the ratio of motor MG1 and MG2 relative to engine speed, torque and power.

From Fig. 2.54, when vehicle velocity is zero, $r_n \to \infty$, along with the increase of vehicle velocity, r_n is reducing to 1 and finally smaller than 1, that leads the power ratio increase rapidly and the system efficiency reduces rapidly. In addition, with the increase of planetary structural parameter ($K_i = R_i/S_i$), α also slowly increases, rotary speed of MG1 and torque of MG2 that relative to the engine will larger and larger. There are also some references define structural parameter as $\rho = S_i/R_i$, this will have no influence on the statics analysis results [23, 24].

Above all, the conclusion is that this kind of hybrid system need a drive motor MG2 with a large torque and MG1 with wide speed range to choose a mechanical point speed ratio to coordinate the drive-line efficiency and motor size.

2.2.3.2 Characteristics Analysis of Compound Power-Split Hybrid System

First, configuration diagram of compound power-split hybrid system is shown in Fig. 2.55.

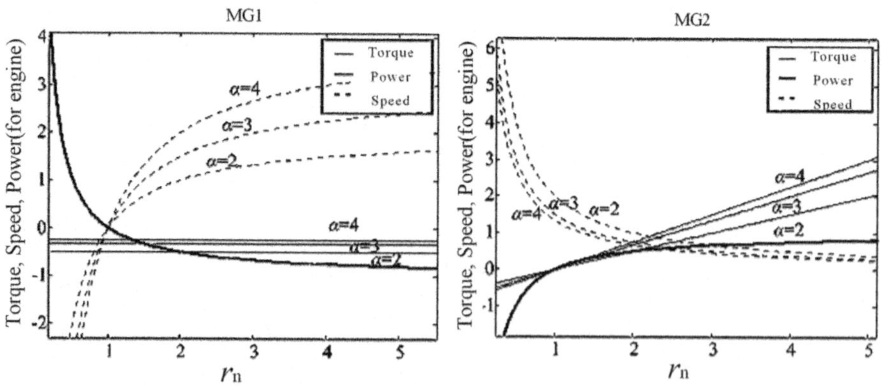

Fig. 2.54 Input power-split mode static characteristics curves of hybrid power system

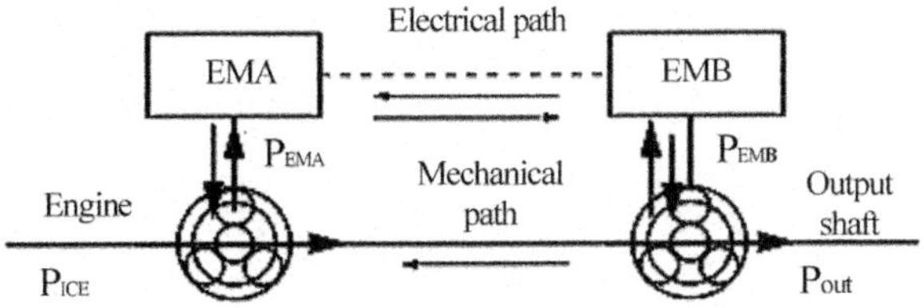

Fig. 2.55 Compound power-split configuration diagram

The static Eq. (2.40) of the compound-split hybrid system can be obtained by Eq. (2.34):

$$\begin{aligned}
\frac{\omega_A}{\omega_i} &= \alpha\left(1-\frac{1}{r_n}\right)\\
\frac{\omega_B}{\omega_i} &= \beta+\frac{\alpha(1-\beta)}{\alpha-1}\cdot\frac{1}{r_n}\\
\frac{T_A}{T_i} &= \frac{\alpha-\beta+\beta(\alpha-1)(r_n-1)}{-\alpha(\alpha-\beta)}\\
\frac{T_B}{T_i} &= \frac{(\alpha-1)(r_n-1)}{(\alpha-\beta)}
\end{aligned} \tag{2.40}$$

If the range of EVT ratio is defined as Φ, then we can get Φ from Eq. (2.41):

$$\phi = \frac{\alpha(\beta-1)}{\beta(\alpha-1)} \tag{2.41}$$

Next, choosing two examples which are the wide ratio range in which $\alpha = 2$, $\beta = 0.1$, $\Phi = 4$ and the narrow ratio range in which, $\alpha = 4$, $\beta = 0.2$, $\Phi = 2$, to explain the static characteristics of compound-split hybrid system. Taking α and β into Eq. (2.40), the static characteristics curves are shown as below:

As can be seen from the static characteristics curves in Fig. 2.56, when velocity of vehicle is zero, $r_n \to \infty$, with the increasing of vehicle velocity r_n will reach the first mechanical point (while $\alpha = 2$, $\beta = 1$, the first mechanical point is $r_n = 2$, while $\alpha = 4$, $\beta = 2$, the first mechanical point is $r_n = 1.5$), between the two mechanical points, the ratio of electric power is smaller, the system efficiency is higher. Similarly, when the ratio of the EVT is greater than the ratio of the second mechanical point, with the increasing of vehicle velocity, the system efficiency will decrease for the reducing of electric power ratio. If the power of motor is negative, it's generating, if the power of motor is positive, it's driving.

The conclusion is that the dual-mode compound power-split hybrid system has advantages to be applied to high speed cruising cycles (in low load); its disadvantage is not suitable for low speed starting (in high load). Especially when the vehicle is starting but engine speed is high, the ratio of electric power in electric path is much bigger than 1, which leads to lower overall efficiency of the EVT.

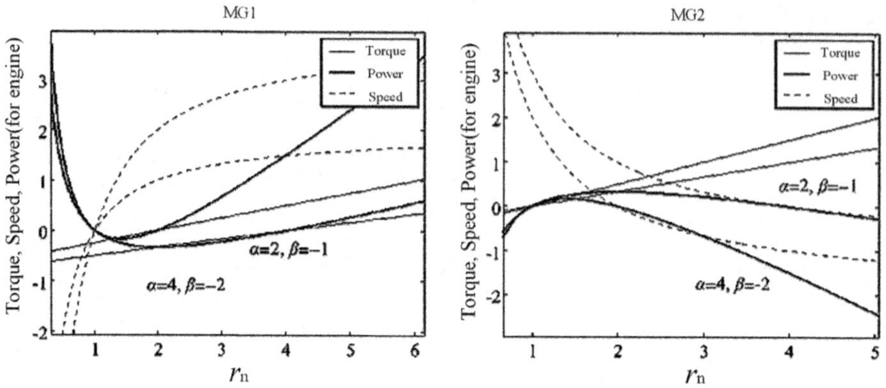

Fig. 2.56 Compound-split static characteristics curves of hybrid power system, when $\alpha = 2$, $\beta = 1$ and $\alpha = 4$, $\beta = 2$

2.2.3.3 Characteristics Analyses of Combinations of Input/Compound Power-Split Hybrid System

Taking the AHS combination power-split mode hybrid system used in America GM Yukon vehicle as example. AHS system is in input power-split mode in low speed (high load), and in compound-split mode in high speed (low load). The static characteristics analysis on AHS system in these two working conditions would be conducted next.

At low speed, the lever model of AHS static characteristics analysis is shown in Fig. 2.57.

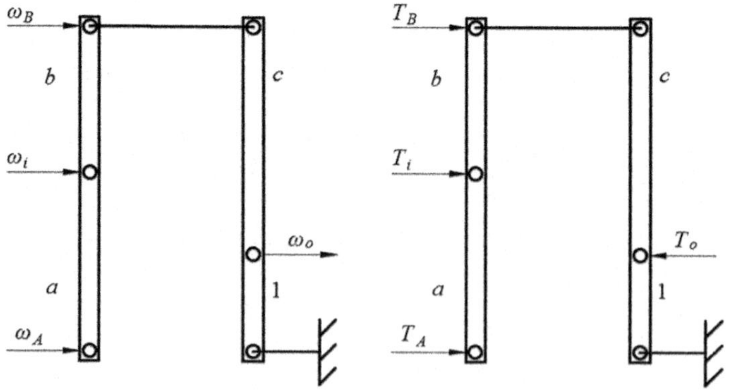

Fig. 2.57 General lever analysis model of AHS system

According to the load balance and relations of speed, the following equation can be obtained:

$$\begin{aligned}&(T_A + T_B + T_i)(c+1) = T_o\\&(a+b)T_A + bT_i = 0\\&\omega_B = (c+1)\omega_o\\&\omega_i = \frac{b}{a+b}\omega_A + \frac{a}{a+b}\omega_B\end{aligned} \tag{2.42}$$

If the ratio of EVT is defined as $R_n \equiv \frac{\omega_i}{\omega_o}$, then Eq. (2.42) can be transferred to Eq. (2.43).

$$\begin{aligned}\frac{\omega_A}{\omega_i} &= \frac{(a+b)}{b} - \frac{a}{b}\frac{(c+1)}{R_n}\\\frac{\omega_B}{\omega_i} &= \frac{(c+1)}{R_n}\\\frac{T_A}{T_i} &= \frac{-b^2}{a+b}(R_n - 1)\\\frac{T_B}{T_i} &= \frac{\frac{b^2}{a+b}\left(\frac{a+b}{b} - \frac{a}{b}\frac{(c+1)}{R_n}\right)}{\frac{(c+1)}{R_n}}\end{aligned} \tag{2.43}$$

At high speed, AHS system is a compound power-split hybrid system consist of two planetary gears. Analyzing its static characteristics need calculate α and β in this system, which is shown as Fig. 2.58.

It's obvious that when the two planetary gears make a compound gear set, n_{R2} and n_{S2} in the left diagram in Fig. 2.58 could get from the speed relations between connect points, which is $n_{R2'} = n_{R2}/(n_{S1} + n_{R1})$, $n_{S2'} = n_{S2}/(n_{S1} + n_{R1})$. To make qualitative analysis, assuming that $n_{R1} = 9$, $n_{S2} = 3$, $n_{R2} = 8$ and $n_{S2} = 4$, correspond to the connect relations of power sources and nodes we get that: $\alpha = 4/5$, $\beta = 8/5$.

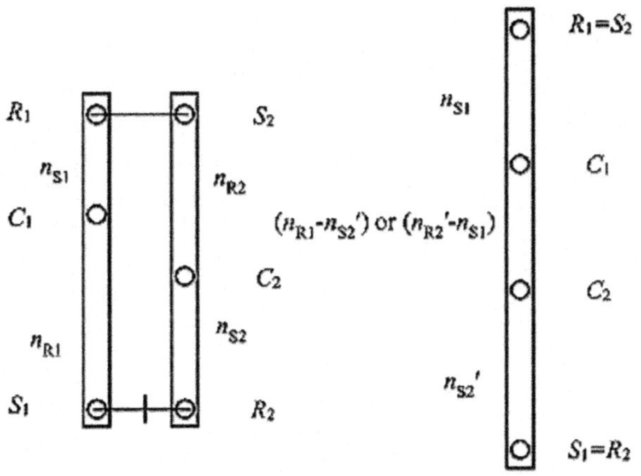

Fig. 2.58 Lever analysis model of compound power-split hybrid system

Similarly, use R_n to take place of r_n in Eqs. (2.42) and (2.44) is shown as below:

$$\begin{aligned}\frac{\omega_A}{\omega_i} &= \alpha + \frac{(1-\alpha)}{R_n}\\ \frac{\omega_B}{\omega_i} &= \beta + \frac{(1-\beta)}{R_n}\\ \frac{T_A}{T_i} &= \frac{\alpha-\beta+\beta(\alpha-1)\left(\frac{\alpha}{\alpha-1}R_n-1\right)}{-\alpha(\alpha-\beta)}\\ \frac{T_B}{T_i} &= \frac{(\alpha-1)\left(\frac{\alpha}{\alpha-1}R_n-1\right)}{\alpha-\beta}\end{aligned} \quad (2.44)$$

The following conclusions can be obtained by substituting $\alpha = 4/5$ and $\beta = 8/5$ into Eqs. (2.43) and (2.44) respectively: for low speed condition, mechanical point is $R_n = 9/4$, for mid-high speed condition, the two mechanical points are $R_{n1} = 9/4$ and $R_{n2} = 3/8$. The statics characteristics curves of this double-mode system (AHS) is conducted from Eqs. (2.43) and (2.44), which are shown as Fig. 2.59.

When the rotary speed of motor MG1 is 0, that is at the mechanical point of input power-split and low speed mode and high speed mode convert to each other. As can be seen from Fig. 2.59, during the process R_n changing from ∞ to 9/4 (in the low speed condition and high load) it takes full advantages of the input-split model: to maintain the power of the vehicle; to avoid the shortcomings of the inefficient transmission system of the compound power-split hybrid system at low speed. During the process R_n changing from 9/4 to 3/8(in the high speed condition and low load), it leverages the advantages of the compound power-split mode EVT to drive the vehicle and improve the fuel economy.

2.2.4 Summary

The basic principle of the planetary hybrid power-split system is introduced through the static characteristic analysis of model of the hybrid system. Then, based on the

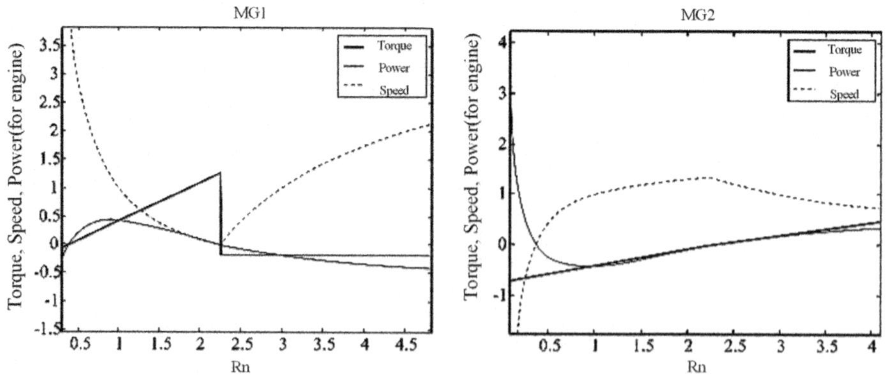

Fig. 2.59 Static characteristics curves of dual-mode combined hybrid system (AHS)

above power splitting modes of the EVT hybrid vehicle, the general model of the series-parallel hybrid system is established by the lever model analysis method. Finally, the characteristics of various configurations and the working characteristics of the system are introduced.

2.3 Differential-Based Hybrid System

In order to provide a new kind of power-split device for series-parallel hybrid vehicle, at the same time, to break the shackles of THS power-split device patent of Toyota corporation, through the research and analysis of the structure and working principle of conventional bevel gear differential, as well as the cooperation to PSD of THS system in Prius hybrid vehicle of Toyota corporation, we obtain the feasibility of using traditional differential as the power-split device for series-parallel hybrid vehicle. This new type of system is called Differential-based Hybrid System (DHS) [25]. In this Section, DHS system is compared with THS system in two aspects: the connection and the mechanical properties. Also, the feasibility of this system is proved theoretically.

2.3.1 Comparison of Powertrain Connection

The connection and the mechanical properties of THS system in Prius hybrid vehicle have already been described in detail in the previous article. So, in this part, the connection of DHS system is mainly introduced.

As shown in Fig. 2.60, differential is regarded as the power-split device for hybrid vehicle. Compared with the structure of system I, system II is connected by a single speed ratio gear between the generator and the left axle. The speed ratio can be determined in accordance with the speed relationship between the rotating speed or the torque relationship between the engines.

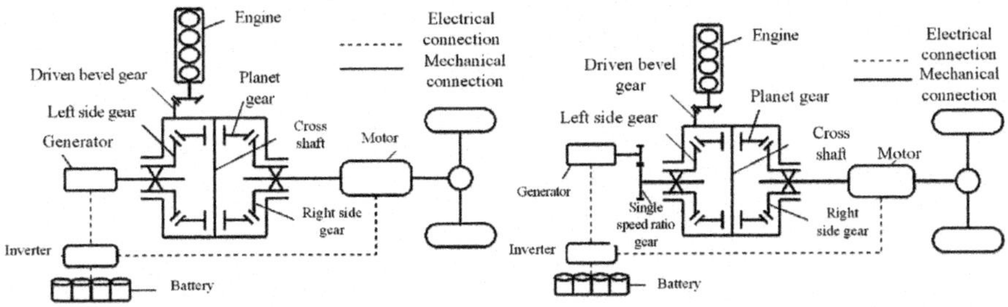

Fig. 2.60 Differential-based hybrid system

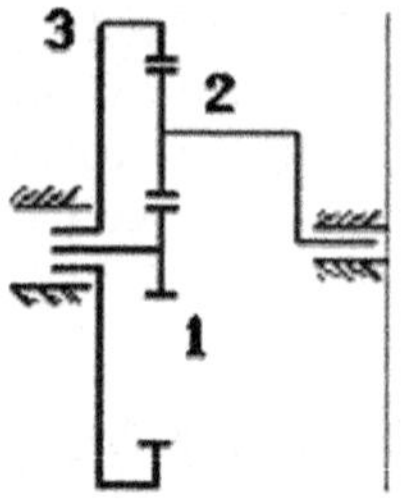

Fig. 2.61 2 K-H planetary structure

The connection relationship of the three-ends of differential and power source is: left side gear is connected with generator; driven bevel gear is connected with engine; right side gear is connected with motor.

Because the differential power-split device and THS-PSD both belong to 2 K-H structure (as shown in Fig. 2.61), i.e. the two forms are the same in structural connection, it is unnecessary to redesign the overall structure, which proves the feasibility in connection. The traditional differential power-split device changes the three-end speed and torque ratio of DHS, but it can be adjusted with a single ratio gear. So, besides the two structural forms described above, the single speed ratio gear may also be arranged between the engine and the driven gear or the right side gear of DHS and drive motor.

2.3.2 *Relationship Between Speed and Torque*

In THS and DHS system, speed relationship between the power sources can both be expressed as:

$$k_1 w_g = k w_m = (1 + k) w_e \tag{2.45}$$

For THS system, $k = z_r/z_s = 2.6$, $k_1 = 1$. For DHS system, $k = 1$, $k_1 = 1$.

As shown in Fig. 2.62, according to Eq. (2.45) and Fig. 2.60, the relationship between the speed of each power source of powertrain structure and the vehicle velocity can be obtained. Among them, the relationship between the speed of the three main power components (generator, drive motor and engine) is:

$$w_y + w_m = 2 \cdot w_e \tag{2.46}$$

The relationship between the driving motor speed and vehicle velocity is:

$$w_m = 35.11 \cdot v \tag{2.47}$$

As shown in Fig. 2.63, according to Eq. (2.46) and Eq. (2.47), the restrictive relationship between the power sources can be obtained.

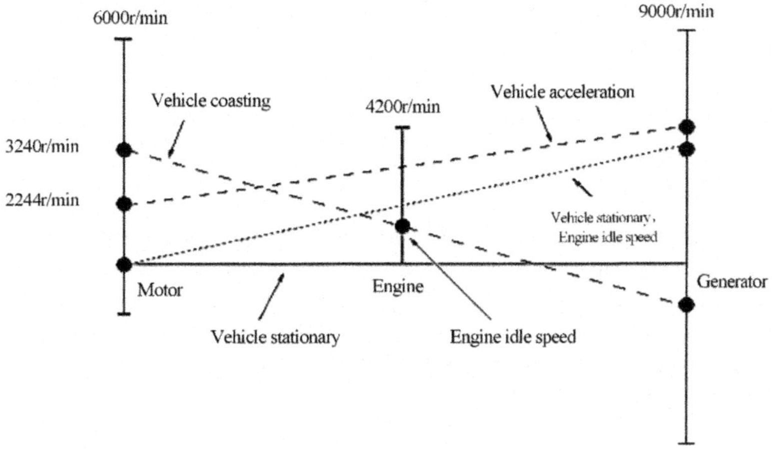

Fig. 2.62 Speed relationships of powertrain components

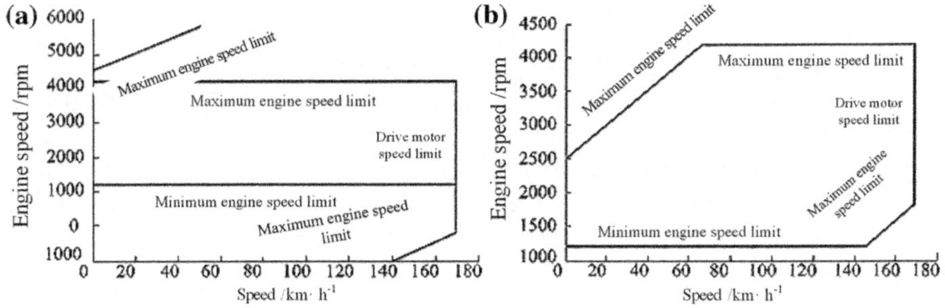

Fig. 2.63 Engine speed limited by powertrain components

Substituting the Eq. (2.45) into the energy conservation equation, the torque relationship of DHS system can be obtained [26–28]:

$$\begin{cases} T_{left\_gear} = T_{right\_gear} = kT_e/(1+k) \\ T_g/k_1 = T_e/(1+k) \end{cases} \tag{2.48}$$

where T_{right_gear} is torque of DHS right side gear. For THS system, $k = z_r/z_s = 2.6$, $k_1 = 1$. For DHS system, $k = 1$, $k_1 = 1$.

Since the size of the motor is proportional to the maximum torque, the first configuration requires a slightly larger generator than the Prius. If the gear ratio of the second configuration is 3.6/2, according to the torque relationship, the size of its generator can be the same with the generator in Prius.

It can be seen from the speed relationship, compared with the THS-PSD, the DHS not only does not limit the best performance of the engine when the speed of generator is high during vehicle is accelerating, it also provides more convenience for powertrain assembly matching. This shows that the differential is feasible as a power coupling device in terms of structure.

2.3.3 Summary

By deeply analyzing the structure and working principle of traditional differential and contrasting it with mature products of Prius THS, it can be seen that the traditional differential as a dynamic coupling device can fully meet the dynamic coupling requirements of hybrid vehicles. Therefore, the differential as a power-split power coupling system for HEV is not only feasible, but also can be a new dynamic coupling system.

2.4 Summary

In this chapter, the typical configurations of planetary hybrid system in the market, including the THS system of Toyota and the AHS system of GM, are introduced. On this basis, the working principle and analysis method of planetary hybrid system are discussed. Characteristic analyses of each mode, including input power-split, compound power-split and combinations of input/compound power-splits hybrid system, are conducted. In the end, the connection and mechanical properties of DHS systems are analyzed, and the feasibility of this DHS is discussed.

References

1. Harada O, Shibata Y, Yamaguchi K. Power output apparatus and method of controlling the same: U.S. Patent 6,067,801. 30 May 2000.
2. Benford HL, Leising MB. The lever analogy: a new tool in transmission analysis. SAE Technical Paper; 1981.
3. Koide T, Matsui H, Nada M. Hybrid vehicle drive system having two motor/generator units and engine starting means: U.S. Patent 5934395. 1999-8-10.
4. Hata H, Matsui H, Ibaraki R. Power output device, hybrid vehicle and method of controlling them. China Patent CN1336879. 20 Feb 2002.
5. Oba H, Yamanaka A, Katsuta H, Kamichi K. Development of a hybrid powertrain system using CVT in a minivan. SAE Technical Paper. 2002-01-0991.
6. Kikuchi Y. Power output unit, automobile having same, and method of controlling power output unit. China Patent: CN101010499. 1 Aug 2007.
7. Hermance D, Abe S. Hybrid vehicles: lessons learned and future prospects. 2006.
8. Muta K, Yamazaki M, Tokieda J. Development of new generation hybrid system THS II—Drastic improvement of power performance and fuel economy. SAE Technical paper: 2004-01-0064.
9. Kimura A, Ando I, Itagaki K. Development of hybrid system for SUV. SAE Technical Paper: 2005-01-0273.

10. Kamichi K, Okasaka K, Tomatsuri M, Matsubara T, Kaya Y, Asada H. Hybrid system development for a high-performance rear drive vehicle. SAE Technical Paper: 2006-01-1338.
11. Grewe TM, Conlon BM, Holmes AG. Defining the general motors 2-mode hybrid transmission. SAE Technical Paper: 2007-01-0273.
12. Schmidt MR. Two-mode, input-split, parallel, hybrid transmission: U.S. Patent 5558588[P]. 24 Sep 1996.
13. Holmes AG, Schmidt MR. Hybrid electric powertrain including a two-mode electrically variable transmission: U.S. Patent 6478705[P].12 Nov 2002.
14. Ai X, Mohr T. An electro-mechanical infinitely variable speed transmission. SAE Technical Paper: 2004-01-0354.
15. Holmes AG, Klemen D, Schmidt MR. Electrically variable transmission with selective input split, compound split, neutral and reverse modes: U.S. Patent 6527658[P]. 4 Mar 2003.
16. Tamai G, Reeves S, Grewe TH. Truck Utility & Functionality in the GM 2-Mode Hybrid. SAE Technical Paper: 2010-01-0826.
17. Schmidt MR. Two-mode, compound-split electro-mechanical vehicular transmission: U.S. Patent 5931757[P]. 3 Aug 1999.
18. Kaehler B. Design criteria, methods of analysis, and evaluation of power split transmissions explained through a two-mode hybrid application. Aachener Kolloquium Fahrzeug- und Motorentechnik 2007.
19. Hendrickson J, Holmes A, Freiman D. General motors front wheel drive two-mode hybrid transmission. Training. 2009;2014:05–22.
20. Meisel J. An analytic foundation for the two-mode hybrid-electric powertrain with a comparison to the single-mode Toyota Prius THS-II powertrain. Training. 2009;2013:09–30.
21. Schmidt MR. Electro-mechanical powertrain: U.S. Patent 5935035. 10 Aug 1999.
22. Bucknor NK, Hendrickson JD, Raghavan M. Electrically variable transmission having three planetary gear sets and three fixed interconnections: U.S. Patent 7238131[P]. 3 Jul 2007.
23. Peizhi T, Bo Y. The calculation of the planetary gear transmission ratio and torque and its power flow chart. Heavy Mach Sci Technol. 2006;19(2):18–20.
24. Xiao M, Sun Y. Comparison between the methods about calculating internal torques and external torques in a planetary transmission. Mach Des Manuf. 2003;(2):114–5.
25. Zeng X, Wang Q, Weihua W, Yuanbin Y, Song D, Jin L, Zhu Q, Yu Y, Wang P. Power coupler for mixed power automobile[P]. China Patent:CN101020410. 22 Aug 2007.
26. Xi B, Aimin D, Feng Xue. Theoretical study and simulation on planetary gear set for hybrid electric vehicle. Automotive Eng. 2006;9:834–9.
27. Chu L. Study on the parametric optimization for a parallel hybrid electric vehicle powertrain. SAE Technical Paper, 2000.
28. Liu, JM; Peng, H; Filipi, Z. Modeling and analysis of the Toyota hybrid system[A]. IEEE/ASME International Conference on Advanced Intelligent Mechatronics[C], 2005.

Chapter 3
Simulation of Hybrid Power-Split System

Due to the complexity of hybrid system, many issues should be considered in the research and development of the hybrid vehicle, such as the choice of components, the determination of the best configuration, the formulation and optimization of vehicle control strategy, etc. Through the simulation technology, reasonable and effective computer models for hybrid vehicle can be established before the technical scheme is determined. Then according to the simulation result, design parameters for each candidate subsystem and hybrid vehicle configuration can be provided. Thus the trail production work of original candidate subsystem and different vehicle configuration can be simplified. After confirming the configuration of hybrid vehicle and each subsystem, the energy distribution strategy can be established quickly and optimized by simulation software.

In a word, as an important technology in the research and development of hybrid vehicle, simulation technology can not only adjust design scheme and optimize design parameters flexibly, but also reduce research costs and shorten development cycle. Many kinds of simulation software for hybrid electric vehicle are used nowadays. Table 3.1 gives the general situation of the simulation software applicable to hybrid vehicle.

Most of the hybrid vehicle simulation softwares are developed in the United States. The development mode mainly has the first development and the second development based on MATLAB/Simulink. Some kinds of software adopt reverse simulation, which calculate the conditions of various components according to the driving cycles. Others adopt forward simulation, which make vehicle components follow the road conditions according to driver's behavior or driving cycles.

The most widely used softwares mentioned above in the hybrid vehicle simulation includes ADVISOR, CRUISE and AMESim. Combined with simulation examples, some applications in the actual project of the three simulation platforms are introduced in the following sections.

X. Zeng and J. Wang, *Analysis and Design of the Power-Split Device for Hybrid Systems*, https://doi.org/10.1007/978-981-10-4272-0_3

Table 3.1 Simulation software for hybrid vehicle

Software name	Application range	Developer	Simulation method
SIMPLEV	EV Series HEV	Idaho National Engineering Laboratory	Reverse simulation
HVEC	EV Series HEV	Lawrence Livermore National Engineering Laboratory	Reverse simulation
V-ELph	Series HEV Parallel HEV	Texas A&M University	Reverse simulation
ADVISOR	EV/HEV Fuel-cell vehicle Conventional vehicle	NREL	Reverse simulation
CRUISE	EV/HEV Fuel-cell vehicle	AVL	Forward simulation
PSAT	EV/HEV Fuel-cell vehicle Conventional vehicle	USACAR/NASA/EPA and DOE	Forward simulation
HEVsim	HEV	Opal-RT	\
AMESim	EV HEV Fuel-cell vehicle Conventional vehicle	LMS Imagine Lab AMESim	Forward simulation

3.1 Hybrid System Development Based on ADVISOR

ADVISOR is a kind of advanced modeling and simulation platform software for automotive system developed by the US department of energy (US DOE) in November 1994, which aims to manage the sub-contracting program of hybrid system, and it was officially named ADVISOR in January 1998. ADVIOSR is a simulation platform based on interactive and graphical modeling Simulink environment.

ADVISOR is based on the modular programming thought and the code is completely public. The related models are all quasi-static models based on experience. It is mainly used for rapid analysis of power performance and fuel economy performance of traditional vehicles, electric vehicles and hybrid vehicles. In addition, ADVISOR is able to solve problems such as whether the vehicle velocity can follow the required cycle velocity, how to make SOC fluctuate reasonably in the whole cycle, and how to distribute the torque of engine. In view of the above advantages listed about ADVISOR, it has been widely used.

The listed hybrid configuration in ADVISOR is very limited, mainly including some configurations used internationally, such as parallel hybrid configuration, series hybrid configuration and Toyota Prius hybrid configuration with planetary. Even for a certain hybrid configuration, there are also many kinds of sub-configuration. For example, there exists different kinds of parallel hybrid configurations according to the different positions of motor. For the planetary series-parallel hybrid system, due to the different ways to connect the planetary mechanism and the power source, there are also many possible types of hybrid configuration.

In this section, the power transmission characteristics and control algorithm development of the DHS system mentioned in the previous chapter are analyzed. According to the analysis result, the DHS system is modeled in Simulink and the model is embedded in ADVISOR, then the system performance can be tested and verified.

3.1.1 Dynamic Modeling of Differential-based Planetary System

1. Fundamental dynamic equation

Power-split device in DHS mentioned above is designed based on the symmetrical differential. Symmetrical differential is a special form of planetary gear structure. It belongs to the 2K-H structure of the planetary gear train. Therefore, the planetary gears of the typical 2K-H structure are analyzed firstly. The structure is shown in Fig. 2.61, where 1 is the sun gear, 2 is the planetary carrier, 3 is the ring gear. The speed relationship of the three is as follows:

$$\frac{\omega_1 - \omega_2}{\omega_3 - \omega_2} = -\frac{z_3}{z_1} \tag{3.1}$$

where $\omega_1, \omega_2, \omega_3$ is the speed of the sun gear, the planetary carrier and the ring gear.

$k = z_3/z_1$ is the gear ratio of the ring gear and the sun gear, so Eq. (3.1) can be simplified as follows:

$$\omega_1 = (1 + k)\omega_2 - k\omega_3 \tag{3.2}$$

Symmetrical bevel gear differential applied to the hybrid system is shown in Fig. 3.1 where sun gear 1 corresponds to the gear of left half-shaft and connects with generator; planetary carrier 2 corresponds to the driven bevel gear and is driven by engine; the ring gear corresponds to the gear of the right half-shaft and outputs torque to the wheel.

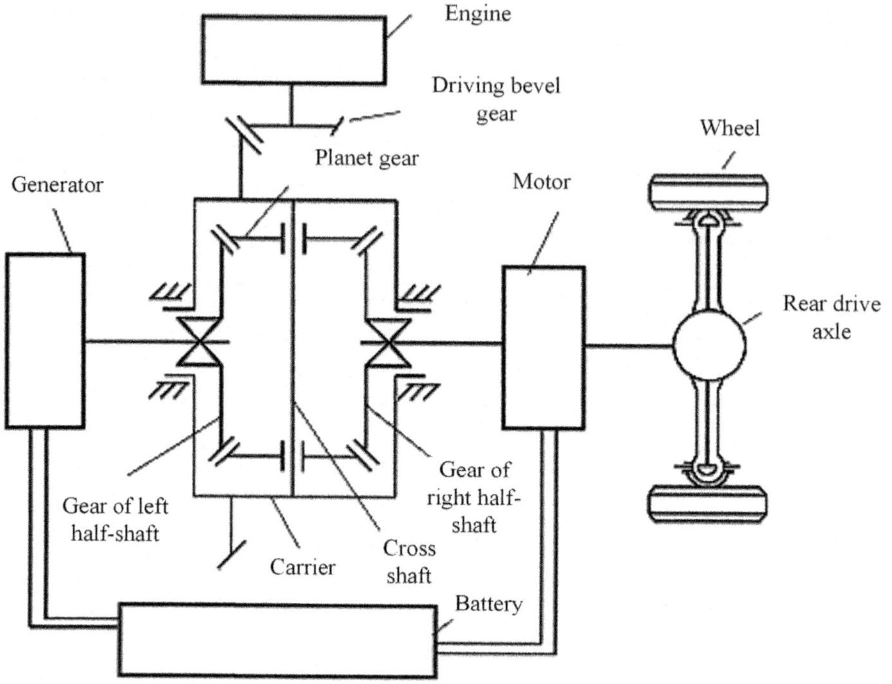

Fig. 3.1 Symmetrical bevel gear differential applied to the hybrid system

The symmetrical bevel gear differential is also a kind of 2K-H planetary gear, but the planetary carrier and the sun gear are distributed vertically, which can also meet the speed relationship of 2K-H planetary gear but $k = 1$. Thus:

$$\omega_1 + \omega_3 = 2\omega_2 \tag{3.3}$$

2. Analysis of power transmission characteristics of DHS

For the convenience of study, it is assumed that the transmission ratio of the engine to the planetary carrier is 1. So the speed relationship can be expressed as follows:

$$\omega_1 = \omega_g, \omega_2 = \omega_e, \omega_3 = \omega_m \tag{3.4}$$

Each port corresponds the torque of each power source, namely:

$$T_1 = T_g, T_2 = T_e, T_3 = T_L - T_m \tag{3.5}$$

where $\omega_g, \omega_e, \omega_m$ is the speed of generator, engine and motor; T_1, T_2, T_3 is the torque of the gear of the left half-shaft, driven bevel gear and the right half-shaft; T_g, T_e, T_m is the generator charging torque, the torque transmitted from engine to driven bevel gear and output torque. For the ring gear, the output is the vector sum of the two torques:

$$T_3 + T_m = T_L \tag{3.6}$$

where T_L is the load torque. If the internal friction loss is ignored, the feature of symmetrical bevel gear differential is: the speed can be different but the torque is the same on half-shaft. That is:

$$T_3 = T_1 \Rightarrow T_g = T_L - T_m \tag{3.7}$$

According to the power balance relationship:

$$T_e\omega_e = (T_L - T_m)\omega_m + T_g\omega_g \tag{3.8}$$

and:

$$2\omega_e = \omega_m + \omega_g \tag{3.9}$$

So the torque of two output shafts is:

$$T_g = \frac{T_e}{2} \tag{3.10}$$

That is, the charging torque of generator is always half of the engine torque, so the motor torque is:

$$T_m = T_L - \frac{T_e}{2} \tag{3.11}$$

From the above analysis, the speed of optimum engine operating point can be determined by adjusting the rotational speed of the generator from the known vehicle velocity (the rotational speed of the motor connected to the wheel with fixed speed ratio); For the torque relationship, when the engine works on the optimum load point, the vehicle velocity and load requirement can be met by regulating load of generator and motor. So the function of electric continuous variable transmission (ECVT) is achieved.

The specific adjustment relationship of output power is as follows: by controlling the charging torque of the generator according to the Eqs. (3.10) and (3.11), it can be determined that the torque requirement of the engine is twice as high as that of the generator. The energy stored in generator is used to meet the road load torque through motor assistance.

3.1.2 Modeling and Control Strategy of DHS

1. Modeling of DHS

Based on the connection relationship and power transmission characteristics of the DHS, the Simulink model is established using the backward modeling style of ADVISOR. The input variables are the power demand, the actual engine input power and the actual generator input power; the output variables are the required power of engine, the required power of generator, the required power of motor and actual output power.

2. Control strategy of DHS

The control strategy of DHS is mainly based on the power transmission characteristics. Under the condition of known required torque T_L, engine is controlled to work at an efficient point (area). Half of the engine output torque is used to generate electricity, the other torque is completed by motor.

The specific method is shown in Fig. 3.2, where *battery charge demand* is the charging demand of battery; *max engine efficiency point* is the optimum engine working point; *pgr_pwr_out_r* is the required power of ring gear; *fc_trq_request* is the required torque of engine; *fc_spd_request* is the required speed of engine.

The input are the required power and torque of power source and SOC. And the balance power of SOC is solved by the fitting equation.

$$P_{ch} = \frac{k \times (SOC_{Tgt} - SOC) \times V_b}{\eta_m} \tag{3.12}$$

where P_{ch} is the required charging power of battery; k is the fitting coefficient; SOC_{Tgt} is the target SOC; V_b is battery voltage; η_m is motor efficient.

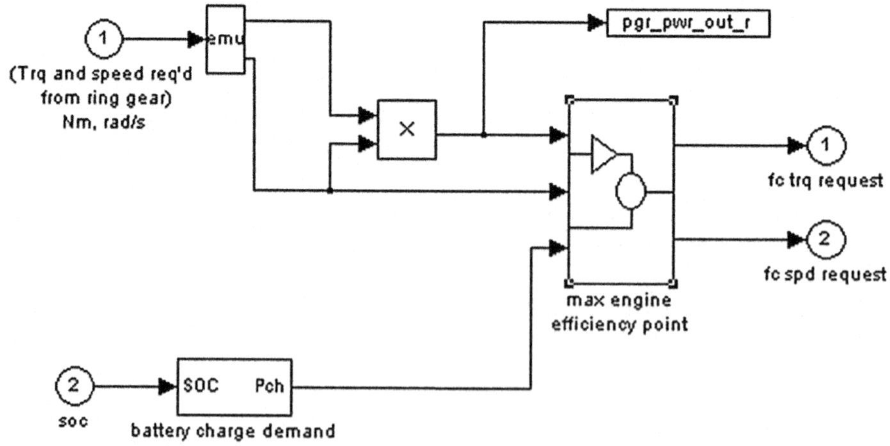

Fig. 3.2 General scheme of control strategy based on differential-based hybrid system

Input charging power P_{ch}, required power $P_r = \omega_L \times T_L$, required speed ω_L (or speed of motor, ring gear) to the control module of engine on/off and optimum operating point. Output the optimum speed point and torque point of engine. The key module consists of two parts.

(1) Engine on/off control

It is judged whether the motor can drive independently by the signals of required power P_r, current vehicle velocity V, SOC and engine temperature T_{fc}. If the following conditions are met at the same time:

$$\begin{cases} P_r < P_{r\_set} \\ V < V_{set} \\ SOC > SOC_{set} \\ T_{fc} > T_{set} \\ Time_{off} > time_{set} \end{cases} \tag{3.13}$$

The engine is allowed to shut down, i.e., the motor drives independently. When any of the above conditions is not satisfied, the engine is engaged.

(2) Optimum engine operating point control

It is judged whether the engine works at the optimum speed point by the sum of required power and charging power. The optimum speed point is determined by the engine efficiency curve (engine performance curve).

Taking into account the moment of inertia of engine, the optimum speed point is also limited by the speed relationship of gear mechanism of DHS.

$$\omega_e = \frac{\omega_g}{(1+k)} + \frac{k}{(1+k)}\omega_m \tag{3.14}$$

The motor speed is known (determined by the velocity in the driving cycle). When the generator reaches the setting maximum speed, the engine speed can increase but the speed should be less than the maximum engine speed. When the generator reaches the minimum speed (reverse), the engine speed can decrease but the speed should be greater than the idling speed. Then the optimum torque value of engine is calculated according to the engine required power and it is limited by the maximum and minimum torque values of the engine.

(3) Vehicle simulation platform

Integrate the above-mentioned differential-based coupling model and control strategy into ADVISOR to establish the vehicle simulation platform. The top model is shown in Fig. 3.3.

In the above figure, the vehicle simulation model mainly includes the following modules: driving cycle, vehicle, wheel and axle, final drive, *fdPSET*, fuel converter, motor/controller, generator/controller, exhaust system, power bus and energy storage.

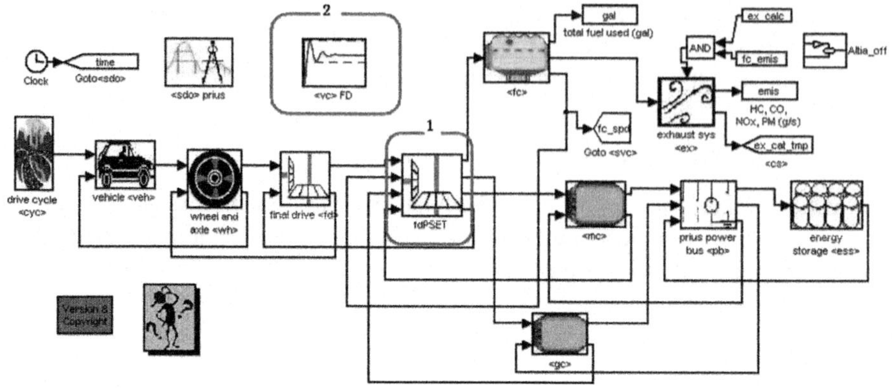

Fig. 3.3 Top model of hybrid vehicle

Mark 1 in above figure is torque coupler module, mark 2 is control strategy module. Based on the simulation platform, the vehicle parameters can be set to analyze how the differential-based coupling device control the speed and torque of each power source. The transmission can be removed to achieve the function of ECVT.

3.1.3 Vehicle Performance Simulation

In order to illustrate the transmission characteristics of differential-based coupling device, the vehicle parameters are set based on the Prius in ADVISOR. The main vehicle parameters are shown in Table 3.2.

1. Fuel consumption simulation analysis

The fuel consumption simulation result is 4.8 L/100 km under NEDC, as shown in Fig. 3.4. 30–40% fuel can be saved compared to the traditional vehicle (the fuel consumption is 7–8 L/100 km). The actual vehicle velocity follows cycle vehicle velocity well and SOC finally maintains near the target value 0.6.

2. Torque relationship simulation analysis

The actual output torque of power source corresponding to the three-terminal of differential-based coupling device are shown in Fig. 3.5. The torque of each power source is controlled normally. The engine is basically controlled to operate in the area of optimum torque and it works smoothly.

In the acceleration process, the insufficient power is provided by the motor. The generator charging torque is directly proportional to the engine output torque and the ratio is 50%, as shown in Fig. 3.6.

Table 3.2 Vehicle parameters of differential-based coupling hybrid vehicle

Vehicle parameters	Vehicle mass	1386 kg
	Tire radius	0.287 m
	Air resistance coefficient	0.30
	Frontal area	1.746 m^2
	Rolling resistance coefficient	0.009
Engine parameters	Maximum power/speed	43 kW/4000 rpm
	Maximum torque/speed	102 Nm/4000 rpm
Motor parameters	Rated power	31 kW
	Maximum speed	6000 rpm
	Rated speed	1000 rpm
	Peak torque	305 Nm
Generator parameters	Rated power	15 kW
	Maximum speed	5500 rpm
	Rated speed	2500 rpm
	Peak torque	55 Nm
NIH battery parameters	Capacity	6 Ah
	Number	40
Powertrain parameters	Final drive ratio	3.939
	Average efficiency	0.85

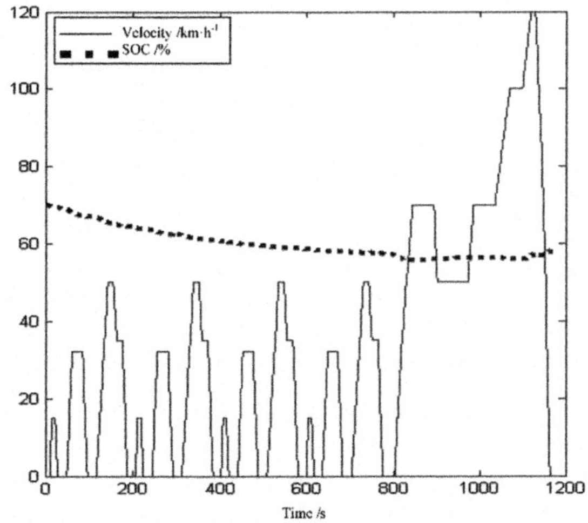

Fig. 3.4 Cycle vehicle velocity and SOC

The simulation result shows that the above-mentioned transmission characteristics of differential-based coupling device is correct and the engine can work in its optimum efficiency point. The motor provides power in acceleration process and generates electricity in deceleration process. And SOC maintains near the target value.

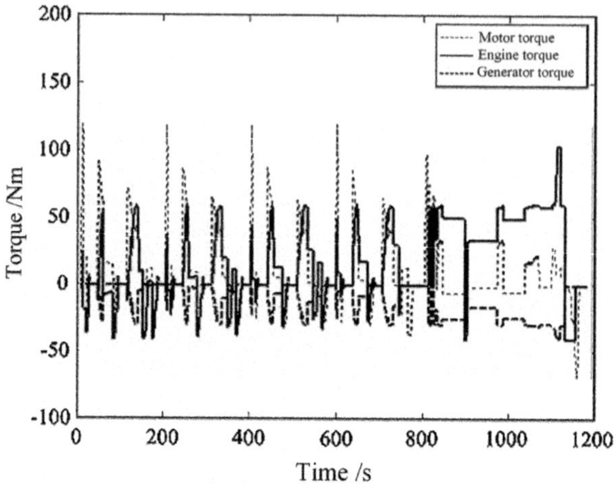

Fig. 3.5 Actual output torque of power source

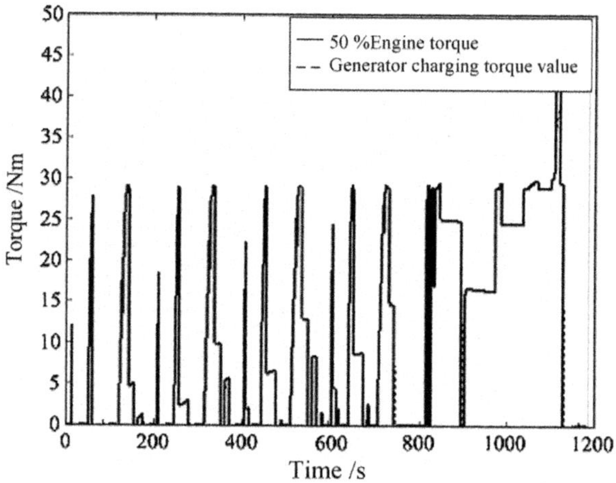

Fig. 3.6 Generator charging torque

3. Speed relationship simulation analysis

The actual speed of power source corresponding to the three terminal of differential-based coupling device are shown in Figs. 3.7 and 3.8.

The motor speed is proportional to the cycle vehicle velocity and the engine speed is controlled near the optimum speed point. At approximately 750 s, the engine is shut down. The reason is that after running a period of time, the engine temperature reaches the warm-up state and other conditions (vehicle speed is low, SOC is high, required power is small) can result in engine off. Then the engine speed decreases to 0. At the same time, the motor speed can follow required speed

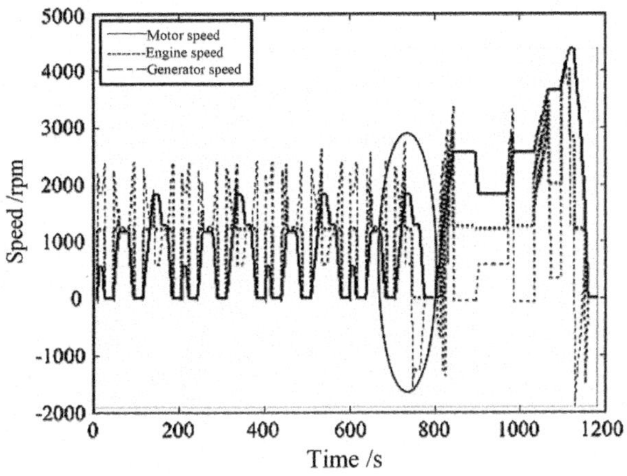

Fig. 3.7 Speed relationship of power source

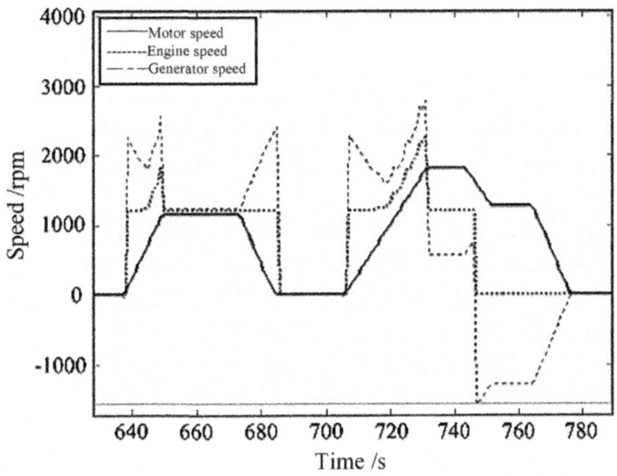

Fig. 3.8 Partial enlarged drawing of speed relationship of power source

by adjusting generator speed. In short, when the engine speed reaches the optimum operating point or shut- down point, motor speed can follow the required speed by adjusting generator speed, which achieves the speed control function of ECVT.

Comparing the speed of engine, motor and generator, the sum of motor speed and generator speed is exactly twice the engine speed, which verifies the speed relationship. The result is shown in Fig. 3.9.

The simulation result shows that the engine can be controlled to work at optimum operating point by regulating the generator speed, which improves the fuel consumption performance in great degree. In the same time, the transmission and clutch can be removed and it can achieve the function of ECVT.

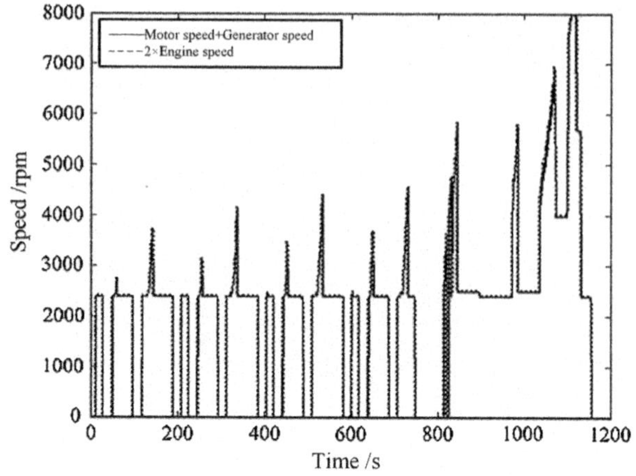

Fig. 3.9 Speed relationship of power source

3.1.4 Summary

In this section, the traditional symmetrical bevel gear differential is used as the power coupling device of hybrid vehicle. The power transmission characteristics are analyzed. The MATLAB/Simulink model of hybrid vehicle and the vehicle control strategy are established. Through secondary development, the vehicle model and control strategy are embedded into ADVISOR, which can effectively simulate the performance of hybrid vehicle. The simulation results of speed, torque and vehicle performance show that it has the advantages of saving energy and removing transmission and clutch for applying the symmetrical bevel gear differential to hybrid vehicle. It can achieve the function of ECVT, verifying the feasibility of this new scheme. Furthermore, it illustrates that simulation test and analysis of the new configuration scheme can be conducted by secondary development.

3.2 Hybrid System Development Based on CRUISE

AVL/CRUISE can easily achieve the simulation analysis of complex vehicle powertrains. It has successfully built a communication bridge between the vehicle manufacturers and part suppliers because of its convenient and generic model components, intuitive data management system, model designing process and software interface based on engineering application development. The main features of the software are summarized as follows:

1. Convenient modeling methods and modular modeling tools allow different project groups to integrate the model easily and quickly. It is quick to build a variety of complex powertrain models, which can be conducted for forward simulation or reverse simulation analysis at the same time;

2. The vehicle performance can be calculated and analyzed: vehicle cycle fuel consumption (for different driving cycles), constant drive fuel consumption (run in all gears and speed), constant drive emissions, maximum climbing degree (with slip), maximum traction force (traction power), maximum acceleration, maximum velocity, full load acceleration (shifting gears from standstill or elasticity), cruising, brake/coast/thrust;
3. Based on the traditional vehicle model, a model of electric vehicle or hybrid vehicle can be established quickly and the vehicle control strategy can be developed through interface with MATLAB (API,DLL,Interface) or C (Black Box);
4. Some new powertrain modes (AT, AMT, DCT, CVT, etc.) and their control strategies can be conveniently analyzed and studied.

In view of the extensive application of AVL/CRUISE, based on AVL/CRUISE and MATLAB/Simulink environment, this section will take a hybrid bus for example to build a forward and modular simulation platform applied for the hybrid vehicle developmental stage.

3.2.1 Vehicle Modeling

Firstly, establish the vehicle model quickly in CRUISE. Then finish standardized and platform-based vehicle control strategy in MATLAB/Simulink environment. The feasibility, generality and convenience of the simulation platform based on CRUISE are verified from the following aspects: parameter matching of vehicle performance, transient controlling and simulation of the main power components and optimization of the performance parameters. The model diagram and energy flow diagram of the hybrid vehicle are shown in Figs. 3.10 and 3.11, respectively.

Users can directly drag modules from the module library to establish a vehicle model in CRUISE and complete parameter settings quickly through proper modification. Then users need to complete mechanical connections and electrical connections among components. The key point is the signal connection between CRUISE and control strategy in MATLAB/Simulink, equivalent to the communication of vehicle and HCU.

3.2.2 Control Strategy Modeling in MATLAB/Simulink

3.2.2.1 Modeling Principles

(1) Control module and vehicle model are independent to each other;
(2) Modeling is clear and easy to read and modify (sub-modular, independence of different driving modes);

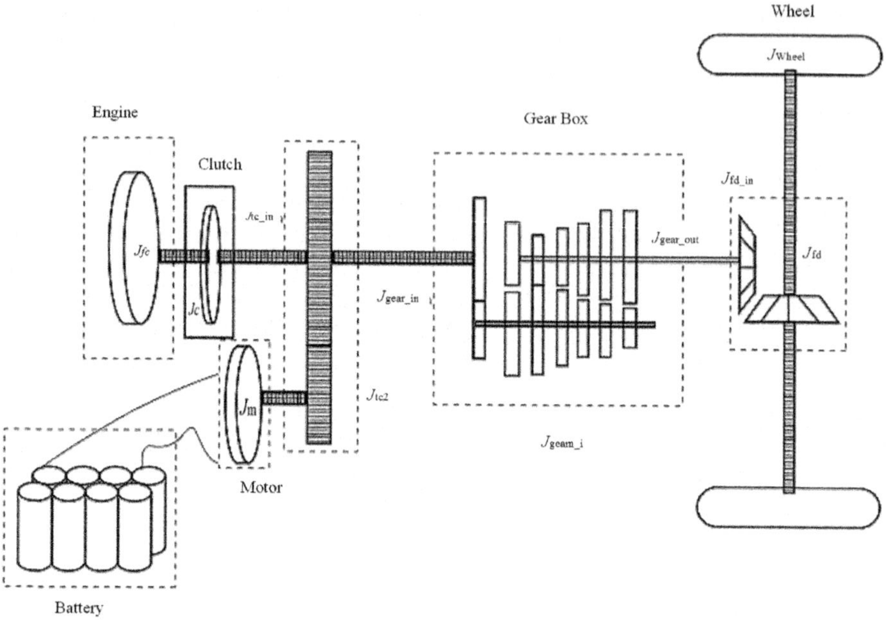

Fig. 3.10 Vehicle physical model

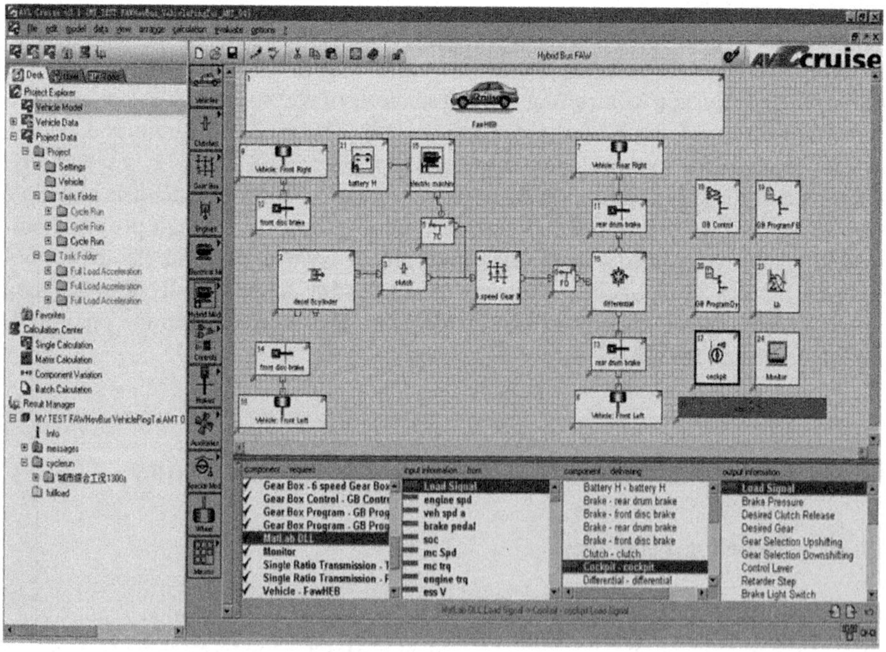

Fig. 3.11 Vehicle model in CRUISE

(3) The control strategy is close to the real vehicle control strategy (the input judgment condition is determined by the actual measured).

3.2.2.2 Drive Mode and Variables

(1) Drive mode settings

To ensure generality, twelve kinds of drive mode are set, taking into account hybrid bus and hybrid car. (including parking mode, pure electric mode,regenerative braking mode, engine drive alone mode, assistance mode, charging mode, engine idling mode, motor starts engine mode, parking charge mode, shifting mode, quick assistance mode, braking mode). The number of drive modes can be added or reduced for special needs of different vehicle.

(2) Input parameters and output parameters settings

Considering the modeling principle (3), the necessary input variables of forward simulation platform are set to approach the real vehicle: acceleration pedal signal, engine speed, velocity, brake pedal signal, SOC, motor speed, actual gear, desired velocity, etc. The output variables include engine load signal, main motor load signal, auxiliary motor load signal, drive mode, desired clutch operating state, required gear, shifting identification, etc. Similarly, for special needs of different vehicles, it can be directly added or deleted input variables and output variables in the reserved input and output ports in Simulink, which ensures the compatibility of the simulation platform.

3.2.2.3 Standardized and Modular Control Strategy Platform

In Fig. 3.12, module 1 is input, module 11 is output, module 3, 4, 5, 6, 7, 8 are torque distribution module, motor temperature controlling module, clutch function module, gear shifting module, driving module and parking module. Standardized control algorithm is easy to debug for improving debugging and simulating speed, and it can provide convenience for learning and communicating with others in the same industry.

3.2.2.4 Control Algorithm Flow Chart of Driving Module in Control Strategy

The control algorithm flow chart of vehicle control strategy is shown in Fig. 3.13. Among them, the driving module is the main module. It is divided into acceleration, deceleration and uniform mode.

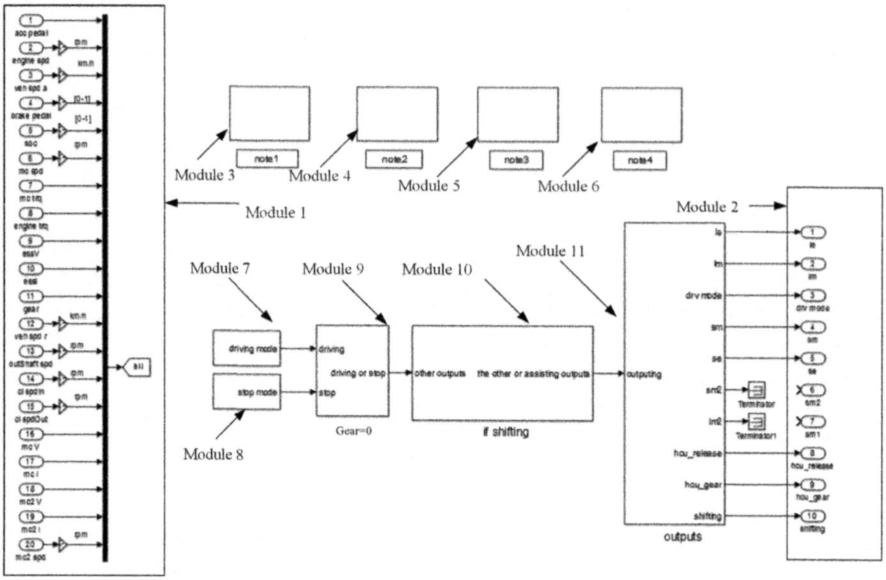

Fig. 3.12 Top-level module of modular vehicle control strategy

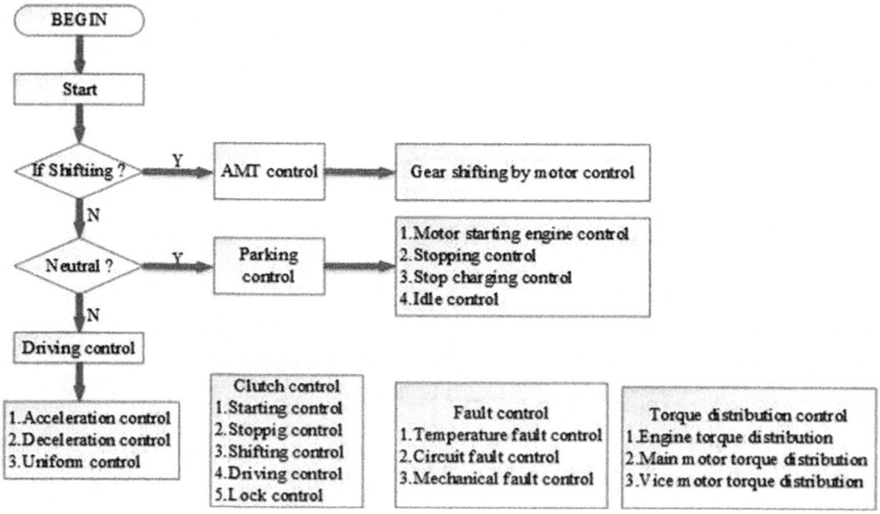

Fig. 3.13 Control algorithm flow chart of vehicle control strategy

3.2.3 Advantages of Forward Simulation Platform Based on CRUISE

For actual project, in the initial stage of vehicle development, what designers need most is a reliable, convenient, fast and standardized simulation platform to carry out relatively accurate forecasting calculation for the concept vehicle. The advantage of CRUISE is that the specific vehicle performance index can be acquired for given vehicle parameters. But CRUISE only provides vehicle model of generality and platform-based, a complete research development platform also requires a common control strategy. CRUISE is the same as other inverse simulation software in terms of accuracy [1]. In practicability, it solves the following important issues:

(1) Forward control. The control strategy of inverse simulation software (such as ADVISOR) is the cycle → wheel → main power components, equivalent to the energy chain of demand power → actual power. There is no driver module, equivalent to no acceleration pedal signal, which is very different from the actual vehicle control. While the forward simulation software (such as CRUISE) has the COCKPIT driver module, meeting the above-mentioned requirement.
(2) Transient control. For the early inverse simulation software, the clutch state includes separation, combination and slipping. However, the vehicle control strategy in Simulink proposed in this book includes four kinds of state: starting, shifting, parking and normal driving. The vehicle transient control can be achieved by co-simulation based on the control strategy and vehicle model in CRUISE. And users can add another two kinds of clutch state (engine out of work, engine into work) to other vehicle models. When the clutch is working in the shifting state, users can conduct active synchronous speed control to engine and motor.
(3) Parameter optimization. The vehicle model in CRUISE is closed. Users can complete the designing model just by modifying the vehicle parameters proper. For the control strategy, users can modify the important parameters through M file, which is very convenient and avoids the shortcomings of poor debugging because of the cross-cutting characteristics and complexity of inverse simulation software.

3.2.4 Simulation Test

Taking bus in urban comprehensive condition as an example, the following three aspects of simulation and analysis are carried out.

3.2.4.1 Simulation of Vehicle and Main Power Components

The simulation results of current velocity, desired velocity and drive mode are shown in Fig. 3.14. It can be seen that the current velocity is in good agreement with the desired velocity and the drive mode is correct at different vehicle velocity (note: the operation mode of hybrid vehicle is closely related to the current cycle condition. Different cycle condition corresponds to different mode. So it's reasonable that not all the drive modes appear in one cycle condition). It can be clearly seen from Fig. 3.15 that when starting the vehicle, the drive mode is the shifting process → motor assistance → engine drive alone. And the speed control of engine and motor in shifting mode is similar to the speed control of real vehicle, which illustrates that the simulation platform established by CRUISE and Simulink is able to observe the control results in real time and verify the rationality of parameters matching [2]. Designers can reset some of the important vehicle parameters to meet the target vehicle performance according to the simulation results. In addition, the simulation platform can also carry out performance tests such as acceleration performance, fuel-economy performance and turning characteristics.

3.2.4.2 Transient Simulation During Shifting Process [3]

In Fig. 3.16, motor load signal and battery current are both 0 in the process of clutch separation. The battery doesn't supply motor energy and the motor speed remains unchanged. In the process of clutch engagement, engine speed decreases and driver steps acceleration pedal according to the acceleration requirement so the engine load signal gradually increases. The simulation result of downshifting process is shown in Fig. 3.17, similar to the upshifting process, so this part will not go into details.

In Fig. 3.18, engine load signal, motor load signal and battery current are 0 in the process of clutch separation. The engine is not engaged in the full clutch

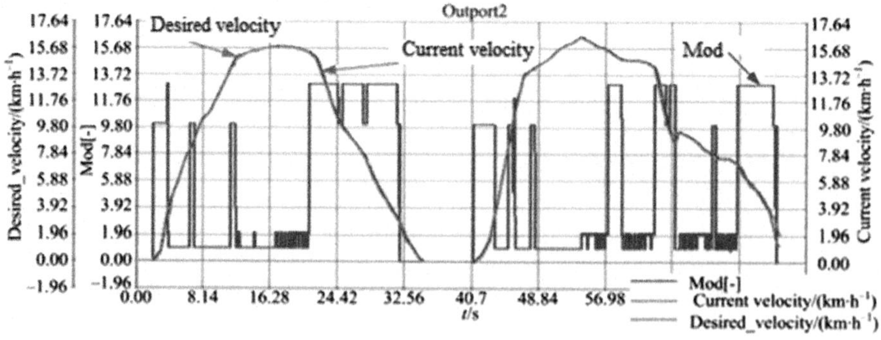

Fig. 3.14 Desired velocity, current velocity and drive mode

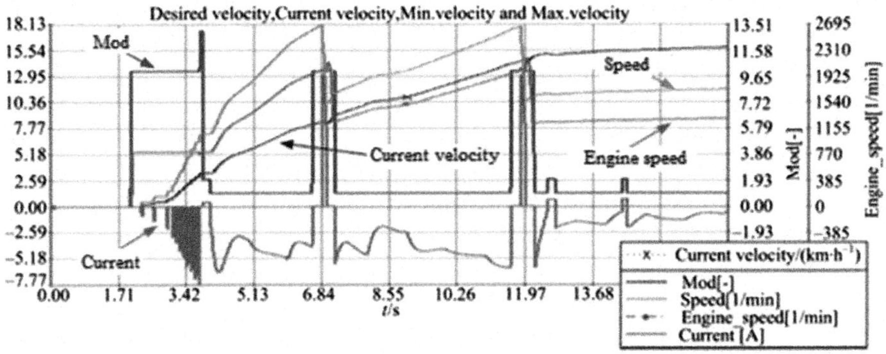

Fig. 3.15 Engine speed, motor speed, current and drive mode

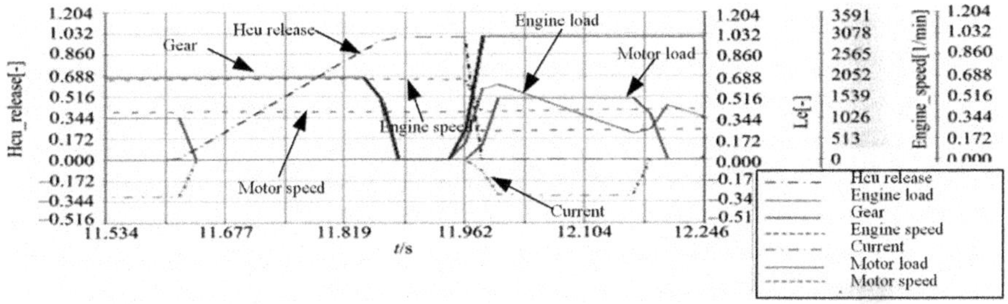

Fig. 3.16 Simulation result of upshifting without active synchronous speed control

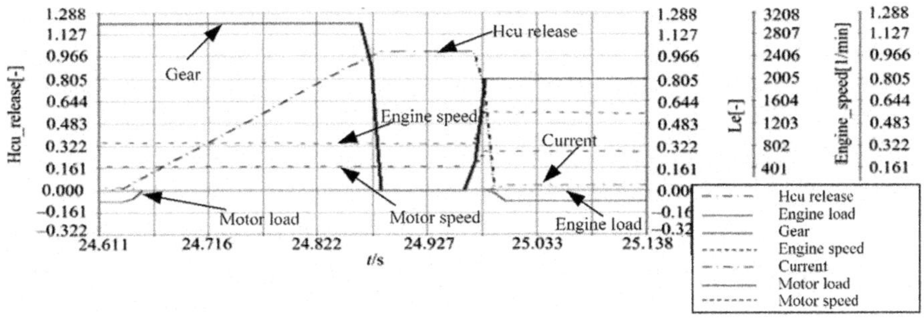

Fig. 3.17 Simulation result of downshifting without active synchronous speed control

separation process so the engine load signal remains 0. Under the upshifting condition, the motor speed needs to decrease according to the target speed, so the motor load signal becomes −1 and the battery current decreases gradually. The active synchronous speed control is not finished until the clutch starts to engage. In the

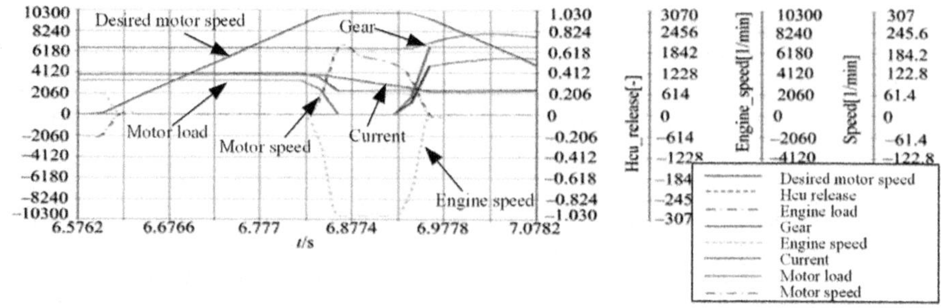

Fig. 3.18 Simulation results of shifting process and main power components in upshifting

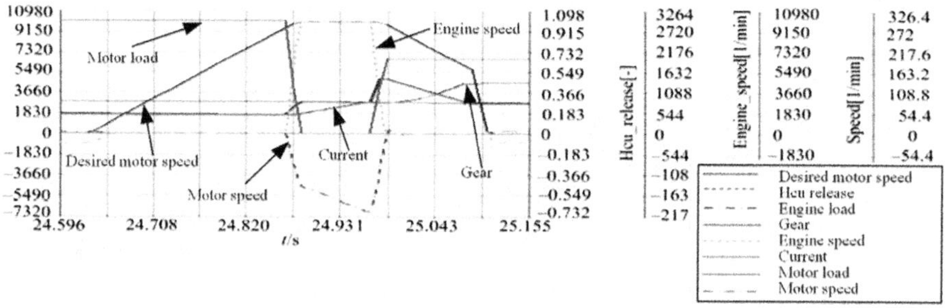

Fig. 3.19 Simulation results of shifting process and main power components in downshifting

process of clutch engagement, the motor doesn't regulate speed so the load signal is 0. The engine works alone. The downshift process analysis is similar to the upshifting process, shown in Fig. 3.19.

The function of analyzing the transient process demonstrates fully that the developed platform can not only carry on the forward simulation, but allow human intervention on the main power components in dynamic process. Users can observe and debug the control results, similar to the real vehicle debugging, which shows the practicality of the forward simulation platform.

3.2.4.3 Optimization of Main Performance Parameters in Control Strategy

The main parameters to be optimized are the upper and lower limits of SOC, the threshold velocity of pure electric driving, the maximum and minimum torque coefficients of engine, etc. SOC is taken as an example in this article. In Fig. 3.20, the four curves coinciding illustrates that the upper and lower limits of SOC have no effect on uncorrected fuel consumption for urban comprehensive condition.

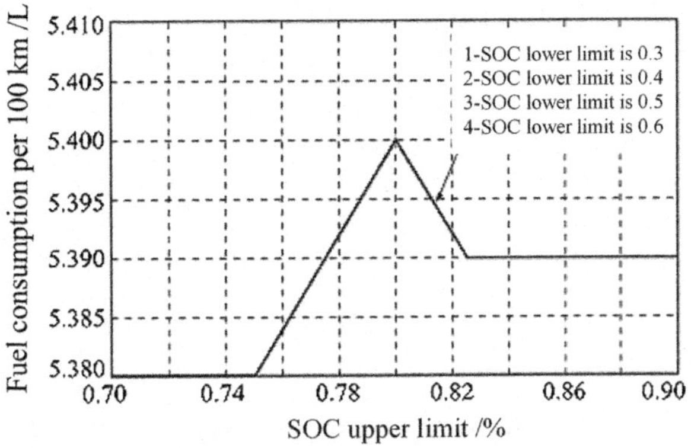

Fig. 3.20 Relationship between upper and lower limits of SOC and fuel consumption per 100 km

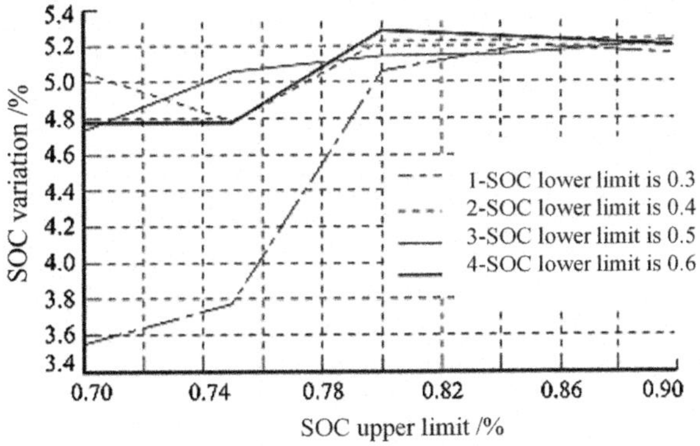

Fig. 3.21 Relationship between upper and lower limits of SOC and SOC variation

In Fig. 3.21, the upper and lower limits of SOC influence SOC variation differently. In the control process, the motor plays an auxiliary role in driving. The SOC level with small change in the whole cycle condition should be chosen from the optimization result and SOC maintains at a high level at the end of simulation. In general, the optimization work can be carried out on some optimization softwares. However, the developed platform in this book can effectively optimize some important parameters, which proves the practicability of the developed platform again.

To sum up, the simulation results give full validation for the pre-established control strategy and control logic. As long as some of the input variables, output variables and vehicle parameters are modified and the component parameters are re-entered in CRUISE, users can finish a complete vehicle model of forward simulation. It is proved that the hybrid vehicle simulation platform based on CRUISE and MATLAB/Simulink is of generality, versatility and practicability.

3.2.5 Summary

Simulation analysis of vehicle performance parameters of a certain hybrid bus under specific cycle condition has fully proved the reasonability, portability and generality of CRUISE. The research-development environment can be standardized to communicate and study easily among the industry, which provides a reliable and efficient forward simulation platform for initial development of hybrid vehicle in the true sense.

3.3 Hybrid System Development Based on AMESim

AMESim is a complex system modeling and simulation platform in the field of multi subjects. It has a set of standard and optimized application libraries, and it also has a large number of models containing 4500 areas. Users can establish a complex model on this single platform, then they can carry out simulation, calculation and analysis on this basis. They can also study steady-state and dynamic performance of any component or system on this platform.

The orientation of AMESim for engineering application makes AMESim an ideal choice for the research and development department in the automotive, hydraulic and aerospace industries. The engineering designers can apply the integrated AMESim application library to design a whole system. AMESim allows engineers to quickly reach the ultimate goal of modeling and simulation: analyzing and optimizing the design, which helps to reduce the development cost and shorten the development cycle.

In this section, a new model of hydraulic hybrid vehicle system based on the AMESim software platform is established firstly, then control strategy based on the optimal working curve is established by Simulink. Finally, the vehicle simulation is carried out. By analyzing the simulation results of dynamic performance, fuel economy and the working point of each power source, the parameter matching method and the control strategy of the key components proposed in this section are tested.

3.3.1 Background

Hydraulic system has the characteristics of large power density, low pollution, high reliability of hydraulic components, high energy cycle efficiency and stable transmission [4–8]. Meanwhile, it has obvious advantage in cost, resulting in its great market application prospect [9–11].

Dual-planetary coupling system has the advantage of double decoupling of the engine torque and speed and can achieve the function of ECVT. Combining the characteristics of the dual-planetary coupling system and hydraulic system, series and parallel hybrid hydraulic system with planetary mechanism is put forward and the city bus is considered as the research object in this section.

The difficulty and emphasis of the vehicle research with above-mentioned system is to distribute the power of the engine, the hydraulic pump and the hydraulic motor effectively and reasonably, so as to meet the power requirements, improve the efficiency of the system and achieve the purpose of fuel economy.

In this section, in view of the current technology status, a kind of parameter matching and control scheme is put forward which is suitable for the hydraulic drive system. In order to make engine and hydraulic pump/motor work in the efficient working area and achieve the best economic goals, by selecting working condition and vehicles first and combining with the established energy management strategy, a parameter matching method called from main to subordination is proposed.

3.3.2 *Principle of the Hydraulic System*

The series and parallel hybrid hydraulic system proposed in this book consists of engine, hydraulic pump/motor A, hydraulic pump/motor B, high-pressure accumulator planetary coupling device, drive axle and control system, as is shown in Fig. 3.22. The engine is connected with the planetary carrier of front planetary row,

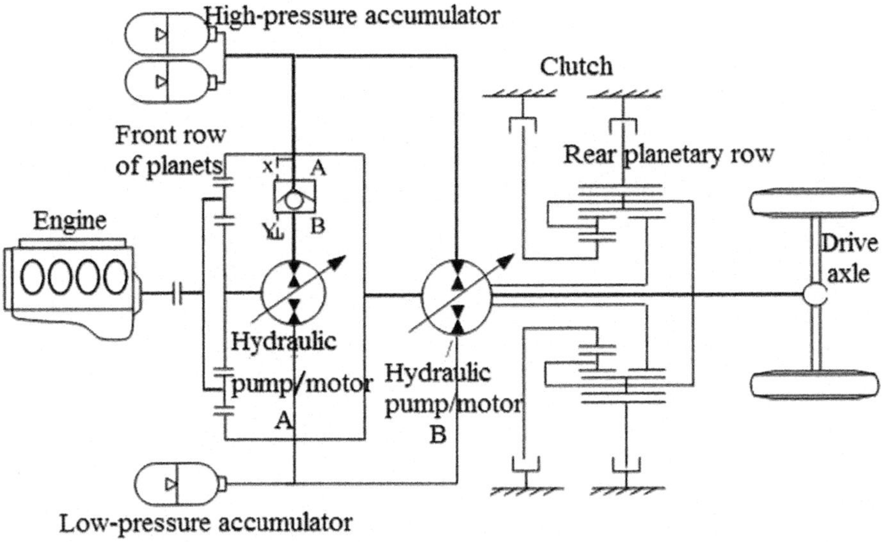

Fig. 3.22 Series and parallel hybrid hydraulic system with planetary mechanism

and hydraulic motor pump A and B are respectively connected to the sun gear of the front and rear planetary row. The ring gear of the front planetary row is connected with the planetary carrier of the rear planetary row in a mechanical way, and is also connected with the driving axle of the vehicle. Through different combination states of the two clutches, hydraulic pump/motor and two planetary rows, the configuration can achieve a variety of operation modes and obtain the function of ECVT.

3.3.3 Modeling of the Hydraulic System

The model in the above picture includes driving cycle, driver model, engine model, car body model, hydraulic drive system and planetary gear hybrid system. As is shown in Fig. 3.23, passing through a one-way valve,the pump/motor A is connected to the accumulator's closing valve and the hydraulic pump/motor B, and then through the hydraulic pump/motor B it's connected to the small sun gear of the rear planetary row. The planetary hybrid system consists of two planetary rows and two clutches. Two modes of high or low speed can be obtained by controlling the different states of the clutch, so as to meet the requirement of high or low speed cruise.

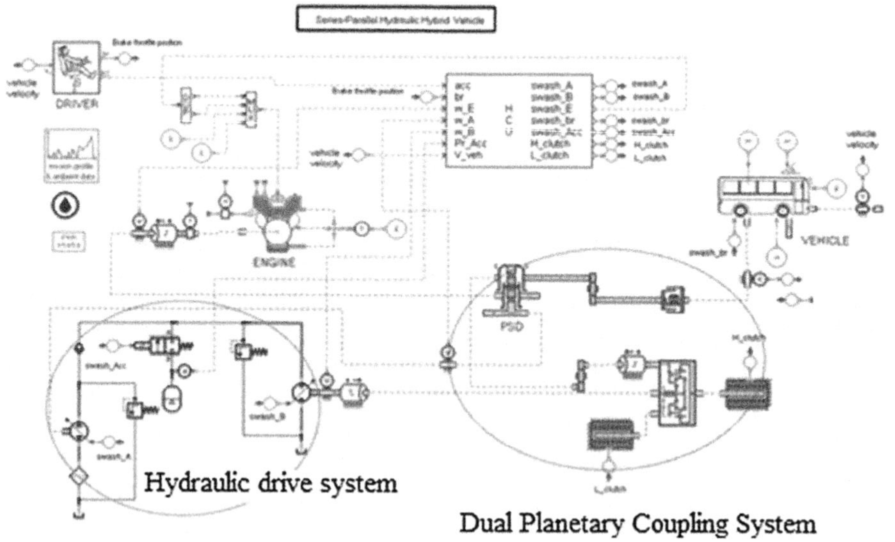

Fig. 3.23 Physical model of the series and parallel hybrid hydraulic system

3.3.3.1 Engine

The effective power of the engine is calculated as follow:

$$P_{\text{eff}} = \frac{T_X \times \omega_e}{1000} \tag{3.15}$$

where P_{eff} is the effective power (kW), T_X is the corrected engine torque (Nm), ω_e is the engine shaft speed (rad/s). The corrected torque and external characteristic of engine are input through data table.

3.3.3.2 Hydraulic Pump/Motor

As the hydraulic pump/motor A and B have a similar mathematical model, so this section will only give the general hydraulic pump/motor model.

(1) The flow q_{pm} of hydraulic pump/motor meets the following equation:

$$q_{pm} = \frac{V_{pm} \cdot n_{\text{pm}} \cdot swash_{pm} \cdot \eta_{\text{v}}}{1000} \tag{3.16}$$

where V_{pm} is displacement of the hydraulic pump/motor (mL/r), n_{pm} is speed of the hydraulic pump/motor (rpm), $swash_{pm}$ is displacement control opening of the hydraulic pump/motor ($-1 \leq \text{swash}_{\text{pm}} \leq 1$), η_v is the volumetric efficiency of hydraulic pump/motor. If the volume effect is not considered, its value is equal to 1.

(2) The output torque of hydraulic pump/motor meets the following equation:

$$T_{Pm} = \frac{(P_{\text{out}} - P_{in}) \cdot V_{pm} \cdot swash_{pm}}{20 \cdot \pi \cdot \eta_{\text{me}}} \tag{3.17}$$

where P_{out} and P_{in} are respectively the output and input pressure (bar) at the two ends of the hydraulic pump/motor, η_{me} is the mechanical efficiency of the hydraulic pump/motor, if the effect of mechanical efficiency is not considered, its value is equal to 1.

3.3.3.3 Accumulator

The equation for the change of the gas volume V_{gas} in the hydraulic accumulator is as follow:

$$V_{\text{gas}} = V_0 \cdot \left(\frac{P_0 + P_{atm}}{P_{gas} + P_{atm}} \right)^n \tag{3.18}$$

where P_{atm} is the atmospheric pressure (bar), P_0 is the initial pressure of the gas (bar), P_{gas} is the gas pressure after its volume has changed (bar).

The change of the gas pressure dP_{gas} always meets the following equation:

$$\mathrm{d}P_{gas} = -\frac{n \cdot (P_{gas} + P_{atm}) \cdot q_{out}}{V_{gas}} \cdot \frac{\rho(0)}{\rho(\text{gas})} \tag{3.19}$$

where $\rho(0)$ is the density of gas under the pressure of 0 at the initial time, $\rho(gas)$ is the density of gas under the pressure at the current time.

3.3.3.4 Planetary Row

The torque relationship between the three components of front planetary row meets the following equation:

$$T_s : T_R : T_{pc} = 1 : k : (-1 - k) \tag{3.20}$$

where T_s, T_R and T_{pc} are the torque of the sun gear, the ring gear and the planetary carrier respectively.

The speed relationship between the three components of front planetary row meets the following equation:

$$(1 + k) \cdot n_{cp} = n_s + k \cdot n_R \tag{3.21}$$

where n_{cp}, n_s and n_R are the speed of the planetary carrier, the sun gear and the ring gear respectively.

Rear planetary row is Ravignaux compound planetary gear, due to the mechanical constraint, the rotational speed relation of this structure meets the following two equations:

$$\omega_r \cdot R_{ring} + \omega_{sr} \cdot S_r = \omega_c \cdot (R_{ring} + S_r) \tag{3.22}$$

$$\omega_r \cdot R_{ring} + \omega_{sf} \cdot (-S_f) = \omega_c[R_{ring} + (-S_f)] \tag{3.23}$$

where ω_R is the speed of ring gear, R_{ring} is the radius of ring gear, ω_{sr} is the speed of sun gear, S_r is the radius of sun gear, ω_C is the speed of planetary carrier; ω_{sf} is the speed of small sun gear; S_f is the radius of small sun gear. The unit of radius is (m), and the unit of speeds is *rpm*.

3.3.3.5 Vehicle

The total vehicle resistance includes slope resistance, rolling resistance and air resistance, as shown in the following equation:

$$\begin{cases} F_{\mathrm{i}} = M_{equ} \times g \times \left(\arctan \frac{\alpha}{100}\right) \\ F_f = M_{equ} \times g \times (f + k + v) \\ F_W = 0.5 \times C_D \times \rho_{air} \times A(v + v_{wind})^2 \end{cases} \tag{3.24}$$

Where F_i, F_f, and F_w are the slope resistance, rolling resistance and air resistance (N) respectively, α is the slope, f is the rolling resistance coefficient, k is the coefficient (1/ms), v is the vehicle speed (m/s), C_D is the air resistance coefficient, A is the frontal area (m^2), ρ_{air} is the air density (kg/m^3), v_{wind} is the wind speed (m/s).

3.3.4 Parameter Matching Method

For this configuration, under the conditions that the matching object, driving cycles and the energy control strategy has been determined, taking better dynamic performance as the premise, the matching idea of minimizing fuel consumption has been put forward. According to the power transmission characteristics of the series-parallel hybrid hydraulic system and its internal energy flow relationship, the matching principle is proposed: according to the principle of 'from main to subordination', and by the sequence of 'total vehicle demand power → engine power → characteristic parameter of planetary row → power, displacement and speed of hydraulic pump/motor A → power, displacement and speed of hydraulic pump/motor B → working pressure and volume of accumulator', the matching of power system parameters can be completed, and will be adjusted or optimized through the follow-up simulation analysis. The specific matching process is shown in Fig. 3.24.

According to the general matching ideas mentioned above, the specific matching process of each component will be introduced in detail.

3.3.4.1 Initial Conditions

Taking a city bus as an example, the parameters of the power system are matched based on the national urban comprehensive driving cycles. The basic parameters of the vehicle are shown in Table 3.3. Its fuel economy needs to be increased at least 35%. Dynamic performance of vehicle is shown in Table 3.4.

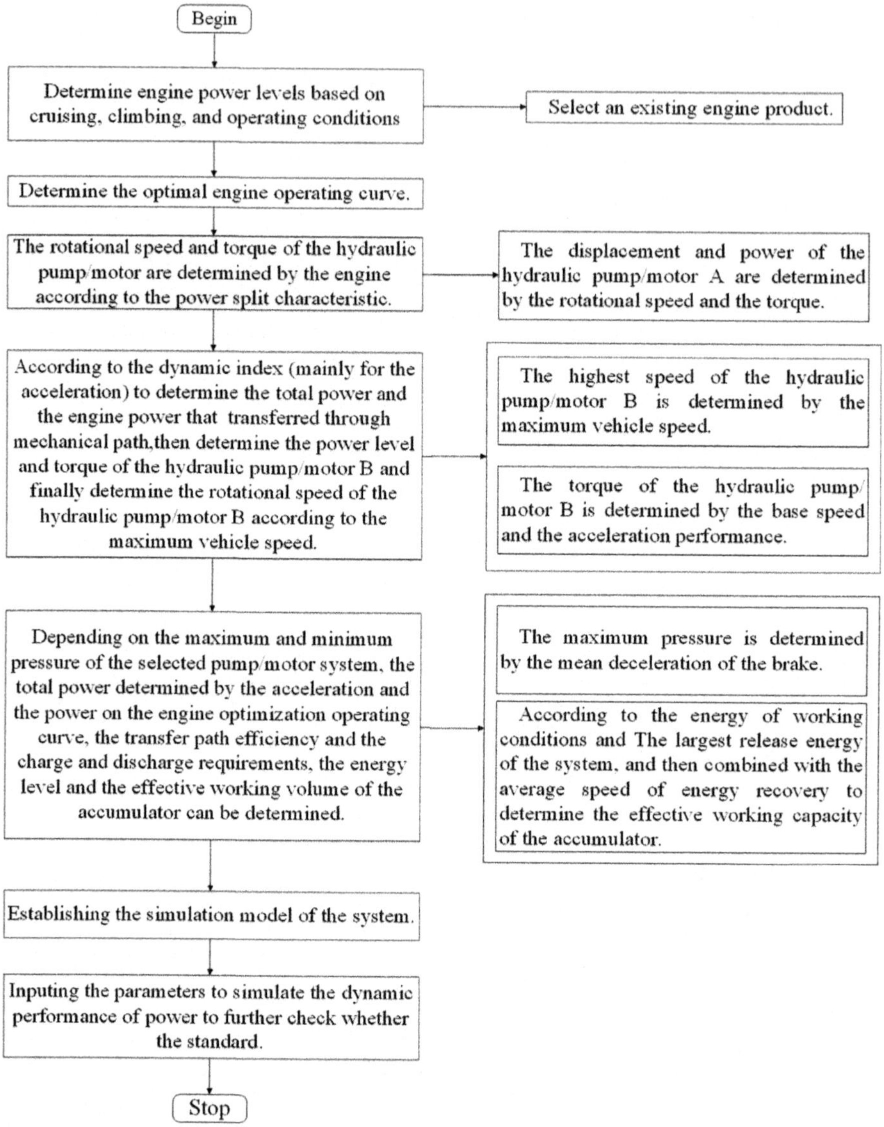

Fig. 3.24 Method and process of general matching

Table 3.3 Parameters of the vehicle

Total weight/kg	Curb weight/kg	Drag coefficient	Rolling radius/mm	Frontal area/m^2
15,000	11,000	0.65	509	5.5

Table 3.4 Index of dynamic performance

Maximum speed/km·h^{-1}	Maximum climbing degree/%	Continuous climbing speed/km·h^{-1} (@4%)	0 ~ 60 km/h acceleration time/s
≥ 80	≥ 25	40	≤ 30

3.3.4.2 Vehicle Power Demand

The total power required by the vehicle must firstly meet the power requirements of the traditional vehicle, on this basis, fuel consumption and emissions should be reduced as far as possible. Generally, the total demand power of the hybrid vehicle is calculated mainly on the basis of the maximum vehicle speed, the maximum climbing degree and the acceleration time. As shown in the following equations: [2, 12]

$$\mathrm{P}_{\max 1} = \frac{v_{\max}}{3600\eta_t}\left(mgf + \frac{C_D A v_{\max}^2}{21.15}\right) \tag{3.25}$$

$$P_{\max 2} = \frac{v_i}{3600\eta_t}\left(mgf \cos\alpha_{\max} + mg \sin\alpha_{\max} + \frac{C_D A v_i^2}{21.15}\right) \tag{3.26}$$

$$P_{\max 3} = \frac{1}{3600T\eta_t}\left(\delta m \frac{v_t^2}{2 \times 3.6} + mgf \frac{v_t}{1.5} T \frac{C_D A v_t^3}{21.15 \times 2.5} T\right) \tag{3.27}$$

$$\begin{aligned} P_{\max 4} &= P_{all}(t)|_{t=t_m} \\ &= \frac{1}{1000\eta_t}\left(\delta m v_m \frac{dv_m}{dt} + mgfv_m + \frac{1}{2} C_D \rho A v_m^3\right) \end{aligned} \tag{3.28}$$

where v_{max}, v_i, v_t, and v_m are the maximum vehicle speed, climbing speed and the speed at the end of accelerating (km/h) respectively, m is the curb weight (kg), g is the acceleration of gravity, f is rolling resistance coefficient, A is the frontal area (m^2), C_D is the air resistance coefficient, η_t is transmission efficiency, δ is the mass conversion coefficient, t_m and T are all accelerating time (s), ρ is the density of air, $\alpha_{\max}$ is the maximum slope (%).

At the same time, $P_{\max} \geq \max(P_{\max 1}, P_{\max 2}, P_{\max 3}, P_{\max 4})$. In conclusion, the total demand power of the hydraulic hybrid vehicle needs to be at least 122 kW.

3.3.4.3 Engine

The power of the engine is mainly used to balance the energy input and output of the hydraulic accumulator, as well as to compensate the energy consumed by the vehicle. So, the engine power is only required to meet the power demand of the vehicle during cruising at a certain high speed on a straight road or meet the power need of the vehicle when cruising on a certain slope with a slightly lower speed. Finally, the engine power can be determined according to the data of the simulation analysis process and the actual product series. The following equations describe respectively the engine power determined by the cruising speed and the certain climbing speed.

$$P_{e1} = \frac{1}{3600\eta_t}\left[Gf + \frac{C_D A v_{cruise}^2}{21.15}\right] v_{cruise} \tag{3.29}$$

$$P_{e2} = \frac{v_i}{3600\eta_t}\left(mgf\ \cos\alpha_i + mg\ \sin\alpha_i + \frac{C_D A v_i^2}{21.15}\right) \tag{3.30}$$

When the vehicle travels at a maximum cruise speed (80 km/h) on a straight road, the engine power should be 55 kW. While the vehicle is traveling at a speed of 40 km/h and at a gradient of 4%, the demand engine power is 88 kW.

In summary, $P_e(\min) \geq \max(P_{e1}, P_{e2})$. Thus, according to the engine power level and the existing products, a 4.76 L diesel engine can be chosen. As shown in Fig. 3.25, the engine reaches its maximum power at 2300 rpm, which is approximately 100 kW. The external characteristic curve and the optimum working curve of the engine are shown in the figure.

The expected engine speed should be between 1000 and 1900 rpm. So, taking into account the efficiency of the engine, ω_{exp} = 1000 rpm is selected firstly.

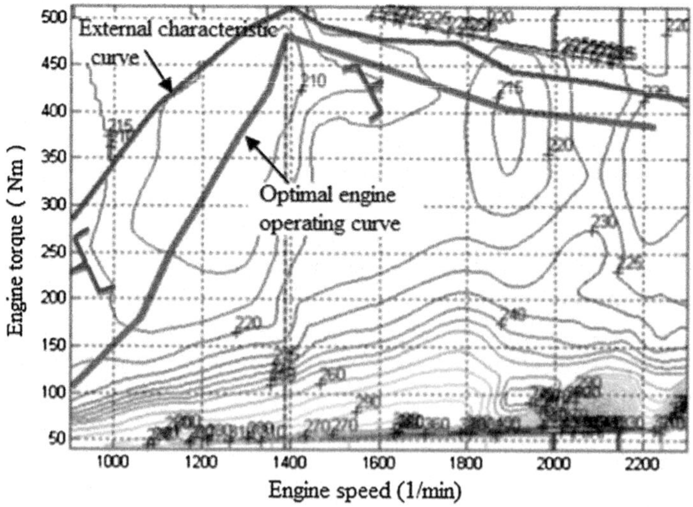

Fig. 3.25 Map diagram of engine

3.3.4.4 Hydraulic Pump/Motor A

The hydraulic pump/motor A is connected with the output shaft of the engine through the front planetary row. By decoupling engine speed and wheel speed, the engine can work more economically. According to the maximum torque of the engine working curve, the theoretical displacement V_p of the hydraulic pump/motor can be obtained from the following equation:

$$V_p = \frac{2\pi T_{emap}}{\Delta p} \eta_{pm} \frac{1}{k+1} \tag{3.31}$$

where T_{emap} is the maximum torque of the engine working curve (Nm), Δp is the pressure difference (MPa), k is the characteristic value of planetary row, η_{pm} is the efficiency of hydraulic system.

The theoretical flow rate of the hydraulic pump can be calculated by the following equation according to the maximum power of the optimum working curve of engine

$$q_p = \frac{60 \cdot P_{emax}}{\Delta p} \cdot \eta_{pm} \tag{3.32}$$

At the same time, the rotational speed relationship of the three components of the front planetary row meets the following equation [5, 13]:

$$(1+k) \cdot n_{cp} = n_s + k \cdot n_R \tag{3.33}$$

where n_{cp}, n_s and n_R are the speed of the planetary carrier, the sun gear and the ring gear respectively.

The speed relationship of the three power sources during normal operation is shown in Fig. 3.26. At first the vehicle speed is zero, hydraulic pump/motor A works at the maximum speed of 380 rpm and the engine works at a minimum operating speed of 1000 rpm. When vehicle starts acceleration, the hydraulic pump/motor A keeps working at the maximum speed, and the engine speed increases to the maximum allowable speed 1900 rpm. Then, the engine speed maintains at the value and the speed of hydraulic pump motor A starts to decrease. Therefore, during the whole acceleration process, the power of the hydraulic pump motor A is the highest when the vehicle speed is 37 km/h.

When n_{Amax} = 3800 rpm, T_{emap} is the torque on the optimal operating curve corresponding to the engine speed of 1900 rpm. So the power of the hydraulic pump/motor A meets the following equation:

$$P_A \geq \frac{T_{emap} \cdot \eta_{Amax}}{(k+1) \cdot 9550} \tag{3.34}$$

Thus, the power of the hydraulic pump/motor A is 42 kW.

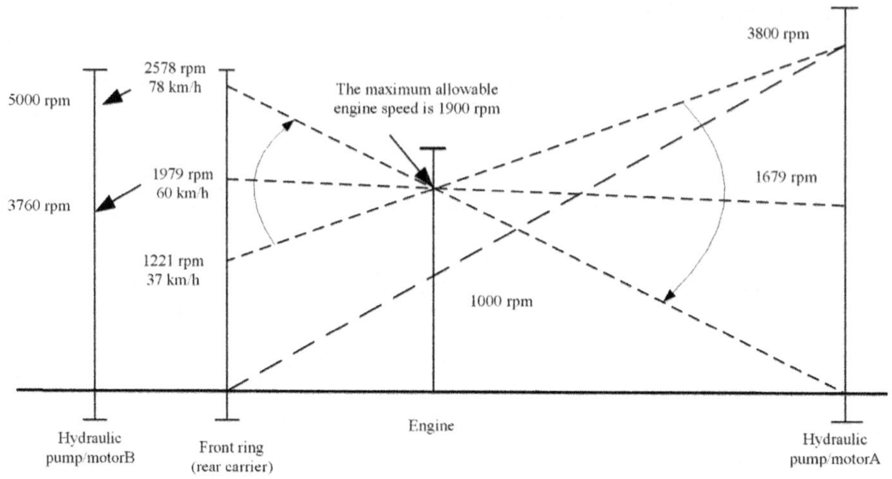

Fig. 3.26 Speed relationship of the three power sources

3.3.4.5 Hydraulic Pump/Motor B

1. Selection of speed

The speed of hydraulic pump/motor B is needed to meet the requirements of the vehicle traveling at the maximum speed, as shown in Eq. (3.35).

$$n_{B\max} \geq i_f \cdot \frac{v_{\max}}{3.6 \cdot R} \cdot \frac{30}{\pi} \cdot i_d \tag{3.35}$$

where i_f is the high speed ratio of rear planetary row, i_f is the ratio of final drive, R is the radius of the wheel.

2. Selection of power

The engine provides stable power during vehicle travel, while the hydraulic components provide the steep peak power. The difference between the total power required by the vehicle and the residual power after the engine overcoming the resistance power is the required power of the hydraulic pump/motor B.

The required power P_d during vehicle travel is calculated as follow:

$$P_d = \frac{1}{3600T\eta_t}\left(\delta m \frac{v_t^2}{2 \times 3.6} + mgf \frac{v_t}{1.5}T + \frac{C_D A v_t^3}{21.15 \times 2.5}T\right) \tag{3.36}$$

During the acceleration process, the residual power P_{e_a} after the engine overcoming the air resistance, rolling resistance and climbing resistance is calculated as follow:

$$P_{e\_a} = \frac{1}{t_m}\int_0^{t_m}(P_{e\_com} - P_r)dt$$
$$= \frac{1}{t_m}\left[\frac{k}{1+k}\cdot\frac{T_e i_d \eta_t}{1000R}\left(\frac{2}{3}v_m t_m\right) - \frac{mgf}{1000}\left(\frac{2}{3}v_m t_m\right) - \frac{C_D \rho A}{2000}\left(\frac{v_m^3 t_m}{2.5}\right)\right] \quad (3.37)$$

where P_r is the power corresponding to the sum of the rolling resistance and air resistance, P_{e_con} is the power that the engine transfers to the wheels through the transmission system and it can be calculated by Eq. (3.38).

$$P_{e\_con} = P_e - P_A = \frac{k}{1+k}\cdot\frac{T_e v i_d \eta_t}{1000R} \quad (3.38)$$

The power of the hydraulic pump/motor B is P_{final_B}, it can be calculated according to the following equation:

$$P_{final\_B} = P_d - P_{e\_a} \quad (3.39)$$

According to P_d and P_{e_a}, the power level of the hydraulic pump/motor B can be obtained. The result is 94 kW.

3. Selection of torque

The torque of the hydraulic pump/motor B can be obtained from the relationship between the power, torque and base speed, as shown in the following equation:

$$T_{B\max 1} = \frac{9550\cdot P_{B\max}}{n_{Bspd}} \quad (3.40)$$

where n_{Bspd} is the base speed determined by average speed of the drive cycle, P_{Bmax} is the maximum power of hydraulic pump/motor B.

Because the hydraulic pump/motor B provides peak power and peak torque during the acceleration process, the hydraulic pump/motor B should meet the requirements of the maximum acceleration of the driving cycles, as shown in the following equation:

$$T_{B\max 2} \geq \frac{1}{2}\cdot\frac{m\cdot a\cdot R}{i_d\cdot i_f} \quad (3.41)$$

where a is the maximum acceleration of the driving cycle.

After the power, torque and rotational speed range of the hydraulic pump/motor B are determined, the flow rate and displacement range can be selected according to the method of matching the hydraulic pump/motor A. The parameters of the hydraulic pump/motor B are then selected in connection with the actual products.

3.3.4.6 Hydraulic Accumulator

The maximum working pressure of the hydraulic accumulator should be no greater than the maximum allowable working pressure of the hydraulic pump/motor. The choice of inflation pressure should ensure that the system energy recovery rate as high as possible. The volume of the accumulator should be calculated by the following equation:

$$E_{Acc} = \frac{p_0 V_0}{n-1}\left[\left(\frac{p_0}{p_2}\right)^{\frac{1-n}{n}} - 1\right] \geq \frac{1}{2} m v_{\max}^2 \tag{3.42}$$

where p_0 is the inflation pressure (MPa), V_0 is the volume of the inflation pressure (mL), p_2 is the maximum working pressure (MPa), n is the gas index.

After the working pressure range and the inflation volume of the hydraulic accumulator are determined, the final parameters can be determined with reference to the actual product.

3.3.4.7 Matching Results

According to the above matching principle, the parameters of the power system of the vehicle can be obtained by calculating and selecting, as shown in Table 3.5.

Table 3.5 Parameters of series-parallel hybrid hydraulic vehicle

Part	Version	Item	Parameter
Engine	Diesel engine	Rated power/kW	105
		Maximum torque/Nm	520
		Displacement/L	4.76
Hydraulic Pump/motor A	Rexroth A6VM28	Maximum displacement/ml·r^{-1}	28.1
		Maximum speed/rpm	5550
		Maximum torque/Nm	179
		Maximum allowable flow/L·min^{-1}	156
Hydraulic Pump/motor B	Rexroth A6VM80	Maximum displacement/ml·r^{-1}	80
		Maximum speed/rpm	3900
		Maximum torque/Nm	509
		Maximum allowable flow/L·min^{-1}	312
Accumulator	American Standard accumulatorMB14/5000	Fluid volume/L	42
		Maximum working pressure/MPa	35
		Minimum working pressure/MPa	21

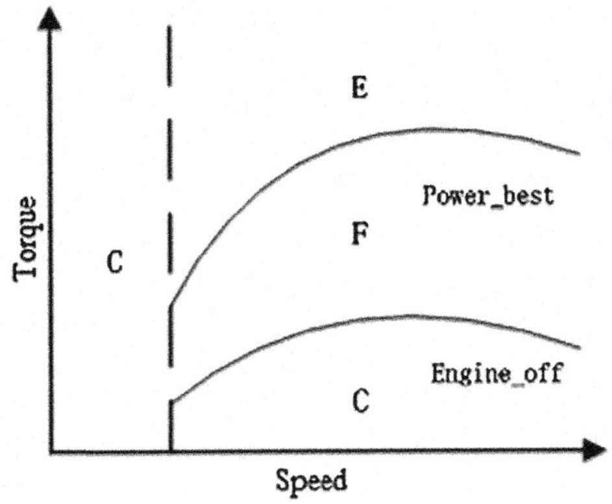

Fig. 3.27 when $SOC > SOC_{low}$

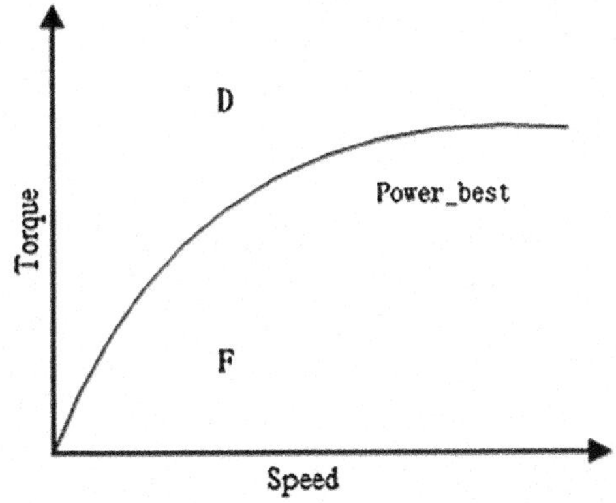

Fig. 3.28 when $SOC < SOC_{low}$

3.3.5 Control Strategy of the Hydraulic System

3.3.5.1 Operation Mode

The operation mode of the system can be determined according to the load rate and the SOC of the hydraulic accumulator. As shown in the Figs. 3.27 and 3.28.

Where the *Power_best* is the best working curve of the engine, *Engine_off* is the working curve after the engine is off.

In the picture above, A–F respectively represents the different vehicle operation mode. The specific meaning is as follows:

A. Regenerative braking, i.e. mechanical and hydraulic co-braking or hydraulic braking separately.
B. Mechanical braking.

C. Hydraulic pump/motor B propels the load alone.
D. Engine propels the load alone.
E. Engine and hydraulic pump/motor B propel the load at the same time.
F. Engine propels the load alone, and the hydraulic pump/motor A is driven to charge the hydraulic accumulator.

Where A operation mode: When the braking torque needed by the vehicle is less than the maximum braking torque that the hydraulic pump/motor A can provide, the hydraulic braking system can be used alone. When the vehicle braking torque is greater than the hydraulic pump/motor A can provide, the mechanical and hydraulic braking system will be used together. B operation mode: The mechanical braking mode is selected when the energy can no longer be stored by emergency braking or hydraulic accumulator.

When the *SOC* of accumulator is greater than SOC_{low}, the operation mode is determined according to the following equation.

$$\begin{cases} C & \omega_{req} < \omega_{str} \ or \ T_{req} < T_{Engine\_off} \\ E & T_{req} > T_{Power\_best} \\ F & T_{req} > T_{Engine\_off} \ and \ T_{Power\_best} \end{cases} \tag{3.43}$$

When the *SOC* of the accumulator is less than SOC_{low}, the operation mode is determined according to the following equation.

$$\begin{cases} D & T_{req} > T_{Power\_best} \\ F & T_{req} < T_{Power\_best} \end{cases} \tag{3.44}$$

where ω_{req} is the demand speed, ω_{str} is the engine starting speed, T_{req} is the demand torque.

3.3.5.2 Control Strategy for Driving

When the vehicle is driven under a cycle condition, the control strategy flow chart is shown in Fig. 3.29.

Firstly, according to the actual vehicle speed V and the target vehicle speed $v_{\_req}$ to calculate the demand power P of the vehicle, torque T_r and vehicle acceleration acc. When driving, $acc \geq 0$. Secondly, the operating mode of the system is determined by the demand power P, the torque T_r, the current engine speed and the SOC value of the hydraulic accumulator. Thirdly, according to the operation mode of the system and the actual SOC value of the hydraulic accumulator, the demand torque, speed, the output torque and speed of the hydraulic pump/motor B, and the gear position of the double planet can be calculated. Finally, convert the working conditions determined in the previous step to a specific command to the actuator.

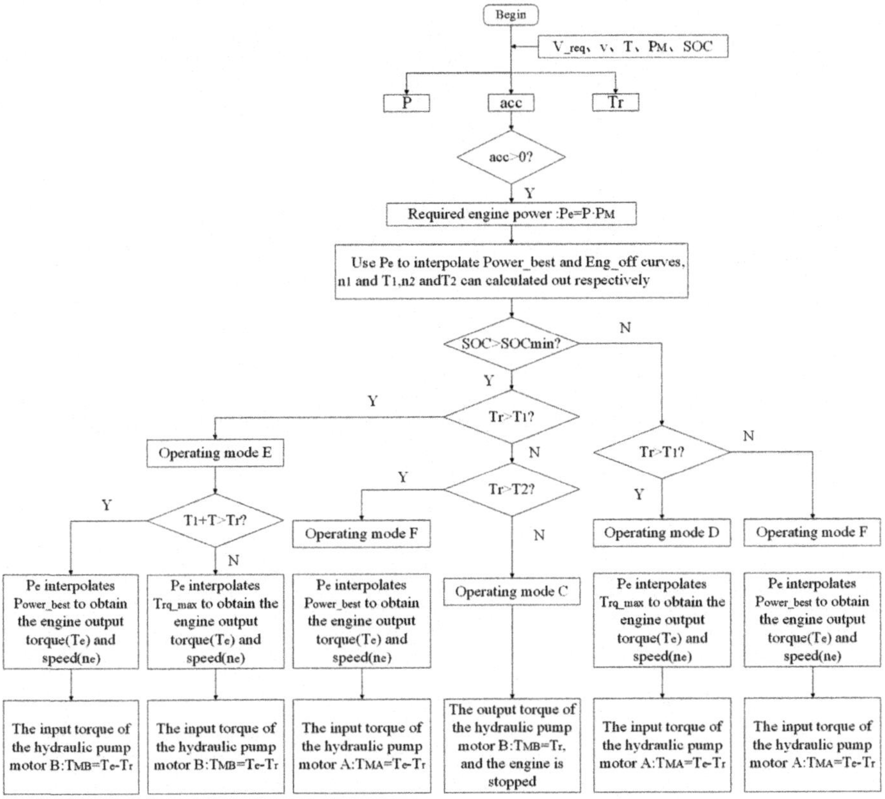

Fig. 3.29 Flow chart of control strategy when driving

3.3.5.3 Control Strategy for Braking

Similarly, the braking control strategy flow chart is shown in Fig. 3.30.

Firstly, according to the actual vehicle speed V and the target vehicle speed $v_{\_req}$ to calculate the demand power P of the vehicle, torque T_r and vehicle acceleration acc. When braking, $acc \leq 0$. Secondly, the operation mode of the system is determined by the requirement of power P, torque T_r and the SOC value of the hydraulic accumulator. Thirdly, according to the operation mode of the system and the actual SOC value of the hydraulic accumulator, the mechanical braking torque, the hydraulic braking torque and the gear position of the double planetary row are calculated respectively. Finally, convert the working conditions determined in the previous step to a specific command to the actuator.

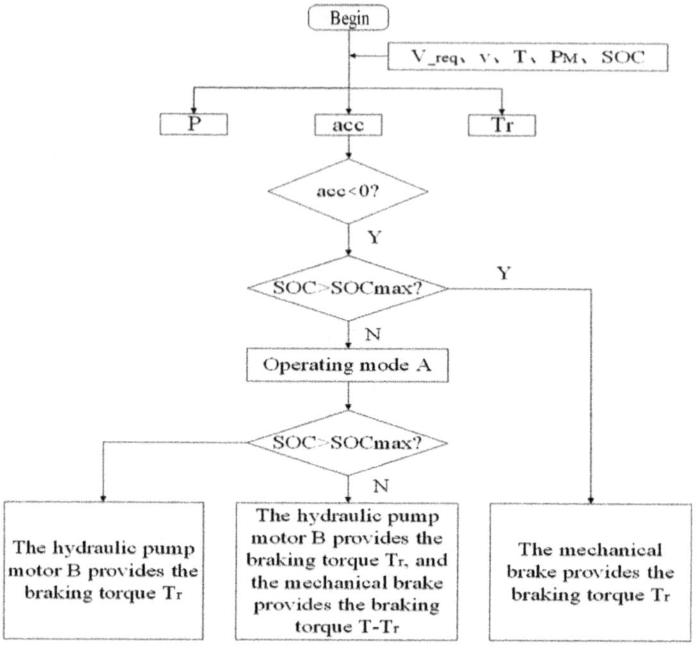

Fig. 3.30 Flow chart of control strategy when braking

3.3.6 *Verification of Parameter Matching Results*

In order to verify the rationality of the results of the above mentioned parameters, a series-parallel hybrid hydraulic vehicle simulation model by MATLAB/Simulink is established. Then, based on the control strategy of minimizing the fuel consumption of the engine, the dynamic performance and fuel consumption simulation under urban comprehensive driving cycles are carried out [14, 15].

3.3.6.1 Verification of Dynamic Performance

Under the condition of full load acceleration, the relationship between the speed and time of the simulation is shown in Fig. 3.31. It can be seen that the acceleration time from stationary to 60 km/h is 26.8 s, which is less than the design target value (28 s). That is, the acceleration performance of the vehicle meets the desired goal.

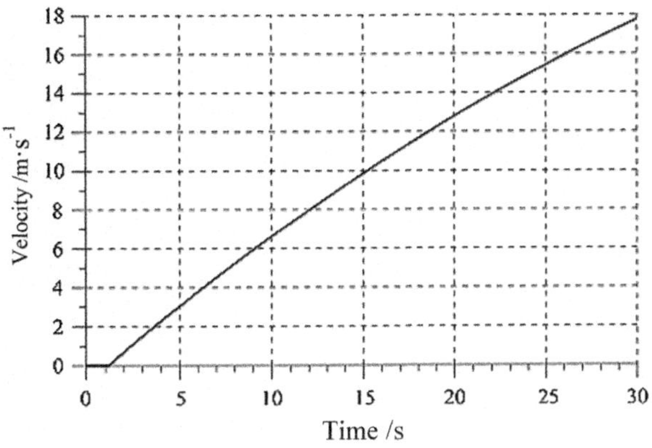

Fig. 3.31 Simulation curve of acceleration performance

3.3.6.2 Verification of Fuel Consumption

Figures 3.32, 3.33 and 3.34 are respectively the operating points of the engine, hydraulic pump/motor A, and hydraulic pump/motor B under the urban comprehensive driving cycle. As it can be seen, the engine is basically working in the optimal operating curve, which is in line with the principle of matching design. The operating points of the hydraulic pump/motor A and the hydraulic pump/motor B are all within their reasonable operating limits.

The simulation results of the fuel consumption of series-parallel hybrid hydraulic vehicles are shown in Table 3.6. It can be known that the fuel economy is improved by 39.4% after reasonable selection of parameters, and it meets the design requirements well.

From the simulation results of dynamic and fuel economy performance, it can be concluded that the matching method, which speed and torque variation of each component in vehicle operation are considered, can not only meet the design requirements, but achieve a better vehicle performance.

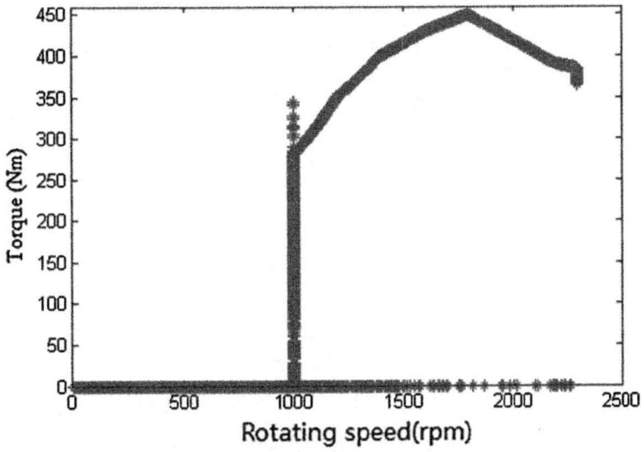

Fig. 3.32 Operating points of the engine

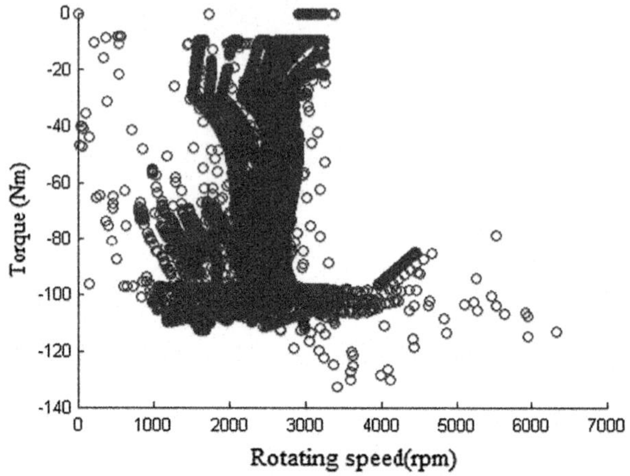

Fig. 3.33 Operating points of hydraulic pump/motor A

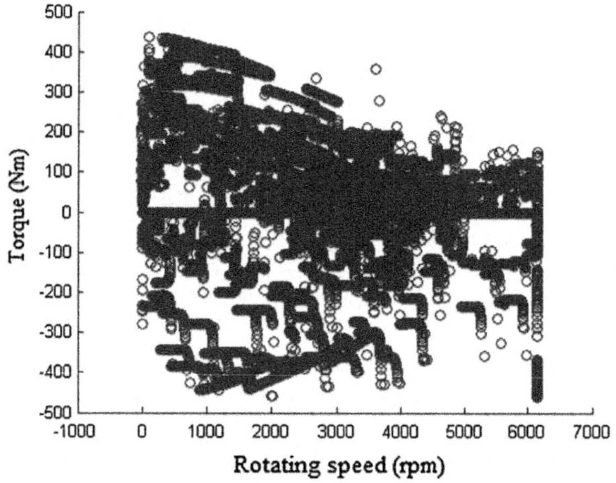

Fig. 3.34 Operating points of hydraulic pump/motor B

Table 3.6 Comparison of fuel economy

Driving cycles	Traditional vehicle/L·100 km^{-1}	Hydraulic vehicle/L·100 km^{-1}	Improvement degree/%
Urban comprehensive	40.0	24.23	39.4

3.3.7 Summary

In this section, the design principle of the key component parameters of a new series-parallel hybrid hydraulic system with planetary mechanism is put forward firstly, and then the energy management strategy is established. Finally, the joint simulation method of Simulink and AMESim verifies the matching results. In short, a kind of matching method and energy management control process suitable for the series-parallel hybrid hydraulic vehicle is formed. The simulation results show that the matching method, based on the configuration characteristics and the established energy management strategy, can effectively regulate the operating state of each power source. Moreover, this method also gives full play to the advantages of the configuration, which improves the energy utilization and reduces the fuel consumption.

3.4 Summary

The above three kinds of software are widely used in the field of hybrid technology and their commonality are: they can all be used for the selection of key components of hybrid vehicles, the determination of the best configuration, and the reasonable formulation or optimization of the vehicle control strategy. Their characteristics are: ADVISOR is the reverse simulation software based on the modular programming thought and the code is completely public. AMESim is the forward simulation software and it has advantages in the establishment of hydraulic hybrid vehicle models. In addition, the steady-state and dynamic performance of hybrid vehicle can be studied on AMESim. CRUISE has convenient universal model components and intuitive data management system, so it is relatively easier to use. Moreover, CRUISE and AMESim also have rich interface, so they can be combined with different commercial software to carry out co-simulation of vehicle performances and they can also be used for the hardware-in-loop testing.

References

1. Yu Y, Zeng X, Wang Q. Validation of hybrid electric vehicle simulation software. J Syst Simul. 2009;21(2):380–4.
2. Zeng X. Study on mechanism of energy saving and method of parameter design for hybrid electric bus. Changchun: Jilin University; 2006.
3. Tong Y. Research on dynamic coordinated control of power-split hybrid electric power system. Beijing: Tsinghua University; 2004.
4. Buchwald P, Christensen H, Larsen H, Pedersen PS. Improvement of city bus fuel economy using a hydraulic hybrid propulsion system—a theoretical and experimental study. SAE Paper 790305, Warrendale, PA, USA, 1979.
5. Yu Y. Optimal design and control of power-split hybrid electric vehicle. Changchun: Jilin University; 2010.
6. Dingel O, Ross J. Model-based assessment of hybrid powertrain solutions. SAE Paper 2011-24-0070.
7. Petric J. A power-split hybrid hydraulic vehicle transmission modeling and comparative analysis. SAE Paper 2010-01-2010.
8. Tavares F, Johri R, Salvi A. Hydraulic hybrid powertrain-in-the-loop integration for analyzing real-world fuel economy and emissions improvements. SAE Paper 2011-01-2275.

9. C Zhao, J Jiang, K Zhao, Harbin Institute of Technology. A study on hydrostatic transmission with secondary regulation and its applications to city bus. Automot Eng. 2001;23(6):423–6.
10. Du J, Wang H, Huang H. Research on peformance of power split hydraulic hydrid powertrain for vehicle. Mech Transm. 2011;35(6):15–8.
11. Li X, Chang S, Han W. Research on performances and matching between engine and accumulator and pump for the vehicle of hydrostatic transmission system. Trans Chin Soc Agric Mach. 2006;37(3):12–6.
12. Yu Z. Automobile theory. Beijing: Mechanical Industry Press; 1998.
13. Du J, Yuan S, Wei C, Zou Y. Performance analysis of dual mode hydro-mechanical transmission. Trans CSAE. 2009;25(4):86–90.
14. Zeng X, Wang Q, Wang W, Chu L. The development of a simulation module for the performance of two-axle drive HEV based on ADVISOR software. Automot Eng. 2003;5:424–7.
15. Zeng X, Wang Q, Li J, Chu L. The development of HEV control strategy module based on ADVISOR2002 software. Automot Eng. 2004;04:394–6.

Chapter 4
Parameter Design Method of Planetary Hybrid Power-Split System

Based on the front introduction of the related theory and simulation technology in the hybrid power system, this chapter proposes a multi-factor integrated parametric design method for a power-split hybrid electric bus (HEB). This method considers the power source efficiency, transmission efficiency, vehicle driving cycle requirements, and primary control strategy. The design method is then applied to a specified city bus. System efficiency is analyzed and the control strategy is determined. Finally, the parametric design method is validated on the co-simulation platform of MATLAB/Simulink and CRUISE. Results show that the fuel economy of the power-split HEB that adopted the proposed parametric design is 36% better than that of the conventional city bus.

4.1 Background

With a global sale of over 10 million [1], Toyota Hybrid System (THS) is undoubtedly the most successful hybrid system in the world. THS consists of one or two planetary gear sets and three power sources, which is also called the power-split system [2].

The power-split system combines the advantages of both series and parallel systems with excellent dynamic performance and fuel economy. The early research on the power-split hybrid system mainly focuses on the analysis and test of some HEV products. An example is the series of studies on the Toyota Prius carried out by the American Argonne Laboratory. This facility conducted a detailed testing and mapping of the Toyota Prius engine [3]. A simulation model of Prius was then built, followed by an experimental test to assess system operation mechanism and control strategy [4]. The performance of the new Prius was also tested and simulated [5]. In recent years, the basic operation characteristics of the power-split system have been extensively studied. A series of studies have been conducted to investigate the transmission efficiency of the planetary gear train and the system efficiency of the

X. Zeng and J. Wang, *Analysis and Design of the Power-Split Device for Hybrid Systems*, https://doi.org/10.1007/978-981-10-4272-0_4

power-split system [6–9]. The concept of separation factor was proposed to determine the most efficient operation points of the THS [10] and the GM power-split systems [11, 12]. Meanwhile, the control strategies of power-split system were also highly developed. Kang et al. [13] developed a control strategy based on transmission efficiency of the dual-mode system [14]. Proposed a motor control strategy to improve the overall system efficiency. Fuzzy logic control strategy [15] and predictive control strategy [16] have also been applied to the power-split system.

Design of the power-split system is another hot research area as it is the foundation of optimal control and optimal sizing. A design methodology for power-split system with a single planetary gear set was proposed based on drivability evaluation and fuel economy optimization [17]. To find out all the suitable compound schemes, a systematic analysis method is presented based upon the characteristics of power-split system [18]. Using the mechanical point characteristics, a design method for the dual-mode power-split transmission was proposed [19]. These researches mainly concentrated on the design of system schemes. In the aspect of parametric design, [20] finished the parametric design of a single mode power-split vehicle by ensuring the vehicle dynamic performance and the rational speed and torque range of the engine [20]. A mathematical model was established to study the relationship between motor capacity and planetary gear characteristic ratio (i.e. ratio of ring gear radius and sun gear radius), and the system was then matched with the goal of minimizing the motor [21]. A three-dimensional operation space was introduced to downsize the motor/generator size based on the full-load operation performance [22]. In order to improve the vehicle operation performance, a geometric method was proposed to determine the gear ratio of the power-split device [23]. Dynamic programming (DP) was also used to optimize the motor size and characteristic ratio [24]. As DP is too time consuming, a near-optimal method called power-weighted efficiency analysis was proposed for rapid sizing, and then was applied to identify the powertrain parameters of 12 possible configurations [25].

Previous studies have made important contributions in terms of proposing configuration analysis methods, design theories and control strategies. Basic parametric design method has been formed, especially with the goal of ensuring the vehicle dynamic performance. However, some issues are still not clarified in current studies. Firstly, the current research on the design of the power-split system mostly focuses on system characteristics while neglecting the influence of the control strategy. In the power-split system, the engine is decoupled from the road load, which makes the design of engine high in free degree. Therefore, engine selection is highly relative to the control strategy because it determines which region the engine works in. Secondly, current studies fail to analyze the driving cycle in detail. For city buses with relatively regular driving cycles, their frequent braking and long acceleration period affect the selection of the electric energy storage. Hence, detailed analysis of driving cycle is important to ensure the vehicle dynamic performance and the full use of regenerative braking. Besides, some optimal parametric design methods optimize both the fuel economy and dynamic performance, while no general guiding rules for the parametric design of the power-split system, especially for the power-split city bus system, are revealed. And the optimal methods are either time-consuming or linearly simplified, which makes them hard to apply in the engineering practice.

In this chapter, the author proposes a multi-factor integrated parametric design method for city bus. The proposed method considers the power source efficiency (i.e. engine, motor, and generator), transmission efficiency, vehicle driving cycle requirements, vehicle operation performance, and primary control strategy. Some general guiding rules for the parametric design of the power-split city bus are revealed in this chapter. The results of this work will contribute to the development of the design theory for a power-split hybrid electric bus (HEB) which is an important part of public transportation. In this chapter, the system configuration of an HEB is described and the parametric design method for the power-split HEB is introduced. The specific design process is conducted based on the parameters of a typical Chinese city bus. Finally, the design method is validated through co-simulation in MATLAB/Simulink and CRUISE.

4.2 System Configuration

The configuration diagram of the power-split system is shown in Fig. 4.1. The system consists of two PG sets. The front PG set is the PSD, and the rear PG set is the reducer given its fixed ring gear. The engine output shaft is connected to the front planetary carrier. Motor/generator (MG) 1 is connected to the front sun gear while MG2 is connected to the rear sun gear. The front ring gear is connected to the rear planetary carrier, which is connected to the final drive. The ultra-capacitor is used as the electric energy storage.

The energy flow of the power-split system is shown in Fig. 4.2. Under general operating conditions, the output power of the engine is split into two portions through the front PG set. One portion is transformed into electric power through MG1 and then transformed back into mechanical power through MG2. The other portion is directly transmitted to the output shaft in the mechanical path.

Neglecting inertia losses and friction, there are two basic equations for the torque and speed of the front PG set, as shown in Eq. (4.1).

$$\begin{cases} T_{S1}=T_{C1}/(1+k_1) \\ (1+k_1)\omega_{C1}=k_1\omega_{R1}+\omega_{S1} \end{cases} \tag{4.1}$$

where T and ω denote the torque and speed, respectively. Subscripts $C1$, $S1$, and $R1$ denote the carrier, sun gear, and ring gear, respectively, of the front PG set. k_1 is the characteristic ratio of the front PG set, that is, the ratio of the ring gear teeth and the sun gear teeth. The two basic equations also apply to the rear PG set, but the speed of the rear ring gear is constantly zero. That is:

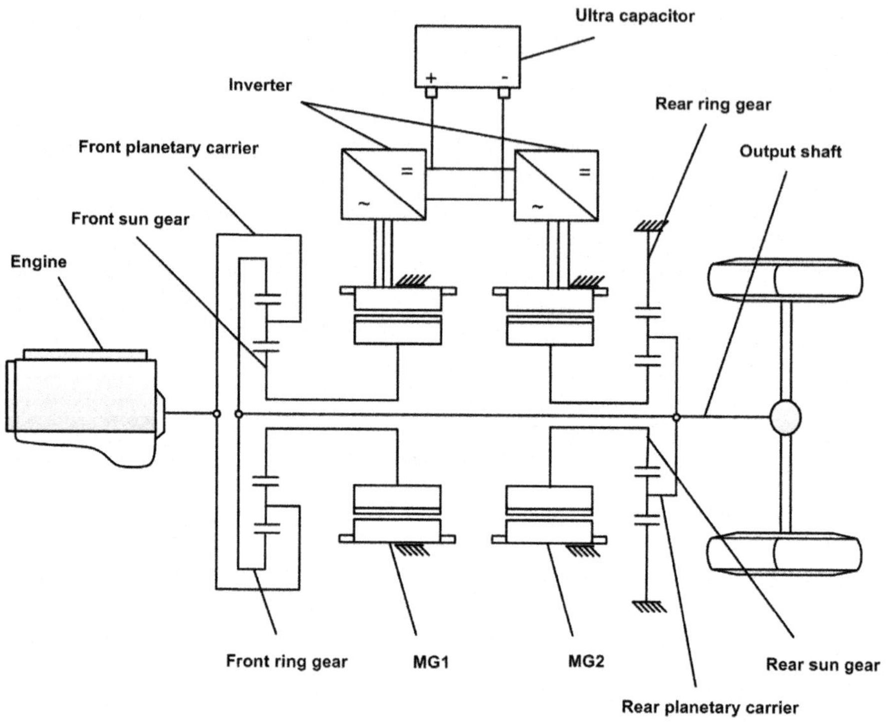

Fig. 4.1 Configuration diagram

Fig. 4.2 Power-split of the system

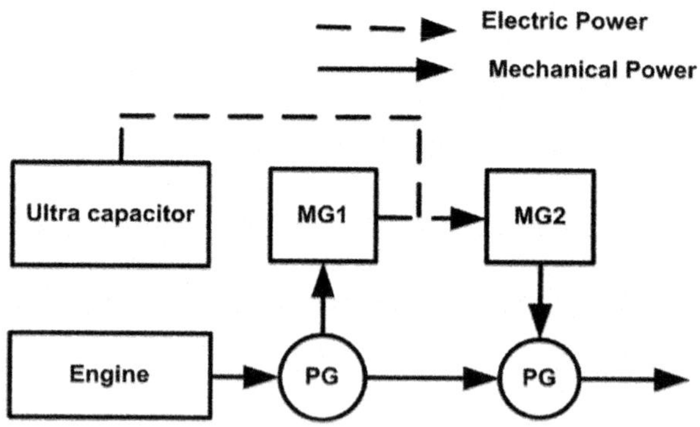

$$\begin{cases} T_{S2} = T_{C2}/(1+k_2) \\ (1+k_2)\omega_{C2} = \omega_{S2} \end{cases} \tag{4.2}$$

where subscripts $C2$ and $S2$ denote the carrier and the sun gear, respectively, of the rear PG set, and k_2 is the characteristic ratio of the rear PG set. The output torque and speed of the power-split system is derived as Eq. (4.3) based on the connection of the PGs and the power sources.

$$\begin{cases} T_{out} = T_e \frac{k_1}{1+k_1} + T_{mg2} \cdot \gamma, \quad \gamma = 1 + k_1 \\ \omega_{out} = \left[\omega_e(1+k_1) - \omega_{mg1}\right]/k_1 = \omega_{mg2}/\gamma \end{cases} \tag{4.3}$$

where subscripts *out*, *e*, *mg*1, and *mg*2 denote the system output shaft, engine, MG1 and MG2, respectively. γ is the transmission ratio of the rear PG set.

4.3 Parameter Design Method

In Eq. (4.1), k_1 is determined by the PSD. Under certain operating requirements, the output of the power-split system (i.e. T_{out} and ω_{out}) can be obtained by the demand torque and the wheel speed according to the mechanical connection, respectively. T_e and ω_e are determined using the primary control strategy when the engine is selected. All the other parameters (i.e. T_{mg2}, ω_{mg1} and ω_{mg2}) in Eqs. (4.1) and (4.2) can then be calculated. The physical meaning of this process is described as follows:

(1) The PSD determines the characteristics of the power-split system (i.e. transmission characteristics).
(2) The output of the core power source engine is determined by the operating requirements and control strategy.
(3) MG1 and MG2 regulate the engine speed and torque, respectively.

Therefore, the parametric design method is as follows: first, a reasonable control strategy is selected; secondly, the core components (i.e. engine and PSD) are determined; thirdly, the parameters of the electric components and the rear PG set are obtained based on the operating requirements. The overall flowchart of the parametric design method is shown in Fig. 4.3.

4.3.1 Engine Selection

The secondary energy storage of the power-split system is the ultra-capacitor, which has high power density but low energy density. Therefore, the engine must provide the required amount of power under steady-state operating conditions to maintain charge sustainability. Considering the driving performance, the steady-state operation requirements mainly include the maximum cruising velocity and the maximum climbing gradient. The demand power of the engine under steady-state operating condition is obtained using Eq. (4.4) [26].

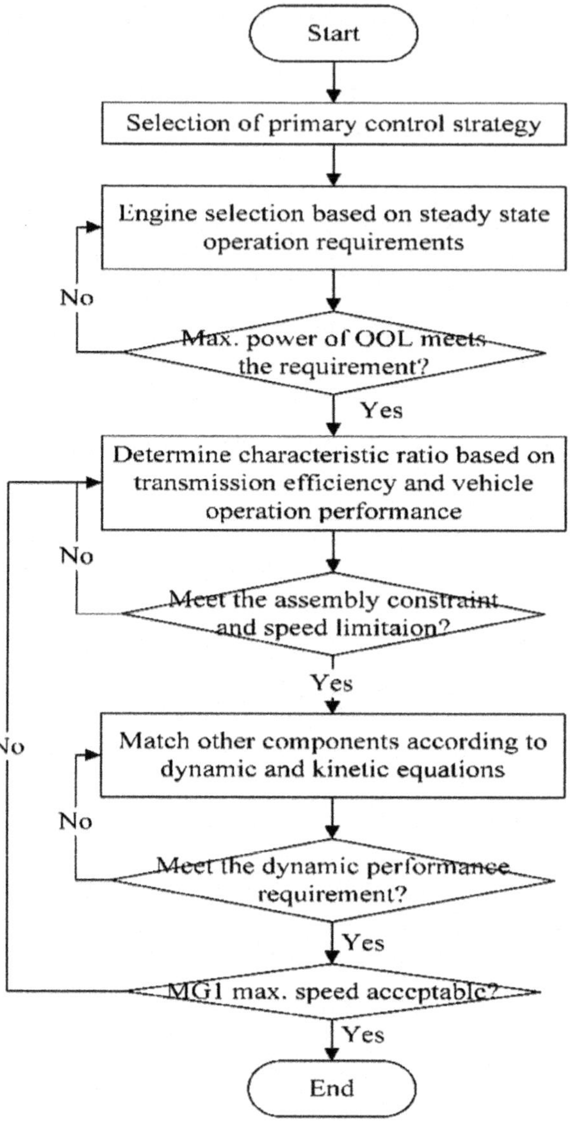

Fig. 4.3 Overall flowchart of the parametric design method

$$P_e = \frac{V_a}{1000\eta_t}\left(Mgf_r + \frac{1}{2}\rho_a C_D A V_a^2 + Mgi\right) \tag{4.4}$$

where P_e is the engine output power (kW), V_a is the vehicle velocity (m/s), η_t is the transmission efficiency, M is the vehicle mass (kg), g is the acceleration of gravity

(m/s^2), f_r is the rolling resistance coefficient, ρ_a is the air density usually set to 1.2258, C_D is the coefficient of air resistance, A is the automotive frontal area (m^2), and i is the climbing gradient.

In addition to the theoretical calculation, we must verify whether the power capacity of the engine is able to satisfy the power demand under most driving cycle conditions. This process must be performed to avoid severe over-discharge of the ultra-capacitor.

As mentioned above, the control strategy affects the parametric design. The engine optimal operation line (OOL) control strategy maximizes the use of the advantages of the power-split system to decouple the road load demand from the engine. Hence, the engine is able to work on the OOL which is compromised of the most efficient operating points of the engine. In this way, the maximum power of the external characteristics of the engine may never be reached during operation. Instead of the rated engine power, the maximum power of the engine OOL determines whether the engine is suitable under the engine OOL control strategy. In this study, the maximum power of the engine OOL should be bigger than P_e.

The OOL is obtained according to the engine map after engine selection. The engine operating speed range is then determined. The following principles can be considered in choosing the speed range:

(1) The engine should reach its maximum power at or before it reaches its maximum speed. Therefore, the maximum speed of the engine should not be too low.
(2) Given that the engine output power consistently increases with the engine speed on the OOL and that MG2 powers the vehicle before the engine starts, the high minimum engine speed results in high electric traction power. Therefore, the minimum engine speed should not be too high.

4.3.2 Characteristics Ratio

The characteristics ratio k_1 of the front PG set determines the power-split characteristics of the hybrid system and therefore determines the transmission efficiency of the system. In the book, the boundaries of k_1 are first determined. An optimal k_1 is then obtained based on the transmission efficiency and the vehicle operation performance of the power-split system. The rear PG set has no influence on the characteristics of the power-split system because it is used as a reducer. Therefore, the rear PG is initially neglected, and MG2 is assumed to be connected to the front ring gear.

First, the condition $k_1 \geq 1.5$ should be met according to the assembly constraint and abutment constraint [27]. Then, the relationship in Eq. (4.5) can be obtained according to Eq. (4.3).

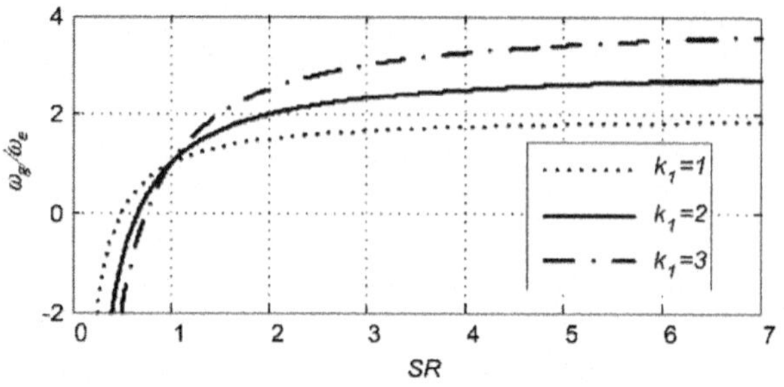

Fig. 4.4 ω_g/ω_e versus *SR*

$$\frac{\omega_{mg1}}{\omega_e} = \frac{\omega_{S1}}{\omega_{C1}} = 1 + k_1 \frac{\omega_{R1}}{\omega_{C1}} = 1 + k_1 - \frac{k_1}{SR} \tag{4.5}$$

where *SR* is the speed ratio of the system defined as $SR = \omega_{C1}/\omega_{R1}$.

The variation in the ratio of the MG1 speed and engine speed versus the speed ratio *SR* is obtained (Fig. 4.4). MG1 speed obviously increases as k_1 increases at a high speed ratio region. Therefore, the condition $k_1 \leq 3$ must be met to avoid high MG1 speed [28].

An optimal k_1 ranging from 1.5 to 3 is then obtained based on the analysis of the transmission efficiency. Transmission efficiency is defined as the ratio of the output power P_{out} and the input power P_{in} of the system, that is, $\eta_t = P_{out}/P_{in}$.

The relationship in Eqs. (4.6) and (4.7) can be obtained according to Eqs. (4.3) and (4.5).

$$\frac{\omega_{S1}}{\omega_{R1}} = (1 + k_1)SR - k_1 \tag{4.6}$$

$$\frac{T_{mg1}}{T_e} = \frac{1}{1 + k_1} \tag{4.7}$$

When the ultra-capacitor is not working, the input power of the system is the engine power transmitted to the front carrier, that is, $P_{in} = P_e = T_e\omega_e$. The output power of the system comprises the electric power and the mechanical power, that is, $P_{out} = P_{mac} + P_{ele}$. The mechanical power is $P_{mac} = T_e k_1/(1 + k_1)\omega_{R1}$ assuming that the mechanical transmission efficiency is 1, as it is much higher than the electric transmission efficiency. The expression of the electric power varies before and after the mechanical point, at which the sun gear speed and the electric power are both zero.

Before the mechanical point, the electric power is

$$P_{ele} = T_{mg1}\omega_{S1}\eta_{mg1}\eta_{mg2} \tag{4.8}$$

where η_{mg1} and η_{mg2} are the efficiencies of MG1 and MG2, respectively. The transmission efficiency before the mechanical point is the same as that in Eq. (4.9).

$$\eta_t = \frac{P_{out}}{P_{in}} = \frac{T_e \omega_{R1} k_1/(1+k_1) + T_{mg1}\omega_{S1}\eta_{mg1}\eta_{mg2}}{T_e\omega_{C1}} = \frac{k_1/(1+k_1) + (T_{mg1}/T_e)(\omega_{S1}/\omega_{R1})\eta_{mg1}\eta_{mg2}}{\omega_{C1}/\omega_{R1}} \tag{4.9}$$

According to Eqs. (4.6) to (4.9), η_{tr} is rewritten as Eq. (4.10).

$$\eta_t = \frac{k_1 + [(1+k_1)SR - k_1]\eta_{mg1}\eta_{mg2}}{SR(1+k_1)} \tag{4.10}$$

The electric power is changed to $P_{ele} = T_{mg1}\omega_{S1}/(\eta_{mg1}\eta_{mg2})$ after the mechanical point. Then, η_t is expressed as Eq. (4.11).

$$\eta_t = \frac{k_1 + [(1+k_1)SR - k_1]/\eta_{mg1}/\eta_{mg2}}{SR(1+k_1)} \tag{4.11}$$

Usually we can assume that the efficiencies of MG1 and MG2 are $\eta_{mg1} = 0.9$ and $\eta_{mg2} = 0.9$, respectively, because both MG1 and MG2 are permanent magnet motors. Hence, the variation in the system transmission efficiency versus *SR* is obtained according to Eqs. (4.10) and (4.11) (Fig. 4.5).

Transmission efficiency increases as k_1 increases in the high speed ratio region, and the maximum efficiency point moves to a high speed ratio as k_1 increases (Fig. 4.5). A large k_1 value is desired because the city bus operates in a high speed ratio region (i.e. low speed region) most of the time.

In reference [23], an optimization method based on geometry technology is proposed. The speed relationship in Eq. (4.1) can be regarded as a space plane, which is called the Characteristic Speed Plane (CSP), in the Cartesian Coordinate System (Fig. 4.6). The CSP represents the operating range of the front PG set. The MSP in Fig. 4.6 is created when each component of the front PG set works at its maximum speed. In order to achieve better vehicle performance, the engine and motors should have wider operation speed range. This means small angle between CSP and MSP is expected to ensure better vehicle performance.

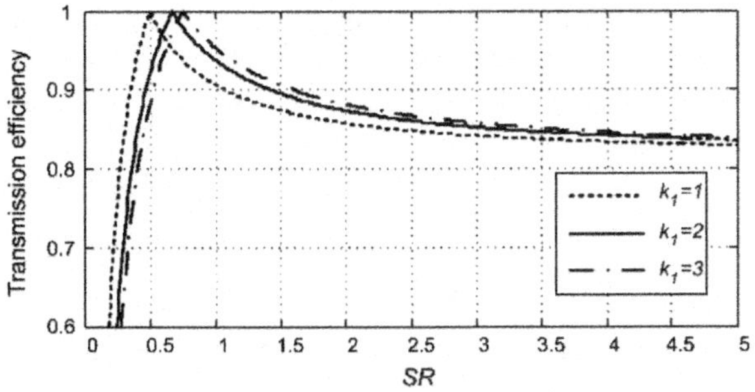

Fig. 4.5 Transmission efficiency versus *SR*

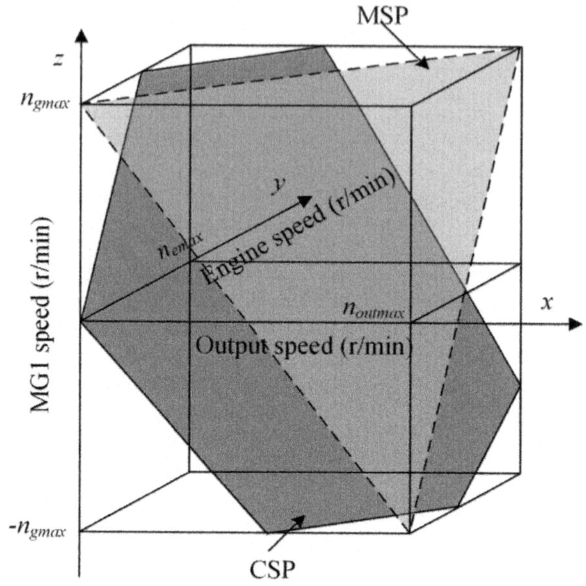

Fig. 4.6 Speed space of the PG set

4.3.3 Other Components

After the parameters of the core components are determined, the other components of the power-split system are identified according to the dynamic and operating requirements. Dynamic performance mainly refers to the maximum velocity, acceleration performance, and grade ability of the vehicle.

After the engine and the front PG set are selected, the demand power of any operation condition can be distributed among the engine and the other power components according to the primary control strategy and the characteristics of the power-split system. While neglecting the pure electric traction ability and the impact of SOC on the traction ability, the engine supplies all the demand power when the demand power is less than the maximum engine power. Otherwise, the engine works at the maximum power point, and the ultra-capacitor supplies the lacking portion of demand power. Based on the engine OOL control strategy, the speed and torque of the engine (i.e. speed and torque of the front planetary carrier) are uniquely determined when the demand power of the engine is determined. In addition, the output shaft speed (i.e. the front ring gear speed) is obtained according to the vehicle speed. Hence, all the other operating parameters are obtained according to the kinetic and dynamic equations of the PG set. The flow chart for calculating the speed and torque of the sun gear is shown in Fig. 4.7.

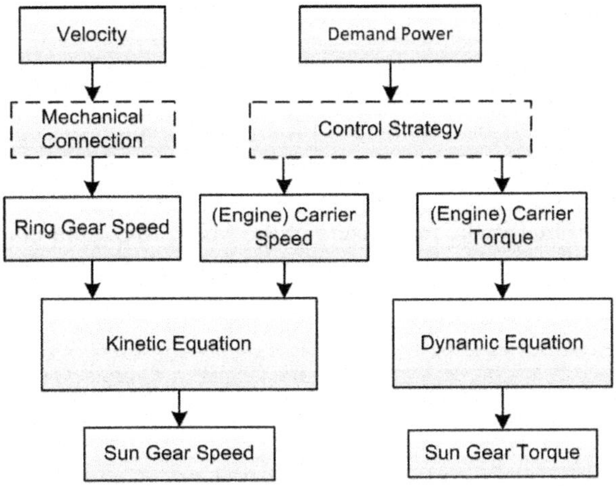

Fig. 4.7 Flow chart of sun gear calculation

Electric power can be calculated with the speed and torque of the sun gear determined above. The mechanical power is calculated in a similar way. The demand power of MG2 is the difference between the demand power and the mechanical power. According to the relationship between the demand power and the engine output power, the output power of MG2 comes from the engine or from both the engine and the ultra-capacitor. The energy utilized by the ultra-capacitor is also obtained by integrating its power over time. The calculation process is shown in Fig. 4.8. At this point, all the operating parameters of the MG1, MG2, and ultra-capacitor are obtained under a certain driving condition.

As the rear PG set is neglected in the analysis above, the MG2 mentioned above is the "*assumed MG2*", the torque and speed of which should be further adjusted. After the "*assumed MG2*" is determined, the characteristics ratio of the rear PG set can be identified according to Eqs. (4.2) and (4.3). The final sizes of all the other components of the power-split system are then determined based on the operating requirements.

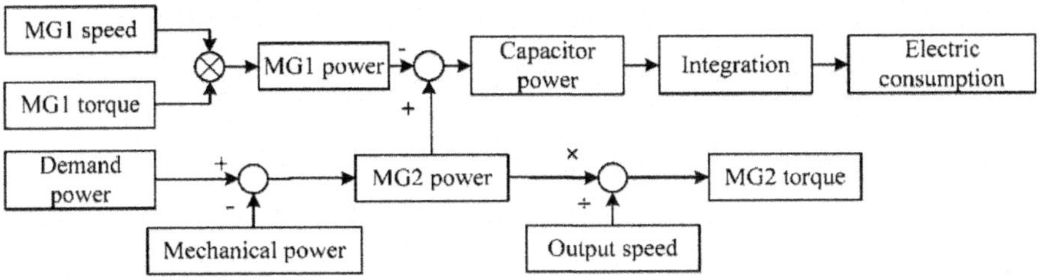

Fig. 4.8 Calculation of other components

4.4 Case Study

The parametric design method is applied based on the parameters of a typical city bus of the world largest bus company. Firstly, the control strategy based on the engine OOL is selected. The engine size is then identified through theoretical calculation. The characteristics ratio of the front PG set is determined based on the system efficiency. Finally, the parameters of the MG1, MG2 and ultra-capacitor are obtained based on the peak values calculated according to the operating requirements.

4.4.1 Initial Condition

The basic parameters of the city bus are shown in Table 4.1. The city bus is 12 m in length and the gross vehicle mass is 16,500 kg. Low rolling resistance tire is used, so the rolling resistance is only 0.0065. The requirements of dynamic performance are shown in Table 4.2. As the bus is operating in the city, the top velocity remains 70 km/h. The acceleration and climbing performances are required to be improved through hybridization.

4.4.2 Control Strategy

This paper intends to adopt the power-following control strategy based on the engine OOL for the power-split city bus. The operation modes of this control strategy consist of the traction mode and the braking mode. The traction mode is

Table 4.1 Basic parameters of the city bus

Items	Parameter
Vehicle gross mass	16,500 kg
Vehicle unladen mass	14,220 kg
Aerodynamic drag coefficient	0.55
Frontal area	6.6 m^2
Rolling resistance coefficient	0.0065
Final ratio	6.17
Rolling radius of tire	510 mm

Table 4.2 Dynamic performance requirements of the city bus

Item	Parameter
Top vehicle speed	70 km/h
Maximum climbing gradient	12%
Acceleration time (0–50 km/h)	<21 s

further divided into the pure electric traction mode and electric-continuously variable transmission (EVT) mode. The braking mode is further divided into mechanical braking, regenerating braking and mixed braking mode [29].

In the traction mode, the engine is off and the MG2 supplies all the traction power when the demand power is less than the minimum power of the engine OOL. The engine works at the maximum power point of the OOL and the MG2 supplies the insufficient power when the demand power is bigger than the maximum power of the engine OOL. Under other conditions, the engine works on the OOL, and the MG2 only converts the electric power coming from the MG1 into mechanical power.

In addition, the engine supplies extra power to charge the ultra-capacitor during driving when SOC is low and the regenerating braking mode will be limited during braking when SOC is high.

4.4.3 Engine Selection

The demand power of the engine is obtained according to Eq. (4.3), as shown in Fig. 4.9. The demand power of the engine is 40 and 92 kW which are calculated according to the maximum vehicle speed and the maximum climbing gradient, respectively. Therefore, the engine power should be more than 92 kW. A product engine rated at 102 kW is selected, considering the accessory power demand. The engine map is shown in Fig. 4.10. As the OOL and the external characteristics of the engine coincide after 1800 rpm, the maximum power of the OOL is also 102 kW, which meets the steady-state operation requirements.

4.4.4 Characteristics Ratio

According to the analysis results of Sect. 4.3.2 of this chapter, k_1 ranges from 1.5 to 3, and a bigger k_1 is desired for the city bus. From this perspective, the closer to 3 k_1

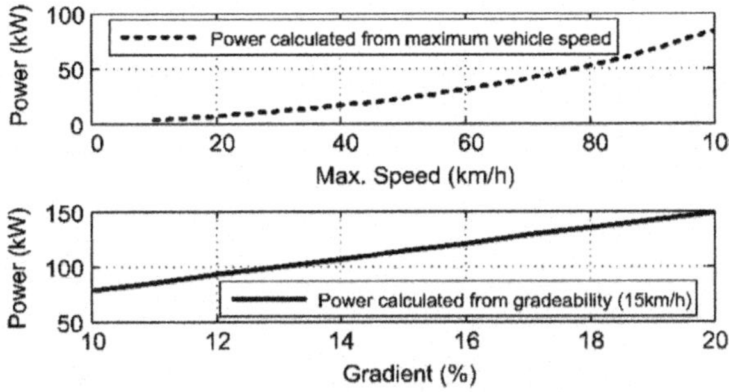

Fig. 4.9 Engine power requirement

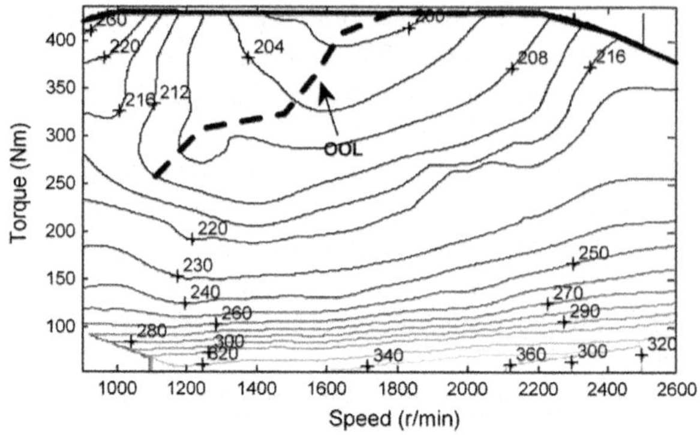

Fig. 4.10 Engine universal characteristics

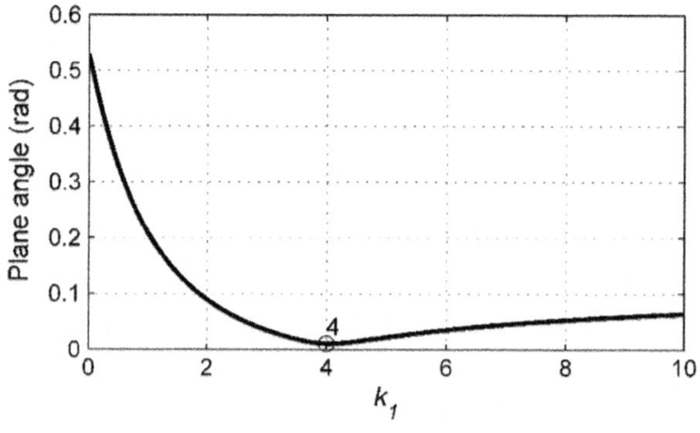

Fig. 4.11 Optimization of characteristics ratio

is, the better the transmission efficiency is. However, it is also known from Sect. 3.2 of this chapter that the maximum MG1 speed will be too high as k_1 comes close to 3. Therefore, the value of k_1 should be finally selected after determination of the maximum MG1 speed. On the other hand, according to the optimization method based on geometry technology, the relationship between k_1 and the angle of CSP and MSP is obtained, as shown in Fig. 4.11. It can be seen that k_1 should be close to 4 to have better vehicle performance and the angle decreases monotonously before 4. Finally, considering the range of k_1, it is tentatively selected that $k_1 = 3$.

4.4.5 Other Components

After the engine and the PSD are identified, the other components remained to be selected include MG1, MG2, ultra-capacitor and rear PG set. As mentioned in Sect. 4.3 of this chapter, the electric components of the power-split system can be selected according to the operating requirements after the engine and PSD are determined. In this section, the peak values of the MG1 and MG2 are extracted from the operating parameters which are calculated according to the requirements of the dynamic performance as shown in Table 4.2. The capacity of the ultra-capacitor is determined mainly based on the driving cycle requirements.

To simplify the parametric design process, it is initially assumed that the MG2 is connected to the front planetary carrier as the rear PG set is utilized to reduce the maximum torque of the MG2, and it has no influence on the characteristics of the power-split system. After the parameters of the "assumed MG2" are determined, the rear PG set is then taken into consideration to guarantee moderate maximum speed and torque of the MG2.

At the end of this section, the parameters of the electric components are determined based on the peak values of all the operating requirements.

4.4.5.1 Ultimate Acceleration Operating Condition

As shown in Fig. 4.12, it is feasible to obtain the ultimate acceleration curve of the vehicle from the curve of a conventional vehicle. Figures 4.13, 4.14 and 4.15 express the demand power, speed and torque of the power sources. Under the ultimate acceleration condition, the vehicle starts at low velocity with low demand

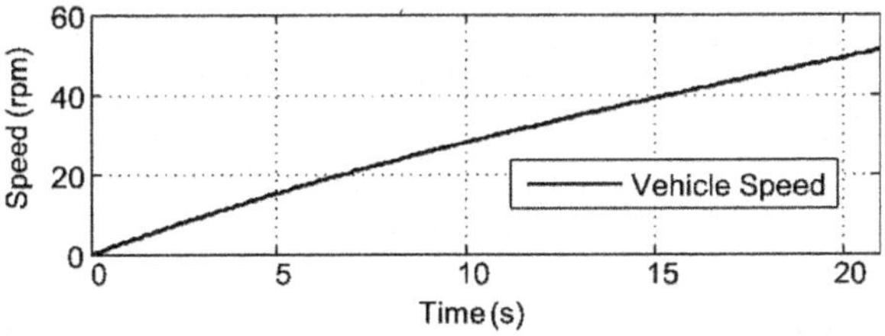

Fig. 4.12 Vehicle speed in acceleration

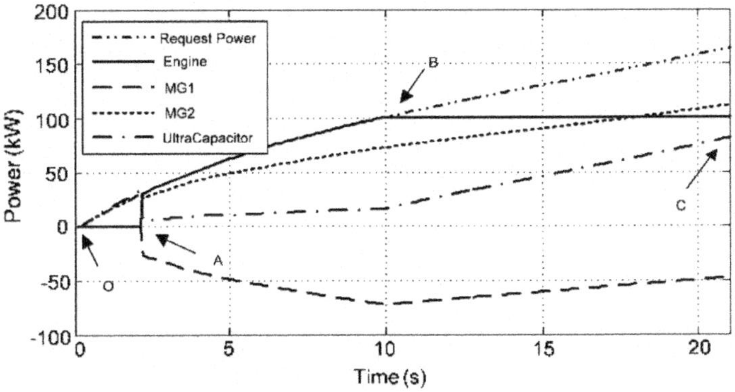

Fig. 4.13 Power of power components in acceleration

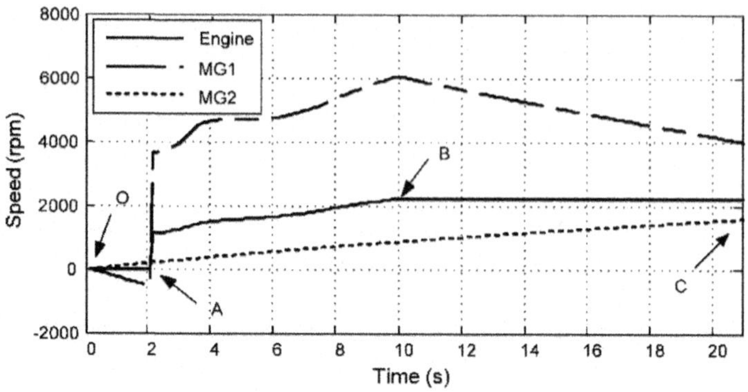

Fig. 4.14 Speed of power components in acceleration

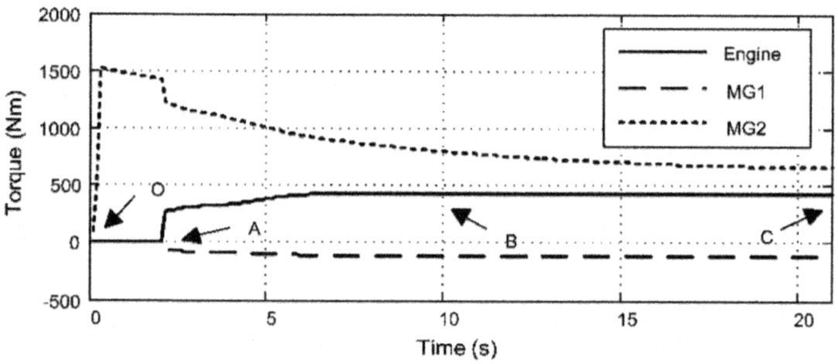

Fig. 4.15 Torque of power components in acceleration

power, which makes it possible for the MG2 to drive the vehicle alone with the engine off as shown in OA stage. When the demand power increases to a certain level, the engine starts, as shown in AB stage. When the engine power reaches its maximum limitation, the output power of the engine remains stable, as shown in BC stage. The energy consumption of the ultra-capacitor is shown in Fig. 4.16. In OA and AB stage, the power demand of ultra-capacitor is relatively low and the

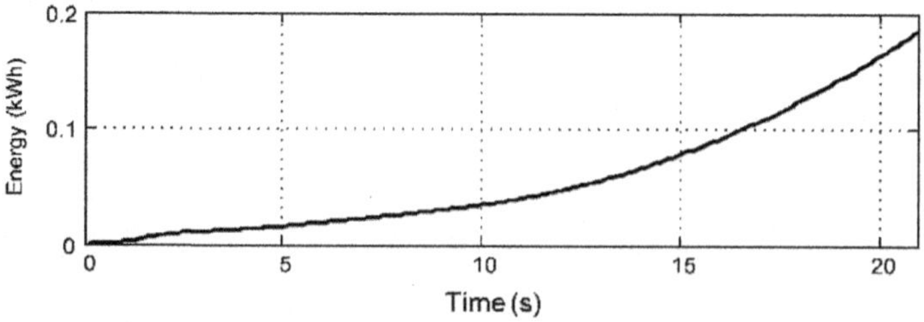

Fig. 4.16 Energy consumption of ultra-capacitor in acceleration

Table 4.3 Peak values in the ultimate acceleration

Component	Max. torque	Max. speed	Max. power
MG1	−113.2 Nm/6018 rpm	6018 rpm/−113.2 Nm	71.3 kW/6018 rpm
MG2	1489 Nm/13.7 rpm	1606 rpm/664.8 Nm	108.5 kW/1606 rpm

energy consumption increases slowly. In BC stage, the demand power of ultra-capacitor increases fast and the energy consumption increases fast too. The final energy consumption of the ultra-capacitor is 0.184 kWh.

The peak points of the three figures are extracted and listed in Table 4.3 In the ultimate acceleration operating condition, each component just works at the peak points for a short time, so it is appropriate to take the peak points as their peak parameters.

4.4.5.2 Maximum Climbing Operating Condition

The demand torque, speed and power of each component in the maximum climbing operating condition are listed in Table 4.4. As the vehicle may be on the slope for a long period of time, it is appropriate to take the values in Table 4.4 as the rated parameters.

4.4.5.3 Maximum Speed Operating Condition

The speed, torque and power of each component in the maximum speed operating condition are listed in Table 4.5.

4.4.5.4 Ultra-Capacitor Selection Based on Driving Cycle

The China Typical Urban driving cycle represents the general operating conditions of a city bus in China. It is significant to conduct the parametric design of the hybrid city bus based on this driving cycle. The speed, acceleration and demand power of the driving cycle is shown in Fig. 4.17. The Statistic characteristics are listed in Table 4.6, including the speed, acceleration, braking deceleration, accelerating power, and the peak and mean values of the braking power.

Table 4.4 Peak values in the climbing

Component	Torque/Nm	Speed/rpm	Power/kW
Engine	430	2005	90
MG1	−113.2	6270	−74
MG2	1489	482	75
Ultra-capacitor	\	\	16

Table 4.5 Values in the maximum speed condition (70 km/h)

Component	Torque/Nm	Speed/rpm	Power/kW
Engine	305	1226	39
MG1	−72	−1630	14
MG2	−81	2246	−17

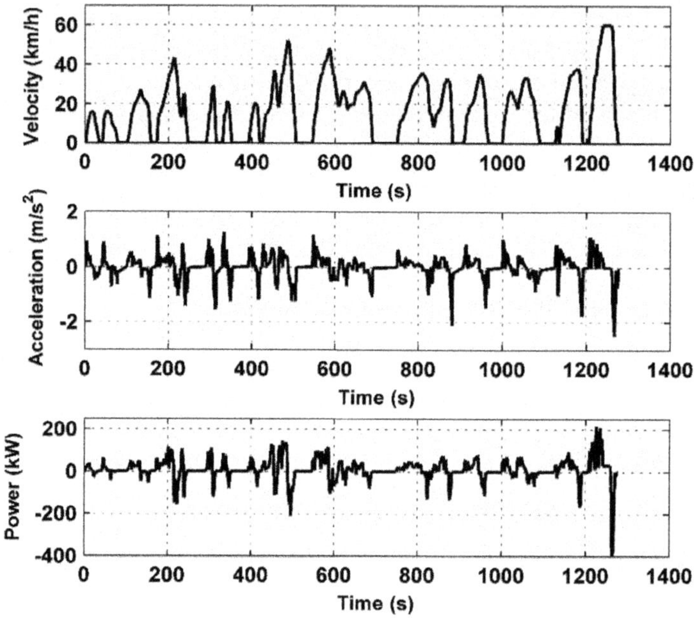

Fig. 4.17 China Typical Urban driving cycle

Table 4.6 Main features of China Typical Urban cycle

Items	Value
Max. velocity	60.3 km/h
Avg. velocity	21.8 km/h
Max. acceleration	1.26 m/s^2
Avg. acceleration	0.31 m/s^2
Max. deceleration	2.49 m/s^2
Avg. deceleration	0.44 m/s^2
Max. acceleration power	217.2 kW
Avg. acceleration power	42.7 kW
Max. brake power	404.1 kW
Avg. brake power	50.2 kW

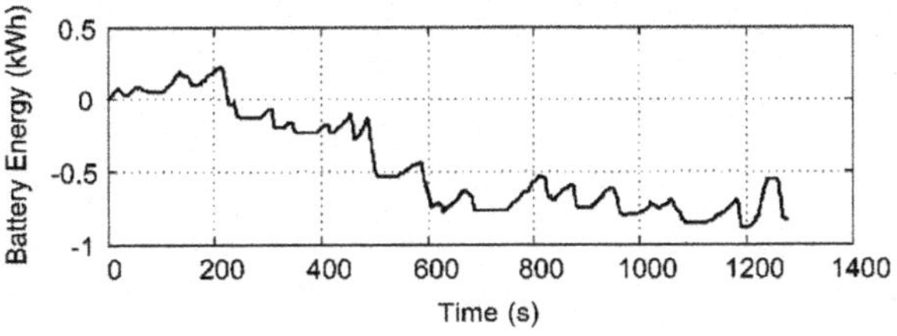

Fig. 4.18 Variation of battery energy in driving cycle

It is obvious that requirements of the dynamic performance are able to cover that of the driving cycle, which makes it unnecessary to take the dynamic performance into consideration in further parametric design. However, it is required that the ultra-capacitor should sustain the charge and recycle as much regenerative energy as possible.

Even if the ideal braking force distribution control strategy is applied, it is impossible to recycle all the regenerative energy. In this paper, it is considered that MG2 works at the peak torque when the braking force is bigger than the maximum braking force of the MG2.

The energy time-varying curve of the ultra-capacitor is shown in Fig. 4.18, defining that the discharging power is positive and charging power negative. The battery energy changes dynamically in the operation cycle. The energy decreases in acceleration period while increases in braking period. Therefore, the change between the final and inertial battery energy may not satisfy the requirement within the operating cycle.

By dividing the driving cycle into 14 kinematic sequences (the start of one idling till the start of the next idling is defined as a kinematic sequence), the energy released and recycled by the ultra-capacitor in each sequence is calculated. Figure 4.19 shows the electric energy variation of the ultra-capacitor at the end of each sequence. It is observed that the ultra-capacitor is mostly on charge under optimal braking force distribution control strategy. Even in some discharging sequences, the energy released is less than 0.1 kWh. To recycle as much regenerative energy as possible, the capacity of the ultra-capacitor is required to be greater than 0.2938 kWh.

Figure 4.20 shows the maximum energy released by ultra-capacitor during each sequence. The ultra-capacitor capacity is required to be greater than 0.3331 kWh to sustain the power support.

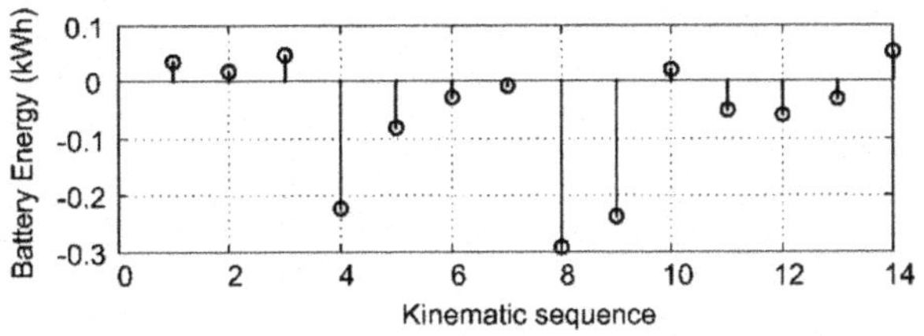

Fig. 4.19 Variation of ultra-capacitor energy in each kinematic sequence

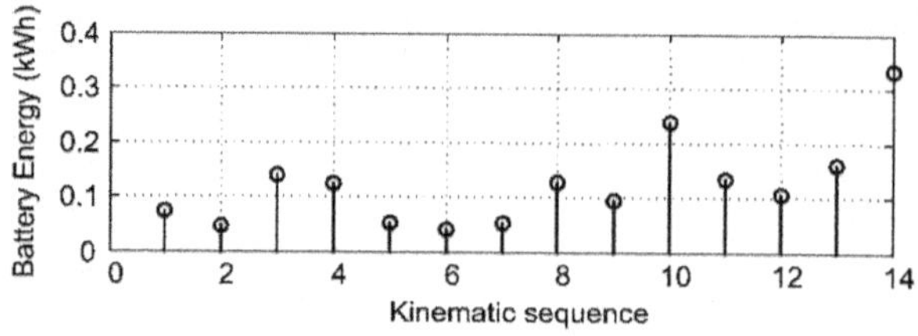

Fig. 4.20 Maximum discharge capacity in every kinematic sequence

Considering all the aspects above and the margin, the ultra-capacitor capacity is determined to 0.5 kWh.

4.4.5.5 Characteristics Ratio of Rear PG Set

Seen from the theoretical calculation above, it is feasible to determine the maximum speed, torque and power of the "assumed MG2" according to the values extracted from the calculation results of the ultimate acceleration and the maximum climbing operating condition.

Considering the impact of the characteristics ratio of the rear PG set on the demand speed and torque of the MG2, the maximum torque of the MG2 decreases and the maximum speed increases as the characteristics ratio of the rear PG set increases (Fig. 4.21). To make a compromise, it is determined that $k_2 = 2$. Consequently, the maximum speed of the MG2 should be greater than 6738 rpm, and the maximum torque should be greater than 496 Nm.

4.4.6 Summary

Meeting all the requirements mentioned above and considering the margin, the electric components are selected as shown in Table 4.7. Figures 4.22 and 4.23 are the efficiency characteristics of the MG1 and MG2. And it can be seen from Table 4.5 that the maximum speed of MG1 is acceptable, so $k_1 = 3$ is finally determined.

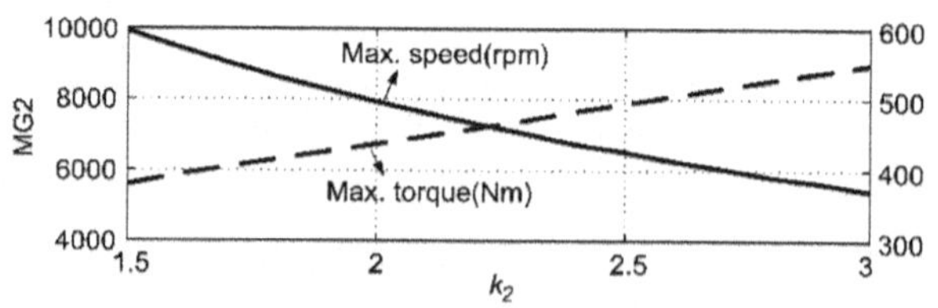

Fig. 4.21 Rear characteristics ratio

Table 4.7 Parameters of electric components

Item	Value
MG2 rated power	110 kW
MG2 rated torque/base speed	500 Nm/2100 rpm
MG2 max. speed	>6750(7000 rpm)
MG1 rated power	75 kW
MG1 torque at max. speed	117 Nm
MG1 max. speed	6400 rpm
Ultra-capacitor capacity	0.5 kWh

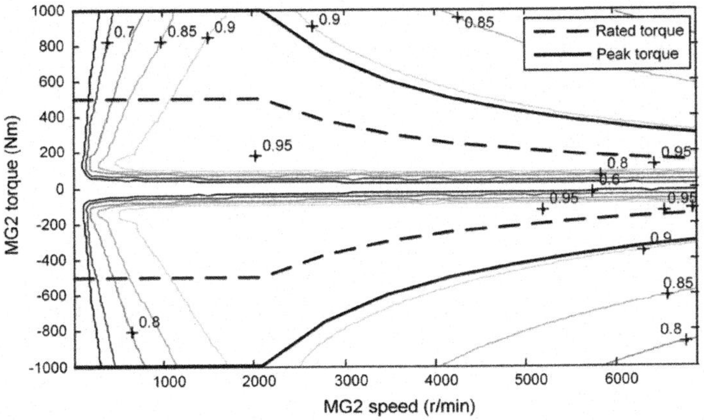

Fig. 4.22 Efficiency characteristics of MG2

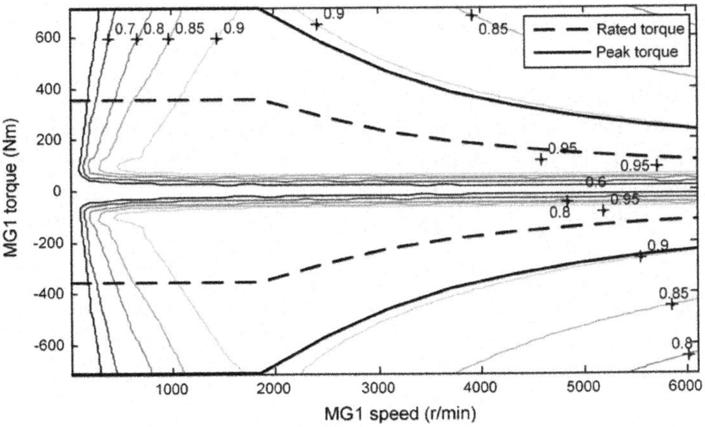

Fig. 4.23 Efficiency characteristics of MG1

4.5 Simulation Verification

Aimed at the verification of the parametric design results, simulation for the power-split hybrid bus with the components determined in the previous section is performed using the co-simulation of CRUISE and MATLAB/Simulink software.

4.5.1 Acceleration Verification

It can be seen from the calculation results in Tables 4.3, 4.4 and 4.5 that acceleration performance requires the most MG2 power. And the ultimate acceleration is an important reference operation condition to finish the parametric design. So it is necessary to verify the acceleration performance to validate the dynamic performance of parametric results.

With the engine OOL control strategy, the vehicle ultimate acceleration curve is obtained by making the engine work at the maximum allowed power and the MG2 works at the peak torque. It is shown that the vehicle reaches the speed of 50 km/h within 15.2 s, which is much better than the requirement of 21 s, as shown in Fig. 4.24. This is because the climbing performance requires higher power than the acceleration performance. Most of the engine points are on the OOL except some points during the engine starting stage, as shown in Fig. 4.25. In conclusion, the parametric design results are able to cover the requirements of the maximum acceleration.

4.5.2 Climbing Verification

In the calculation, climbing performance requires the most MG1 power. And climbing performance is important to expand the application area of the power-split city bus especially in some mountainous region. Also, the maximum climbing is another important reference operation condition to finish the parametric design. So the climbing performance is verified in this section.

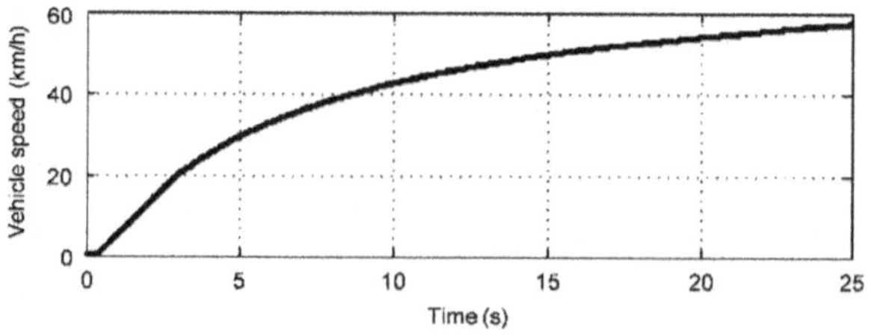

Fig. 4.24 Simulation vehicle speed in acceleration

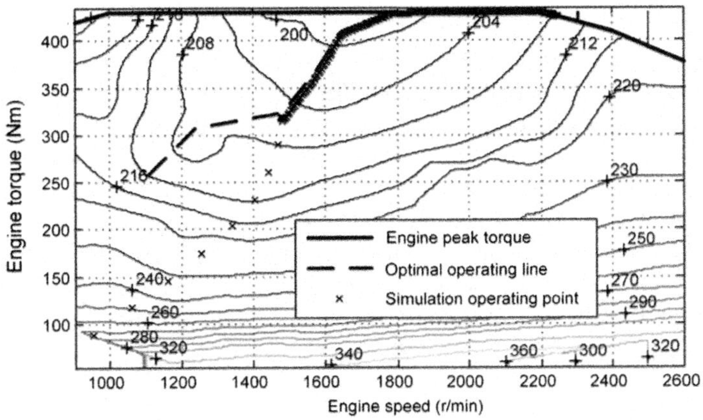

Fig. 4.25 Engine operating points in simulation

Fig. 4.26 Climbing operating condition

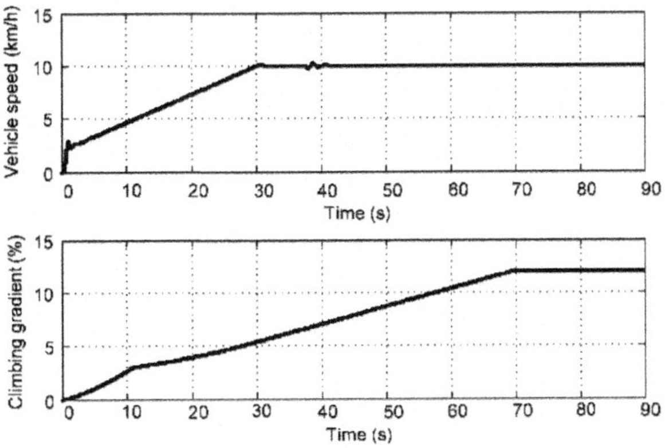

The climbing operating condition is set as Fig. 4.26. When the bus is driving on the slope of 12%, the speed, torque and power of each power source are listed in Table 4.8. In theoretical calculation, the motor average efficiency is lower than that in reality, which makes the output power of each component in simulation less than that of the theoretical calculation. The parametric design results meet the requirements of gradeability in simulation.

Table 4.8 Simulation values in the climbing condition

Component	Torque/Nm	Speed/rpm	Power/kW
Engine	420	1720	76
MG1	−90	5950	56
MG2	497	940	49

4.5.3 Fuel Economy Verification

Fuel economy is the most concentrated indicator of hybrid electric system. And the improvement of fuel economy can be used to validate the analysis of transmission efficiency. Finally, a fuel economy simulation for the power-split hybrid bus is performed under the China Typical Urban driving cycle. In the simulation the bus speed follows the cycle well and the SOC keeps balanced as shown in Fig. 4.27. And it can be seen from Fig. 4.28 that the engine operating points are mostly around the OOL except that some dynamic points deviate from the OOL during mode shifting. It is obvious that the fuel economy is remarkably improved using the design results, as shown in Table 4.9.

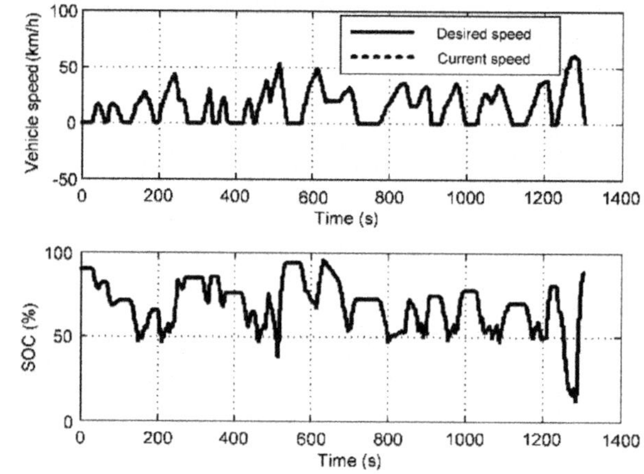

Fig. 4.27 Vehicle speed and SOC in simulation

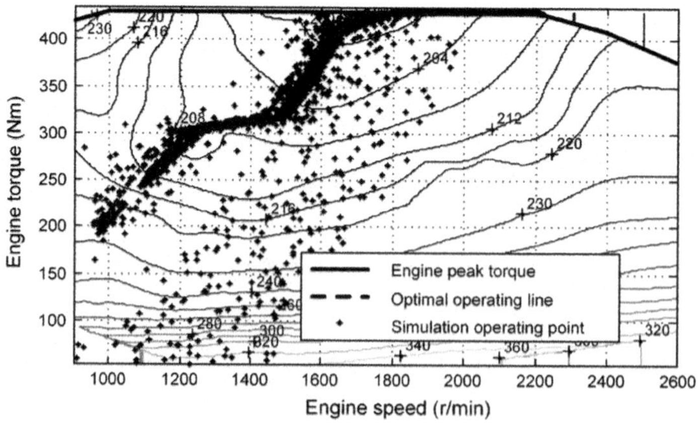

Fig. 4.28 Engine operating points in simulation

Table 4.9 Fuel economy of power-split hybrid bus and conventional bus

Bus type	Fuel economy	Improvement
Conventional bus	35 L/100 km	\
Power-split hybrid bus	22.39 L/100 km	36%

4.5.4 Summary

The simulation results verified that the power capacity of the engine and the energy capacity of the ultra-capacitor are able to maintain the SOC balance in the driving cycle, and the parametric design results of the MG1 and MG2 from theoretical calculation are able to meet the requirements of the dynamic performance. In addition, the favorable fuel economy validates the analysis of the transmission efficiency of the power-split system and the determination of the characteristics ratio of the PSD. Above all, the simulation results have verified the reasonability of the parametric design method.

4.6 Summary

This chapter proposes a multi-factor integrated parametric design method for a novel power-split HEB. Firstly, the engine should supply steady-state power demand to maintain charge sustainability of HEB, which means the demand engine power can be calculated by considering the top vehicle speed and maximum climbing gradient. This is a general parametric design principle for HEV. However, control strategy should also be taken into consideration in the engine selection as the engine of power-split system is decoupled from the road load. Secondly, the transmission efficiency and vehicle operation performance are analyzed to determine the characteristics ration of the PSD. Besides, the characteristics ratio should meet the assembly constraint. After the main power source and the PSD are determined, main parameters of MG1, MG2 and real PG set are obtained by calculating the kinetic and dynamic equations of the power-split system, based on dynamic operating cycle and typical driving cycle. Finally, the capacity of the ultra-capacitor is design by analyzing the energy variation in the driving cycle. The proposed method is verified through simulation, which reveals favorable results.

Above all, the proposed method considers the driving cycle requirements, control strategy, and transmission efficiency of the power-split hybrid electric system. Except some common parametric design principles used for HEV, some specific design principles are raised for the power-split HEB:

(1) MG1 should stabilize and adjust the operation of the engine. In extreme cases, MG1 should provide sufficient torque at its maximum speed to balance the maximum torque of the engine.

(2) To ensure the dynamic performance, the rated torque of MG2 should meet the maximum climbing requirements, and the peak torque should meet the ultimate acceleration requirements.
(3) To improve both the dynamic performance and the fuel economy, ultra-capacitor should supply sufficient energy and recover regenerative braking energy as much as possible in every kinematic sequence of the driving cycle.

In conclusion, the parametric design method proposed in this work meets the requirements for dynamic vehicle performance and obtains favorable fuel economy. The proposed method can thus provide an improved understanding of design theory for power-split hybrid electric systems with two PG sets, especially for the development of power-split HEBs.

References

1. Toyota. Worldwide sales of Toyota hybrids top 10 million units. 2017. http://www.toyota-global.com/innovation/environmental_technology/hv10million/.
2. Hutchinson T, Burgess S, Herrmann G. Current hybrid-electric powertrain architectures: Applying empirical design data to life cycle assessment and whole-life cost analysis. Appl Energy. 2014;119:314–29.
3. Duoba M, Ng H, Larsen R. In-situ mapping and analysis of the Toyota Prius HEV engine. SAE Technical Paper. 2000.
4. An F, Rousseau A. Integration of a modal energy and emissions model into a PNGV vehicle simulation model, PSAT. SAE Technical Paper. 2001.
5. Rousseau A, Kwon J, Sharer P, Pagerit S, Duoba M. Integrating data, performing quality assurance, and validating the vehicle model for the 2004 Prius using PSAT. 2006.
6. Chen C. Power flow and efficiency analysis of epicyclic gear transmission with split power. Mech Mach Theory. 2013;59:96–106.
7. Chen C, Angeles J. Virtual-power flow and mechanical gear-mesh power losses of epicyclic gear trains. J Mech Des. 2007;129:107–13.
8. Chen C, Liang TT. Theoretic study of efficiency of two-dofs of epicyclic gear transmission via virtual power. J Mech Des. 2011;133:031007.
9. Gupta A, Ramanarayanan C. Analysis of circulating power within hybrid electric vehicle transmissions. Mech Mach Theory. 2013;64:131–43.
10. Meisel J. An analytic foundation for the Toyota Prius THS-II powertrain with a comparison to a strong parallel hybrid-electric powertrain. SAE 2006 World Congress & Exhibition. 2006.
11. Meisel J. An analytic foundation for the two-mode hybrid-electric powertrain with a comparison to the single-mode Toyota Prius THS-II powertrain. SAE World Congress & Exhibition. 2009.
12. Meisel J. Kinematic study of the GM front-wheel drive two-mode transmission and the Toyota Hybrid System THS-II transmission. SAE Int J Engine 2011;4:1020–34.
13. Kang J, Choi W, Kim H. Development of a control strategy based on the transmission efficiency with mechanical loss for a dual mode power split-type hybrid electric vehicle. Int J Automot Tech. 2012;13:825–33.
14. Kim J, Kim N, Hwang S, Hori Y, Kim H. Motor control of input-split hybrid electric vehicles. Int J Automot Tech. 2009;10:733–42.
15. Abdelsalam AA, Cui S. A fuzzy logic global power management strategy for hybrid electric vehicles based on a permanent magnet electric variable transmission. Energies. 2012;5:1175–98.

16. Johannesson L, Pettersson S, Egardt B. Predictive energy management of a 4QT series-parallel hybrid electric bus. Control Eng Pract. 2009;17:1440–53.
17. Kim J, Kang J, Kim Y, Kim T, Min B, Kim H. Design of power split transmission: design of dual mode power split transmission. Int J Automot Tech. 2010;11:565–71.
18. Wang W, Song R, Guo M, Liu S. Analysis on compound-split configuration of power-split hybrid electric vehicle. Mech Mach Theory. 2014;78:272–88.
19. Li C.-T, Zhang X, Peng H. Design of power-split hybrid vehicles with a single planetary gear. 2012.
20. Yu Y, Gao Y, Peng H, Wang Q. Parametric design of power-split HEV drive train. In: Vehicle Power and Propulsion Conference; 2009. VPPC'09. IEEE. IEEE; 2009. p. 1058–63.
21. Feng Z, Yingke W, Xuhui W, Li Z. Optimal design of planetary gear in multi-mode hybrid drive system. In: Transportation Electrification Asia-Pacific (ITEC Asia-Pacific); 2014 IEEE Conference and Expo. IEEE; 2014. p. 1–5.
22. Ahn K, Papalambros PY. Design optimization of motor/generator full-load characteristics in two-mode hybrid vehicles. SAE Technical Paper. 2009.
23. Li Y, Kar N. Advanced design approach of power split device of plug-in hybrid electric vehicles using dynamic programming. In: Vehicle Power and Propulsion Conference (VPPC); 2011 IEEE. IEEE; 2011. p. 1–6.
24. Liu J, Peng H. A systematic design approach for two planetary gear split hybrid vehicles. Veh Syst Dyn. 2010;48:1395–412.
25. Zhang X, Peng H, Sun J. A Near-Optimal Power Management Strategy for Rapid Component Sizing of Multimode Power Split Hybrid Vehicles. IEEE Trans Control Syst Tech. 2015;23:609–17.
26. Cole DE. Elementary vehicle dynamics. Department of Mechanical Engineering, The University of Michigan; 1972.
27. Liu HY, Zhang ZQ, Zhang MQ. Mechanisms and machine theory[M] (Chinese edition). Beijing: China Machine Press; 2007.
28. Liu J. Modeling, configuration and control optimization of power-split hybrid vehicles[D]. The University of Michigan; 2007.
29. Mansour C, Clodic D. Dynamic modeling of the electro-mechanical configuration of the Toyota Hybrid System series/parallel power train[J]. Int J Automot Technol. 2012;13(1): 143–166.

Chapter 5
Dynamic Coordinated Control Method of Planetary Hybrid Power-Split System

Parameter-matching methods and optimal control strategies of the top-selling hybrid electric vehicle (HEV), namely, power-split HEV, are widely studied. In particular, extant research on control strategy focuses on the steady-state energy management strategy to obtain better fuel economy. However, given that multi-power sources are highly coupled in power-split HEVs and influence one another during mode shifting, conducting research on dynamic coordination control strategy (DCCS) to achieve riding comfort is also important. This chapter proposes a predictive-model-based DCCS. First, the dynamic model of the objective power-split HEV is built and the mode shifting process is analyzed based on the developed model to determine the reason for the system shock generated. Engine torque estimation algorithm is then designed according to the principle of the nonlinear observer, and the prediction model of the degree of shock is established based on the theory of model predictive control. Finally, the DCCS with adaptation for a complex driving cycle is realized by combining the feedback control and the predictive model. The presented DCCS is validated on the co-simulation platform of AMESim and Simulink. Results show that the shock during mode shifting is well controlled, thereby improving riding comfort.

5.1 Background

With the increasing concerns toward environmental pollution and energy crisis, the development of vehicles with hybrid and pure electric powertrains has received significant attention. The most successful hybrid system is undoubtedly the Toyota Hybrid System (THS) [1–3]. Since the launch of the first Prius in 1997, Toyota has sold over 6 million hybrid electric vehicles (HEVs) with the THS. THS consists of one or two planetary gear sets and three power sources, namely, engine, generator, and motor. The engine output power of the THS is split into two paths: mechanical and electric. This kind of hybrid system is called the power-split system [4].

X. Zeng and J. Wang, *Analysis and Design of the Power-Split Device for Hybrid Systems*, https://doi.org/10.1007/978-981-10-4272-0_5

As a kind of series-parallel hybrid electric system, the power-split system has both the advantages of series and parallel systems with excellent dynamic performance and fuel economy [5]. In this regard, the analysis method, design theory, and control strategy of such a system are significantly related to exploring and developing the conditions of this advanced device. An early study on the power-split system includes massive vehicle tests to analyze its operation principle and basic control strategy [6]. In the power-split system, engine speed is adjusted by the generator, and engine torque is tuned by the motor. Hence, the engine in such a system is decoupled from the road, providing convenience to the optimal control [7]. The existence of electric power in the particular system, however, affects transmission efficiency [8, 9]. This effect is the reason the characteristics of transmission efficiency are extensively analyzed to further accomplish the design method [10] and optimal energy management strategy (EMS) [11, 12] for obtaining better fuel economy.

The three power sources are highly coupled in the power-split system, with each power source having its own dynamic response characteristics. The start or stop of any power source during mode shifting may lead to the fluctuation of the output torque, which seriously influences riding comfort. Therefore, the dynamic coordination control strategy (DCCS) for attaining good riding comfort is also important in this research field. State-feedback control [13] and sliding mode control [14] are important control methods for dynamic control strategies. However, existing research on DCCS focuses on the gear shift control of the parallel hybrid system, which usually utilizes PID and feedback [15, 16] to manipulate motor torque [17, 18] or engine speed [19]. A control strategy without a clutch is proposed to shorten the time of gear shifting [20], and a linear quadratic regulator (LQR) is applied to optimize the position control algorithm [21]. Some previous studies on dynamic coordination to improve the power-split hybrid system only used feedback and gain to realize the close-loop motor control [22]. However, among current research, the DCCS based on feedback and gain not only requires an extensive amount of experimental data, but also extra sensors and actuators. Such precondition leads to the adverse adaption of the strategy to complex driving cycle and the increase of total cost. Therefore, a newly developed DCCS that can efficiently adapt to a complex driving cycle without any extra sensors and actuators has obvious technological superiority. Moreover, this advanced control strategy conforms to the application and promotion of the power-split hybrid system. This research proposes a predictive model-based DCCS based on the rapidly developed model predictive control (MPC) technology. MPC is based on rolling optimization, which avoids massive computation and is suitable for solving the optimization problem of nonlinear systems. Previous studies on predictive control utilized GPS to obtain further information on driving cycle and to realize EMS optimization [23]. Within the development of MPC technology, the optimization of HEV EMS based on MPC recently obtained the approximate global optimal solution [24, 25]. These studies also employed the MPC to optimize the EMS of the power-split hybrid system [26, 27]. However, the application of the MPC is still constrained because of its poor real-time performance. Based on this instance, the current study removes the online optimization component of the MPC algorithm that influences real-time performance. A dynamic coordination strategy with predictive model and feedback control is subsequently designed. This strategy has good real-time performance, can satisfactorily adapt to a complex driving cycle, and can eliminate the need for massive experimental data.

The engine torque signal in the CAN bus of an actual vehicle is obtained with the table look-up method, which cannot reflect the dynamic characteristics of the engine. Hence, such approach cannot satisfy the requirement of DCCS. Forming an exact engine torque estimation algorithm is another important component of the DCCS. Model-based observer is widely used for optimal state estimation [28]. A mode-dependent estimator with fast sampling frequency is proposed for the nonuniformly sampled nonlinear system [29]. The design method of the state estimator for discrete-time Takagi-Sugeno fuzzy system is proposed, and the solving method of the estimator parameters is determined [30]. A model-based observer for state and stress estimation is designed and validated by experiment [31]. Among extant research for engine torque estimation, the algorithm based on Hammerstein nonlinear model cannot indicate the exact dynamic characteristics of the engine [32]. The algorithm based on neural network has massive computation load and poor real-time performance, and is therefore not suitable for DCCS [33]. The algorithm based on the least square method vector machine [34] and the sliding-mode observer [35, 36] relies on massive experimental data. Current torque estimation algorithms cannot be used directly for the DCCS. Therefore, based on the principle of nonlinear observer [37] and model-based engine torque observer [38], this chapter designs a nonlinear engine torque estimation algorithm for the power-split system, which obtains good precision and avoids massive experimental data.

The following are the significance of the work in this chapter: (1) The reason for the transmission shock that occurred during the mode shifting process is determined for the power-split system based on the analysis of the dynamic model; (2) The established nonlinear engine torque estimator is fit for transient control and is easy to apply in engineering practice; (3) This chapter creatively applies the model-based prediction technology for dynamic coordination, which ensures the control strategy with good adaptation for complex driving cycles.

The rest of the chapter is organized as follows. Following the Introduction, Sect. 2 constructs the dynamic model of the power-split system and establishes the engine torque estimation algorithm to realize exact engine torque estimation without extra sensors. Section 3 establishes the predictive model and dynamic coordination strategy. Section 4 validates the engine torque estimation algorithm and dynamic coordination strategy by co-simulating AMESim and Simulink. Section 5 presents the conclusion of the chapter.

5.2 Dynamic Analysis

The research object of this chapter is a dual-planetary power-split hybrid electric bus (HEB). The configuration of the HEB is displayed in Fig. 4.1. This section first forms the dynamic model of the power-split HEB, and then solves the relationship between the degree of shock and the torque of the power sources based on the system dynamic model. Next, the reason for the system shock generated under steady-state EMS is analyzed, and the engine torque estimation algorithm is finally established.

5.2.1 Dynamic Model

The system dynamic model is constructed by dividing the power-split system into three parts, namely, front planetary gear set, rear planetary gear set and system output. The free-body diagram of each part is then developed to explore the system dynamics.

The free-body diagram of the front planetary gear set is illustrated in Fig. 5.1, which shows that the engine outputs the torque through PG1 with the adjustment of MG1. PG1 is highlighted in the dashed line box. The pinion gears are assumed to be an ideal force transfer mechanism because of their small mass. The internal forces of the gears, which neglect the coulomb and viscous friction, are shown in Fig. 5.1. Applying Eula's law derives the basic dynamic equations:

$$I_{r1} \cdot \dot{\omega}_{r1} = F_1 \cdot R_1 - T_{r1} \tag{5.1}$$

$$I_{c1} \cdot \dot{\omega}_{c1} = T_{c1} - F_1 \cdot S_1 - F_1 \cdot R_1 \tag{5.2}$$

$$I_{s1} \cdot \dot{\omega}_{s1} = F_1 \cdot S_1 - T_{s1} \tag{5.3}$$

where T, ω and I denote torque, speed, and inertia, respectively; $r1$, $c1$ and $s1$ represent the front ring gear, planetary carrier, and sun gear, respectively; R_1 and S_1

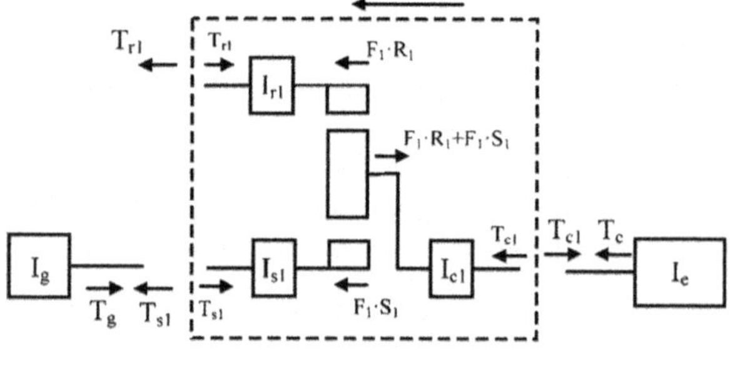

Fig. 5.1 Free-body diagram of front planetary gear set

are the radiuses of the front ring gear and sun gear; and F_1 is the internal force of PG1.

According to the connection relationship of the engine, MG1, and PG1, the governing equations of the engine and MG1 can be obtained by assuming that moving to the left denotes positive motion, as presented in Eqs. (5.4) and (5.5).

$$I_e \cdot \dot{\omega}_e = T_e - T_{c1} \tag{5.4}$$

$$I_g \cdot \dot{\omega}_g = T_{s1} - T_g \tag{5.5}$$

where e and g denote the engine and MG1 respectively.

Speed relationship can be derived from the connection of PG1 and the power sources, that is, $\omega_e = \omega_{c1}$ and $\omega_g = \omega_{s1}$. According to Eqs. (5.2) to (5.5), the following can be obtained:

$$(I_e + I_{c1}) \cdot \dot{\omega}_e = T_e - F_1 \cdot R_1 - F_1 \cdot S_1 \tag{5.6}$$

$$\left(I_g + I_{s1}\right) \cdot \dot{\omega}_g = F_1 \cdot S_1 - T_g. \tag{5.7}$$

Finally, according to the internal meshing relationship of PG1, the speed of the internal components of PG1 should satisfy the following relationship:

$$\omega_g + \omega_{r_1} \cdot k_1 = (k_1 + 1) \cdot \omega_e \tag{5.8}$$

The free-body diagram of the rear planetary gear set is illustrated in Fig. 5.2, which indicates that MG2 outputs the torque through PG2 (highlighted in the dashed line box).

The following relationship can be obtained by using the derivation similar to that of the dynamic equations of PG1:

$$(I_m + I_{s2}) \cdot \dot{\omega}_m = T_m - F_2 \cdot S_2 \tag{5.9}$$

$$I_{c2} \cdot \dot{\omega}_{c2} = F_2 \cdot S_2 + F_2 \cdot R_2 - T_{c2} \tag{5.10}$$

where m, $s2$ and $c2$ represent MG2, rear sun gear, and rear planetary carrier, respectively; $R2$ and S_2 are the radiuses of rear ring gear and sun gear respectively.

Fig. 5.2 Free body diagram of rear planetary gear set

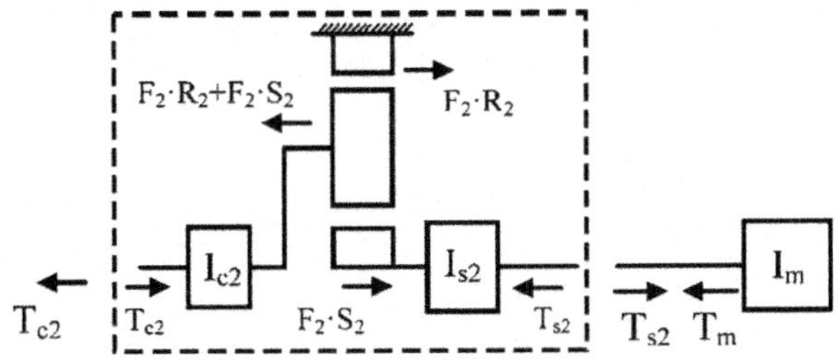

Similarly, the components of PG2 are constrained by the speed relationship. Given that the rear ring gear is fixed to the transmission house, of which the speed is constantly zero, the speed relationship of PG2 components is as following:

$$\omega_m = (k_2 + 1)\omega_{c2} \tag{5.11}$$

where k_2 is the characteristics parameter of PG2.

According to Eqs. (5.9) to (5.11), the relationship between the output torque of PG2 (T_{c2}) and the input torque (T_m) is obtained, as shown in Eq. (5.12):

$$T_{c2} = T_m \cdot (1 + k_2) - \left[(I_m + I_{s2})(1 + k_2)^2 - I_{c2}\right] \cdot \dot{\omega}_{c2} \tag{5.12}$$

Figure 5.3 is the free-body diagram of the system output, which indicates that the output torques of PG1 and PG2 are coupled and transmitted to the final drive (FD) to overcome the driving resistance of the bus.

Considering only the longitude dynamics and neglecting the wheel slip lead to the following equation:

$$(\dot{\omega}_{r1}/i_o)R_t^2 \cdot m = (T_{c2} + T_{r1})i_o - T_f \tag{5.13}$$

where i_o is the FD ratio, R_t is the wheel radius, m is the vehicle mass, and T_f is the driving resistance of the bus. On a straight and flat road, the following relationship exists:

$$T_f = T_{fb} + mgf_rR_t + 0.5\,\rho AC_D\left(\frac{\omega_{r1}}{i_o}\right)^2 R_t^3 \tag{5.14}$$

where T_{fb} is the braking torque, f_r is the rolling resistance coefficient, A is the frontal area, and C_D is the air drag coefficient.

According to Eqs. (5.1), (5.12), (5.13), and (5.14) and to the connection of front ring gear and rear planetary carrier, the following relationship is derived:

$$\begin{aligned} &\dot{\omega}_{r1}\left[R_t^2 \cdot m/i_o + I_{r1}i_o + (I_m + I_{s2})(1 + k_2)^2 i_o - I_{c2}i_o\right] \\ &= [T_m(1 + k_2) + F_1R_1]i_o - T_{fb} - mgf_rR_t - 0.5\,\rho AC_D(\omega_{r1}/i_o)^2 R_t^3 \end{aligned} \tag{5.15}$$

Finally, the dynamic model of the dual-planetary power-split system is formed with Eqs. (5.6), (5.7), (5.8), and (5.15).

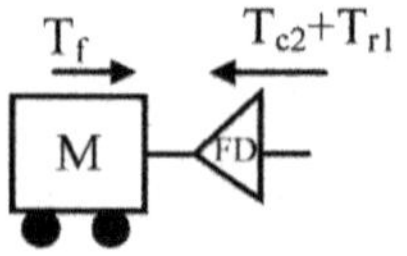

Fig. 5.3 Free body diagram of system output

5.2.2 Shock Analysis

The longitude degree of shock (i.e., jerk) is defined as the gradient of the acceleration (a) and is assumed to be positive with the same direction of the vehicle speed, as shown in Eq. (5.16),

$$j = \frac{da}{dt} = \frac{d^2 v}{dt^2} \tag{5.16}$$

where v is vehicle velocity, which can be expressed as $v = \omega_{r1}/i_o \cdot R_t$. Eq. (5.16) is then reformed as Eq. (5.17):

$$j = \frac{\ddot{\omega}_{r1}}{i_o} R_t. \tag{5.17}$$

According to the dynamic equations of the power-split system, the acceleration of the system output can be obtained:

$$\begin{aligned} \ddot{\omega}_{r1}\left[\frac{I'_v}{k_1^2 i_o I'_g} + \frac{(1+1/k_1)^2 I'_v}{i_o I'_e} + 1\right] = & -\frac{1}{k_1 I'_g}\dot{T}_g + (1+k_2)\left[\frac{(1+1/k_1)^2}{I'_e} + \frac{1}{k_1^2 I'_g}\right]\dot{T}_m \\ & + \frac{1+1/k_1}{I'_e}\dot{T}_e - \left[\frac{(1+1/k_1)^2}{i_o I'_e} + \frac{1}{k_1^2 i_o I'_g}\right]\dot{C} \end{aligned} \tag{5.18}$$

where $I'_v = R_t^2 m/i_o + I_{r1} i_o + (I_m + I_{s2})(1+k_2)^2 i_o - I_{c2} i_o$, $I'_e = I_e + I_{c1}$, $I'_g = I_{s1} + I_g$, $C = T_{fb} + mgf_r R_t + 0.5\,\rho A C_D \left(\frac{\omega_r}{i_o}\right)^2 R_t^3$.

When the operation mode changes from electric vehicle (EV) to electronic-continuously variable transmission (EVT), the engine is started and adjusted to the required speed. Under steady-state EMS, the system output torque is supplied by MG2 in EV mode, while in EVT mode, the output torque comes from both the engine and MG2, as shown in Eqs. (5.19) and (5.20) respectively.

$$T_{\text{out\_EV}} = T_m(1+k_2) \tag{5.19}$$

$$T_{\text{out\_HEV}} = \frac{T_e k_1}{(1+k_1)} + T_m(1+k_2) \tag{5.20}$$

Within the EVT mode, the engine torque is calculated according to the current engine speed and steady-state EMS, that is $T_e = f(\omega_e)$. Meanwhile, MG2 supplies insufficient torque to satisfy driver demand. In this process, engine speed is controlled by MG1 through the PID control method. The large target torque of MG1 is calculated according to the principle of PID control when EV changes to EVT because engine speed is significantly lower than target speed when the mode is shifted. The large MG1 torque accelerates the engine, ensuring a rapid engine start. However, with the fast response of MG1, a large $\dot{T}_g$ is generated. Meanwhile, because of poor efficiency in the low-speed region, the engine only manages to output just enough torque to assist in starting the engine. Equation (5.20) reveals that the output torque of MG2 gradually decreases with the slow increase of engine torque; specifically, both $\dot{T}_m$ and $\dot{T}_e$ are insufficient. In addition, braking torque does not exist in the driving mode, and the driving resistance nearly changes during the transient process, which implies that $\dot{C} \approx 0$. In this mode shifting process, a negative jerk occurs because of the rapid response of MG1, as specified in Eqs. (5.17) and (5.18). When EVT changes to EV, the output torque partially and completely comes from MG2 before and after the mode is shifted, according to Eqs. (5.19) and (5.20). Therefore, the output torque of MG2 is managed for its rapid increase to satisfy the required torque of steady-state EMS when the mode is shifted. However, the engine torque does not suddenly disappear, but decreases instead at a specific response speed. In particular, the response speed of MG2 is remarkably larger than that of the engine, that is, $|\dot{T}_m| > |\dot{T}_e|$, which induces a positive jerk in this process. The torque changing ratio of each power source during a specific mode shifting process is summarized as an example to determine the system shock, as shown in Table 5.1.

The above analysis signifies that neglecting the inertia moments and response speed of power sources in steady-state EMS is the essential reason for system shock during mode shifting. When EV changes to EVT, the rapid response of MG1 produces a negative jerk. Contrarily, when EVT changes to EV, the uncoordinated response speed of the engine and MG2 causes a positive jerk.

5.2.3 Engine Torque Estimation

According to the preceding analyses, the control objective of the DCCS is to realize smooth mode shifting by coordinating the torque changing ratio of power source. As mentioned earlier, engine torque estimation is the basis of DCCS. Therefore, engine toque estimation algorithm is established in this section.

The relationship of the engine torque, MG1 torque, MG2 torque, and engine speed is obtained according to the system dynamic model, as shown in Eq. (5.21):

Table 5.1 Mode changing process

Item	$\frac{1}{k_1 I'_g}\dot{T}_g$	$(1+k_2)\left[\frac{(1+1/k_1)^2}{I'_e} + \frac{1}{k_1^2 I'_g}\right]\dot{T}_m$	$\frac{1+1/k_1}{I'_e}\dot{T}_e$	$\dot{C}$	j
EV changes to EVT	12,990	3946	148	≈0	Negative
EVT changes to EV	6709	37,480	−927	≈0	Positive

$$\left[\frac{i_o k_1}{(1+k_1)} + \frac{I'_v}{k_1(1+k_1)I'_g}\right] T_e = \left[(1+1/k_1)I'_v + \frac{I'_e i_o}{(1+1/k_1)} + \frac{I'_v I'_e}{k_1(1+k_1)I'_g}\right] \dot{\omega}_e - i_o T_m - \frac{I'_v}{k_1 I'_g} T_g + C \tag{5.21}$$

Motor torque is estimated in the motor control unit, and engine speed can be easily acquired with the already installed speed sensor. Thus, T_m, T_g, and ω_e can be directly obtained in the CAN bus of the vehicle. Meanwhile, C can be calculated according to vehicle speed and brake pressure. Therefore, in the power-split system, engine torque can be determined along with the already-known state variables according to Eq. (5.21). The dynamic equation is discretized to be applied in the control strategy as Eq. (5.22):

$$\left[\frac{i_o k_1}{(1+k_1)} + \frac{I'_v}{k_1(1+k_1)I'_g}\right] T_e(\mathrm{n}) = \left[(1+1/k_1)I'_v + \frac{I'_e i_o}{(1+1/k_1)} + \frac{I'_v I'_e}{k_1(1+k_1)I'_g}\right] [\omega_e(\mathrm{n}) - \omega_e(\mathrm{n}-1)] - i_o T_m(\mathrm{n}) - \frac{i'_v}{k_1 I'_g} T_g(\mathrm{n}) + C \tag{5.22}$$

where n denotes the current sample point, and n−1 denotes the previous sample point.

The above analysis suggests that, based on the characteristics of the highly coupled power-split system, the proposed engine torque estimation algorithm utilizes the available signals in the vehicle to estimate the unknown engine torque, taking into consideration the inertia moments and acceleration of each component. However, current research aimed at the gear shifting process of the parallel HEV is unable to use the proposed algorithm because the separation of the clutch decouples the engine from the transmission. Evidently, the engine torque estimation algorithm for the power-split hybrid system fully considers the system characteristics, thus avoiding massive computation load, extra sensors, and several calibrations, as well as ensuring good real-time performance. The estimation algorithm also acquires good nonlinear performance because the engine torque estimation algorithm is derived from the nonlinear dynamic model of the power-split system.

5.3 Dynamic Coordination Control Strategy

Based on the above discussion, shock occurs because engine response speed is significantly slower than MG1 when the engine is started. Moreover, due to the poor instantaneous control characteristics of the engine, the torque changing ratio of

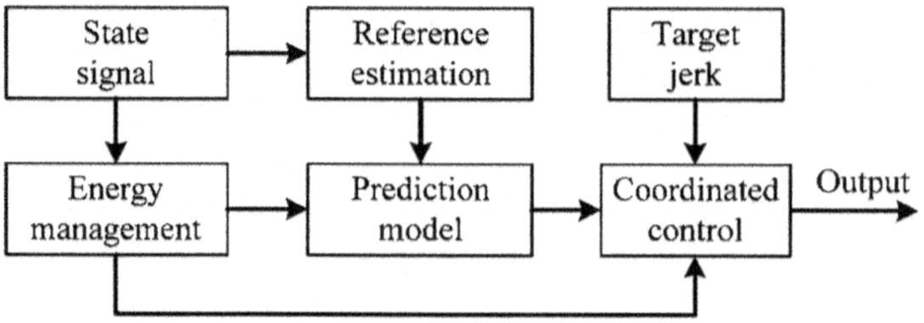

Fig. 5.4 Scheme of the DCCS

MG1 should be limited, and MG2 should offer an appropriate compensation torque to realize smooth mode shifting. When the engine shuts down, the increase of MG2 torque is uncoordinated with the decrease of engine torque. In this event, DCCS should coordinate the torque changing ratio of the two power sources.

The control scheme of DCCS is shown in Fig. 5.4. The required torque changing ratios of MG1 and MG2 are first predicted according to the specific mode shifting process, satisfying the requirement of riding comfort. The output of steady-state EMS is then adjusted according to the predictive value, that is, the dynamic coordination.

Reference estimation (i.e., engine torque estimation) was introduced in Sect. 2.3 of this chapter. The present section mainly aims to discuss the coordination control methods of MG1 and MG2, which are the core components of DCCS.

5.3.1 *Limitation of MG1 Torque*

The relationship between torque changing ratio of MG1 and degree of shock is obtained according to Eqs. (5.6), (5.7), (5.8), (5.15), and (5.17), as shown in Eq. (5.23):

$$\dot{T}_g = \left[(1+k_2)\left(\frac{(1+1/k_1)^2}{I'_e} + \frac{1}{k_1^2 I'_g}\right)\dot{T}_m + \frac{1+1/k_1}{I'_e}\dot{T}_e - \ddot{\omega}_r\left(\frac{I'_v}{k_1^2 i_o I'_g} + \frac{(1+1/k_1)^2 I'_v}{i_o I'_e} + 1\right) - \left(\frac{(1+1/k_1)^2}{i_o I'_e} + \frac{1}{k_1^2 i_o I'_g}\dot{C}\right)\right] k_1 I'_g \tag{5.23}$$

No braking torque exists and the driving resistant force changes little, which means $\dot{C} \approx 0$. Thus, Eq. (5.24) can be obtained by iterating Eqs. (5.17) to (5.23) and by discretion.

$$\Delta T_g = U_m[T_m(\mathrm{n}+1) - T_m(\mathrm{n})] + U_e[T_e(\mathrm{n}+1) - T_e(\mathrm{n})] - U_j j \tag{5.24}$$

where ΔT_g denotes the variation of MG1 torque; n + 1 denotes the next sample point; and U_m, U_e and U_j are presented by Eqs. (5.25), (5.26), and (5.27), respectively.

$$U_m = (1 + k_2)\left[\frac{(1 + 1/k_1)^2}{I'_e} + \frac{1}{k_1^2 I'_g}\right] k_1 I'_g \tag{5.25}$$

$$U_e = (k_1 + 1) I'_g / I'_e \tag{5.26}$$

$$U_j = \left[\frac{I'_v}{k_1^2 i_o I'_g} + \frac{(1 + 1/k_1)^2 I'_v}{i_o I'_e} + 1\right] \frac{k_1 I'_g i_o}{R_t}. \tag{5.27}$$

In Eq. (5.24), T_m(n + 1) and T_e(n + 1) represent the predictive MG2 and engine torque, which are obtained from the outputs of steady-state EMS with the first-order inertia element, respectively, as shown in Eqs. (5.28) and (5.29).

$$T_m(\mathrm{n} + 1) = T_{m\_ctr} \frac{1}{\tau_m s + 1} \tag{5.28}$$

$$T_e(\mathrm{n} + 1) = T_{e\_ctr} \frac{1}{\tau_e s + 1} \tag{5.29}$$

where T_{m_ctr} and T_{e_ctr} denote the control signals of MG2 and engine, which are the outputs of steady-state EMS; τ_m and τ_e are the time constants of MG2 and engine; and s is the Lagrange factor.

In Eq. (5.24), T_m(n) is the current MG2 torque obtained in the CAN bus, and T_e(n) is the estimated engine torque. With the limitation of maximum allowable shock $|j_{\max}|$, the variation of MG1 torque (ΔT_g) is then restricted between the top $\Delta T_{g\_\max}$ and bottom $\Delta T_{g\_\min}$ thresholds, as indicated in Eq. (5.30).

$$\begin{cases} |\Delta T_g|_{\min} = U_m[T_m(n+1) - T_m(\mathrm{n})] + U_e[T_e(\mathrm{n}+1) - T_e(\mathrm{n})] - U_j j_{\max} \\ |\Delta T_g|_{\max} = U_m[T_m(n+1) - T_m(\mathrm{n})] + U_e[T_e(\mathrm{n}+1) - T_e(\mathrm{n})] - U_j j_{\max} \end{cases} \tag{5.30}$$

The allowable range of variation of MG1 torque is obtained with Formula. (5.30). If the difference between the control signal and the current torque of MG1 is larger than the top limitation, the output signal is adjusted to T_g(n) + $\Delta T_{g\_\max}$. If, however, the difference is less than the bottom limitation, the output signal is adjusted to T_g(n) + $\Delta T_{g\_\min}$. If the difference is in the limitation range, the control signal is a direct output.

5.3.2 Compensation of MG2

Similarly, the relationship between MG2 torque and degree of shock is obtained according to Eqs. (5.6), (5.7), (5.8), (5.15) and (5.17), as shown in Eq. (5.31):

$$\dot{T}_m = \frac{-j\frac{i_o}{R_t}\left[\frac{I_v'}{k_1^2 i_o I_g'} + \frac{(1+1/k_1)^2 I_v'}{i_o I_e'} + 1\right] + \frac{1}{k_1 I_g'}\dot{T}_g - \frac{1+1/k_1}{I'}\dot{T}_e + \left[\frac{(1+1/k_1)^2}{i_o I_g'} + \frac{1}{k_1^2 i_o I_g'}\right]\dot{C}}{(1+k_2)\left[\frac{(1+1/k_1)^2}{I_e'} + \frac{1}{k_1^2 I_g'}\right]} \tag{5.31}$$

The discrete equation of Eq. (5.31) can be obtained as follows:

$$\Delta T_m = U_g'[T_g(\mathrm{n}+1) - T_g(\mathrm{n})] - U_e'[T_e(\mathrm{n}+1) - T_e(\mathrm{n})] - jU_j' \tag{5.32}$$

where ΔT_m is the variation of MG2 torque; $T_g(\mathrm{n}+1) = T_{g\_ctr}/(\tau_g s + 1)$; T_{g_ctr} is the control signal of MG1, which is the output of steady-state EMS; τ_g is the time constant of MG1. U_j', U_g' and U_e' are presented by Eqs. (5.33) to (5.35), respectively:

$$U_j' = \left\{\frac{i_o}{R_t}\left[\frac{I_v'}{k_1^2 i_o I_g'} + \frac{(1+1/k_1)^2 I_v'}{i_o I_e'} + 1\right]\right\} \Big/ \left\{(1+k_2)\left[\frac{(1+1/k_1)^2}{I_e'} + \frac{1}{k_1^2 I_g'}\right]\right\} \tag{5.33}$$

$$U_g' = \frac{1}{k_1 I_g'} \Big/ \left\{(1+k_2)\left[\frac{(1+1/k_1)^2}{I_e'} + \frac{1}{k_1^2 I_g'}\right]\right\} \tag{5.34}$$

$$U_e' = \frac{1+1/k_1}{I_e'} \Big/ \left\{(1+k_2)\left[\frac{(1+1/k_1)^2}{I_e'} + \frac{1}{k_1^2 I_g'}\right]\right\} \tag{5.35}$$

In Eq. (5.32), $T_g(\mathrm{n})$ is the current MG1 torque obtained in the CAN bus, and $T_e(\mathrm{n})$ is the estimated engine torque. With the limitation of the maximum allowable shock $|j_{\max}|$, the variation of MG2 torque is then restricted between the top $\Delta T_{m\_\max}$ and bottom $\Delta T_{m\_\min}$ thresholds, as indicated in Eq. (5.36):

$$\begin{cases} |\Delta T_m|_{\min} = U_g'\left[T_g(n+1) - T_g(\mathrm{n})\right] + U_e'[T_e(\mathrm{n}+1) - T_e(\mathrm{n})] - j_{\max} U_j' \\ |\Delta T_m|_{\max} = U_g'\left[T_g(n+1) - T_g(\mathrm{n})\right] + U_e'[T_e(\mathrm{n}+1) - T_e(\mathrm{n})] + j_{\max} U_j' \end{cases} \tag{5.36}$$

With the range of MG2 torque variation obtained, if the difference between control signal and current torque of MG2 is less than the bottom limitation, then the output signal is adjusted to $T_m(\mathrm{n}) + \Delta T_{m\_\min}$. If the difference is larger than the top limitation, the output is adjusted to $T_m(\mathrm{n}) + \Delta T_{m\_\max}$. If the difference is in the limitation range, the control signal is directly obtained.

5.4 Simulation Verification

To validate DCCS, the co-simulation of AMESim and Simulink is conducted based on the parameters of a typical Chinese city bus (Table 5.2).

In DCCS, the power sources are simplified as first-order elements. The time constants and inertia moments of the power sources are obtained by bench test, as shown in Table 5.3.

The diagram of the co-simulation platform is shown in Fig. 5.5. The AMESim vehicle model is in the left dash-dot line box. In this model, the components in the dashed line boxes are the core components, namely, the engine, MG1, MG2, PG1, PG2, ultra-capacitor, and the bus. Meanwhile, the components in the solid-line boxes are the inertia moments of the core components. The rest of the components are signal receivers and transmitters. Placed inside the right dash-dot line box is the top-level structure of the control strategy, which consists of parameter estimation, mode switching module, energy management, and dynamic coordination modules.

Simulations with only steady-state EMS and the DCCS-integrated EMS are carried out. In the simulation without DCCS, the control strategy is produced through port A; otherwise, it is induced via port B, as illustrated in Fig. 5.5.

5.4.1 Simulation in Chinese Bus Driving Cycle

Chinese Bus Driving Cycle (CBDC) represents the average driving condition of a Chinese city bus. Therefore, CBDC is chosen as the driving cycle to validate DCCS. The simulation vehicle speed of the DCCS-integrated EMS is depicted in Fig. 5.6. The figure indicates that the DCCS-integrated EMS satisfies the requirement of dynamic performance of CBDC.

Table 5.2 Bus parameters

Item	Parameter
Gross mass (kg)	16,500
Curb mass (kg)	14,220
Air resistance coefficient	0.55
Frontal area (m^2)	6.6
Rolling resistance coefficient	0.0065
Final ratio	6.17
Wheel rolling radius (mm)	510

Table 5.3 Power source parameters

Item	Engine	MG1	MG2
Peak torque (kW)	167	90	100
Inertia moment ($kg{\cdot}m^2$)	1.35	0.33	0.60
Time constant (s)	0.3	0.01	0.01

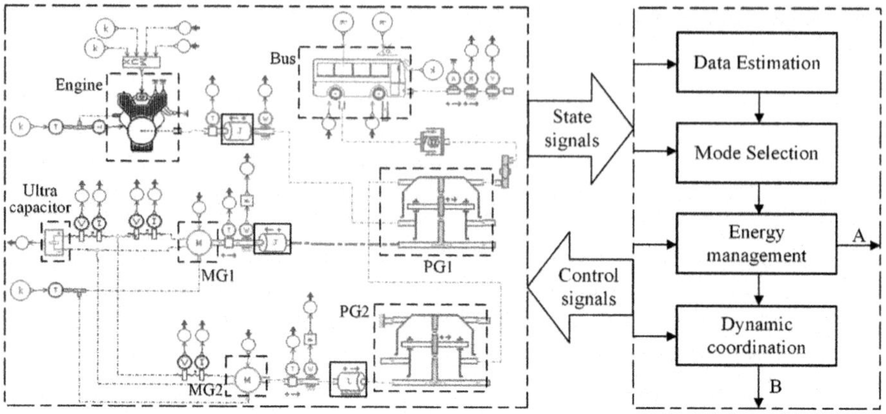

Fig. 5.5 Diagram of co-simulation platform

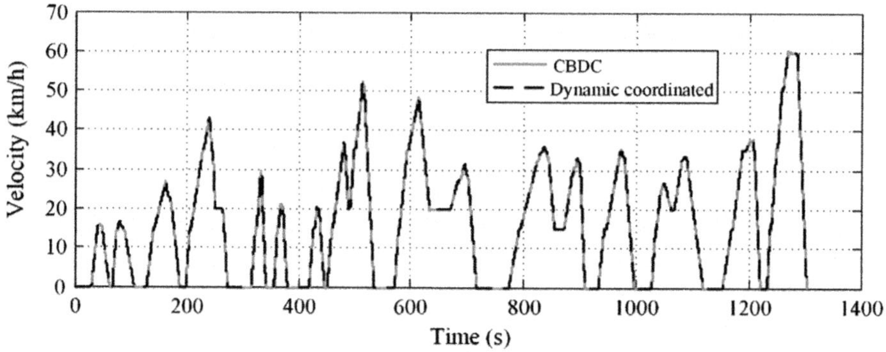

Fig. 5.6 Simulation vehicle speed

5.4.2 Jerk Observation and Prediction

In Sect. 5.2.2 of this chapter, the relationship between degree of shock and torque changing ratios of the power sources is obtained based on the dynamic model, as signified in Eqs. (5.17) and (5.18). Therefore, the shock calculated ("Jerk calculated" in Fig. 5.8) from the acceleration in the AMESim model is compared with that observed ("Jerk observed" in Fig. 5.8) in the Simulink model to verify the system dynamic model. In Fig. 5.7, operation modes 1 and 2 represent the EV and EVT modes respectively. In particular, the figure manifests that the calculated jerk and the observed jerk are nearly the same in both EV and EVT modes, as well as in the mode shifting process. This observation verifies the correction of the system dynamic model and certifies the rationality of the assumption and simplification presented above.

Fig. 5.7 Jerk observation

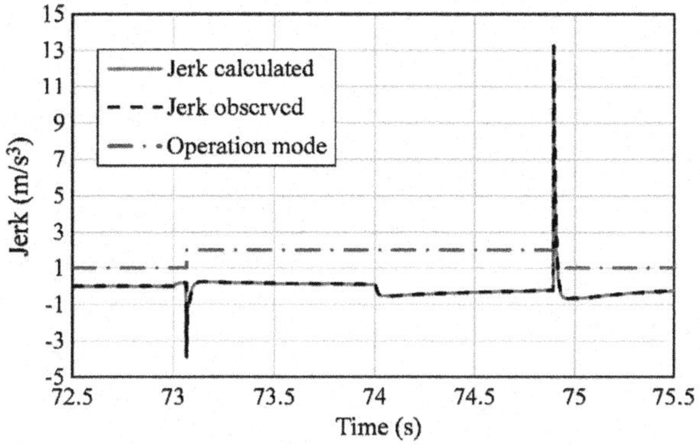

Fig. 5.8 Jerk prediction

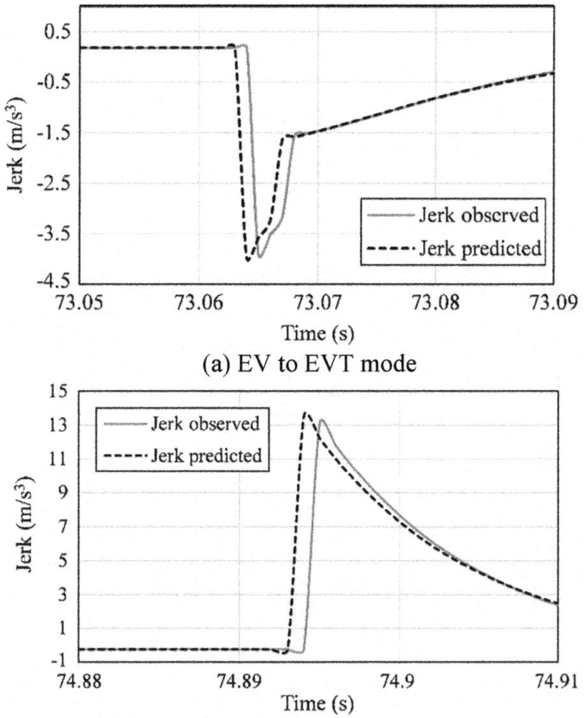

(a) EV to EVT mode

(b) EV to EVT mode

Discrete predictive model is then validated based on the verification of the system dynamic model. As shown in Fig. 5.8, in the mode shifting processes of EV and EVT, the predictive jerk is one simulation step size earlier than the observed jerk, and the values of the two jerks are close. Therefore, the discrete predictive model is validated, which serves as the basis of the DCCS and guarantees real-time performance.

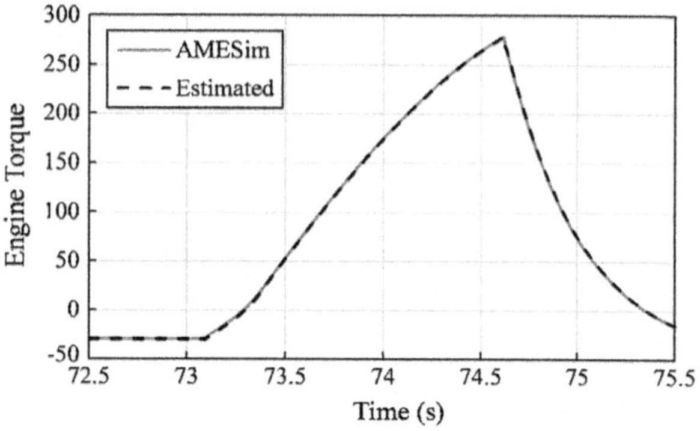

Fig. 5.9 Engine torque estimation

5.4.3 *Engine Torque Estimation*

The estimated engine torque is compared with the engine torque in the AMESim in Fig. 5.9. The comparative results reveal that the estimated value accurately matches the simulation result. Errors between the two engine torques do not exceed 3% under most conditions, conforming with the requirement of the engineering application.

5.4.4 *Dynamic Coordination*

After validating the system dynamic model and engine torque estimation algorithm, this section sets the mode shifting process of EV and EVT as an example to further discuss the simulation results of DCCS.

5.4.4.1 Mode Shifting from EV to EVT

The system jerk in the mode shifting process from EV to EVT under the control strategy with and without DCCS is shown in Fig. 5.10. The results of the analyses

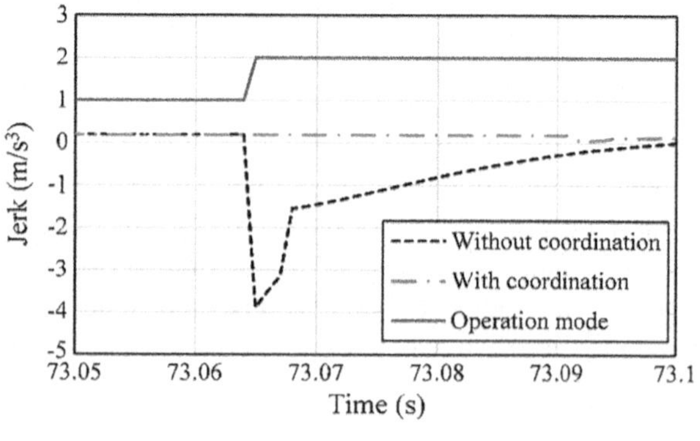

Fig. 5.10 Jerk comparison (EV to EVT)

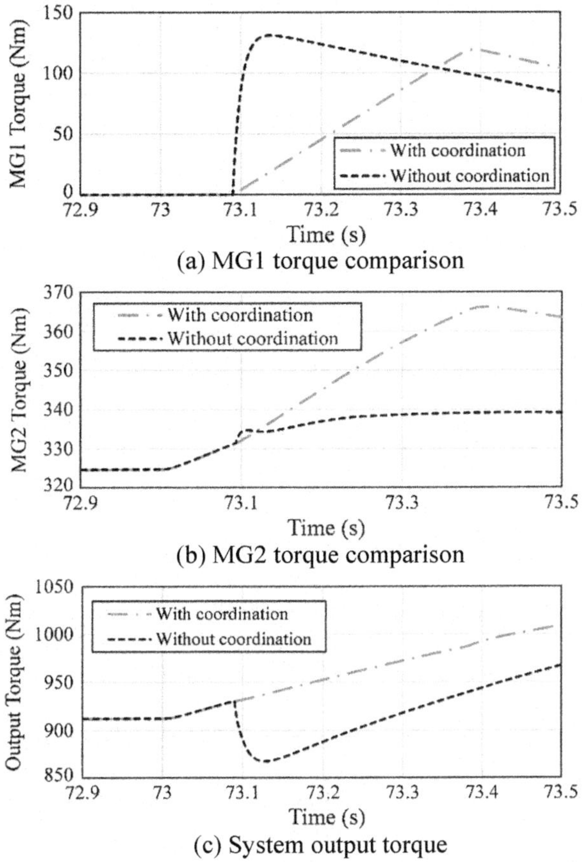

Fig. 5.11 Dynamic coordination (EV to EVT)

verify that, with the integrated DCCS, the system jerk is effectively controlled and smooth mode shifting is realized. Figure 5.11 displays the MG1, MG2 and system output torques in the mode shifting process. Without dynamic coordination, MG1 torque rapidly increases (Fig. 5.11a), while MG2 cannot supply adequate torque (Fig. 5.11b) to compensate for the decrease of system output torque (Fig. 5.11c). In this event, negative jerk, which is nearly 4 m/s^3, occurs. Meanwhile, with the dynamic coordination strategy, the increase speed of MG1 torque is effectively limited (Fig. 5.11a), prompting the smooth increase of MG2 torque (Fig. 5.11b) to compensate for the reduced system output torque (Fig. 5.11c). Thus, the jerk is limited to less than 0.2 m/s^3. Conclusively, riding comfort during the mode shifting process from EV to EVT is obviously improved by the DCCS.

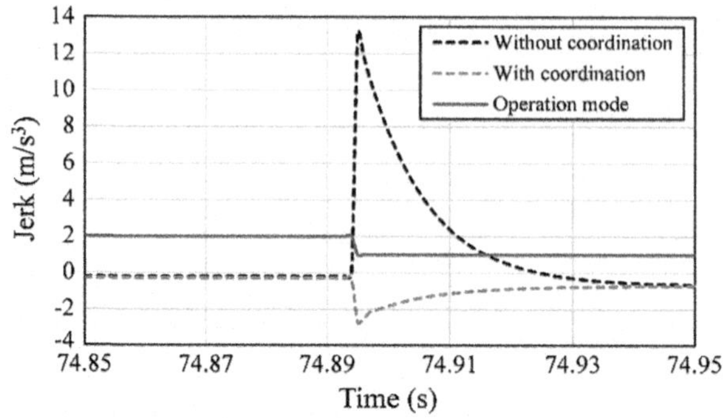

Fig. 5.12 Jerk comparison (EVT to EV)

5.4.4.2 Mode Shifting from EVT to EV

The system jerk in the mode shifting process from EVT to EV under control strategy with and without DCCS is compared, as shown in Fig. 5.12. Without dynamic coordination, a large positive jerk over 12 m/s^3 occurs. Such a large shock in the transmission not only affects riding comfort, but may also cause traffic accidents in heavy traffic. Meanwhile, with DCCS, system jerk is effectively limited to less than 3 m/s^3. Figure 5.13 illustrates that steady-state EMS increases MG2 torque sharply (Fig. 5.13a). This increase leads to a sharp increase of the system output torque (Fig. 5.13b). With DCCS, the decrease of MG2 torque is limited before the mode shifting process (Fig. 5.13a), which indicates that active compensation exists. In addition, the response speed of the engine is considered. Then, in the mode shifting process, the changing ratio of MG2 torque changes slightly without any sharp increase. Therefore, the system output torque only decreases slightly in the mode shifting process and decreases smoothly on the trend (Fig. 5.13b).

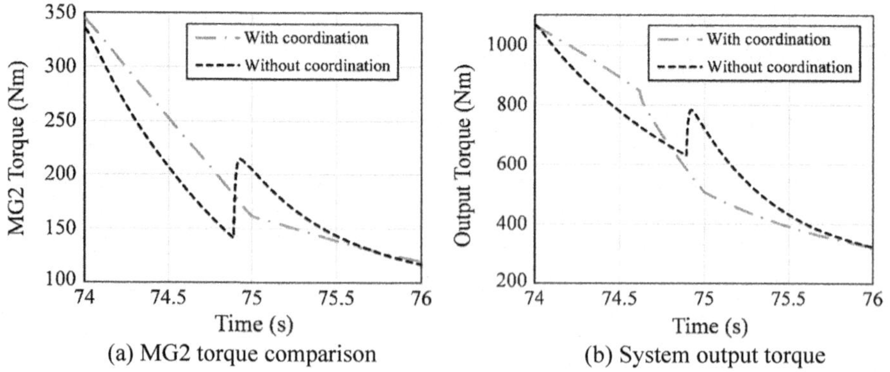

Fig. 5.13 Dynamic coordination (EVT to EV)

In conclusion, dynamic coordination strategy involves the smooth change of the system output torque and obvious improvement of the riding comfort regardless of whether EV changes to EVT or EVT changes to EV. Moreover, the dynamic coordination strategy determines whether the limitation or compensation of the power source is necessary in different mode shifting processes. If so, the amount is also determined according to the specific driving condition and mode shifting process. The simulation results verify that the dynamic coordination strategy is adapted for complex driving conditions.

5.5 Summary

This chapter develops the dynamic model of the dual-planetary power-split HEB and analyzes the reason for the occurrence of system jerk during the mode shifting process under steady-state EMS. Based on this analysis, the engine torque estimation algorithm is established according to the principle of the nonlinear observer and the model-based torque estimator. Finally, a predictive-model-based dynamic coordination strategy is designed based on the theory of MPC. This study also co-simulated AMESim and Simulink to validate the system dynamic model, engine torque estimation algorithm, and DCCS. The results of the simulations indicate that the system jerk is nearly eliminated when the mode is shifted from EV to EVT. Meanwhile, the system jerk is limited to less than 3 m/s^3 from 12 m/s^3 when the mode is shifted from EVT to EV. The results specify that the proposed DCCS can reduce the system jerk and ensure smooth mode shifting in different mode shifting processes.

Compared with the slope limitation method generally applied in engineering, the newly developed DCCS realizes real-time estimation of the slope threshold according to the target system jerk range. When the power sources respond at an exceedingly fast pace, the output torque is limited by the top threshold. When the power source does not supply enough torque, the output torque is constrained by the bottom threshold and adequate torque is supplied for compensation. The real-time estimation of dual thresholds in the proposed DCCS ensures good adaptation for complex driving cycles, which signifies that the control effect is consistent and the excessive limitation is avoided to guarantee the swiftness of mode shifting.

The proposed dynamic coordination strategy effectively decreases system jerk and improves riding comfort and safe driving. This control strategy helps to complete the control theory of the power-split hybrid electric system and conforms to promoting the application of such system.

References

1. Matsubara T, Yaguchi H, Takaoka T, Jinno K. Development of new hybrid system for compact class vehicles. SAE Technical Paper 2009.
2. Kimura A, Ando I, Itagaki K. Development of hybrid system for SUV. SAE Technical Paper 2005.
3. Kamichi K, Okasaka K, Tomatsuri M, Matsubara T, Kaya Y, Asada H. Hybrid system development for a high-performance rear drive vehicle. SAE Technical Paper 2006.
4. Hutchinson T, Burgess S, Herrmann G. Current hybrid-electric powertrain architectures: Applying empirical design data to life cycle assessment and whole-life cost analysis. Appl Energy. 2014;119:314–29.
5. Çağatay Bayindir K, Gözüküçük MA, Teke A. A comprehensive overview of hybrid electric vehicle: powertrain configurations, powertrain control techniques and electronic control units. Energy Convers Manag. 2011;52:1305–13.
6. Duoba M, Ng H, Larsen R. In-situ mapping and analysis of the toyota prius HEV engine. SAE Technical Paper 2000.
7. Duoba M, Lohse-Busch H, Carlson R, Bohn T, Gurski S. Analysis of power-split HEV control strategies using data from several vehicles. SAE Technical Paper 2007.
8. Kim J, Kim N, Hwang S, Hori Y, Kim H. Motor control of input-split hybrid electric vehicles. Int J Automot Technol. 2009;10:733–42.
9. Meisel J. An analytic foundation for the two-mode hybrid-electric powertrain with a comparison to the single-mode Toyota Prius THS-II powertrain. SAE Technical Paper 2009.
10. Wang W, Song R, Guo M, Liu S. Analysis on compound-split configuration of power-split hybrid electric vehicle. Mechanism Mach Theor. 2014;272–88.
11. Liu J, Peng H. Modeling and control of a power-split hybrid vehicle. IEEE Trans Control Syst Technol. 2008;16:1242–51.
12. Kim N, Cha S, Peng H. Optimal control of hybrid electric vehicles based on Pontryagin's minimum principle. IEEE Trans Control Syst Technol. 2011;19:1279–87.
13. Zhang H, Zhang X, Wang J. Robust gain-scheduling energy-to-peak control of vehicle lateral dynamics stabilisation, Veh Syst Dyn. 2014;309–40.
14. Zhang H, Liu X, Wang J, Karimi HR. Robust gain-scheduling energy-to-peak control of vehicle lateral dynamics stabilisation. Int J Adv Manuf Technol. 2014;73:1095–104.
15. Gao B, Liang Q, Xiang Y, Guo L, Chen H. Gear ratio optimization and shift control of 2-speed I-AMT in electric vehicle. Mech Syst Signal Process. 2015;50–51:615–31.
16. Phillips AM, Jankovic M, Bailey KE. Vehicle system controller design for a hybrid electric vehicle. In Proceedings of the 2000 IEEE International Conference on Control Applications. 2000. p. 297–302.
17. Baraszu RC, Cikanek SR. Torque fill-in for an automated shift manual transmission in a parallel hybrid electric vehicle. American Control Conference IEEE 2002. Vol. 2, p. 1431–36.
18. Jo H, Park Y, Lee J, Lee H. A development of an advanced shift control algorithm for a hybrid vehicles with automated manual transmission. Int J Heavy Veh Syst. 2000;7:281–98.
19. Hwang HS, Yang DH, Choi HK, Kim HS, Hwang SH. Torque control of engine clutch to improve the driving quality of hybrid electric vehicles. Int J Automot Technol. 2011;12:763–8.
20. He H, Liu Z, Zhu L, Liu X. Dynamic coordinated shifting control of automated mechanical transmissions without a clutch in a plug-in hybrid electric vehicle. Energies. 2012;5:3094–109.
21. Zhong Z, Kong G, Yu Z, Xin X, Chen X. Shifting control of an automated mechanical transmission without using the clutch. Int J Automot Technol. 2012;13:487–96.
22. Moriya K, Ito Y, Inaguma Y, Sato E. Design of the surge control method for the electric vehicle powertrain. SAE Technical Paper 2002.
23. Johannesson L, Pettersson S, Egardt B. Predictive energy management of a 4QT series-parallel hybrid electric bus. Control Eng Pract. 2009;17:1440–53.

24. Kermani S, Delprat S, Guerra T, Trigui R, Jeanneret B. Predictive energy management for hybrid vehicle. Control Eng Pract. 2012;20:408–20.
25. Payri F, Guardiola C, Pla B, Blanco-Rodriguez D. A stochastic method for the energy management in hybrid electric vehicles. Control Eng Pract. 2014;29:257–65.
26. Borhan HA, Vahidi A, Phillips AM, Kuang ML, Kolmanovsky IV. Predictive energy management of a power-split hybrid electric vehicle. American Control Conference, ACC 09, IEEE, 2009. p. 3970–76.
27. Borhan H, Vahidi A, Phillips AM, Kuang ML, Kolmanovsky IV, Di Cairano S. MPC-based energy management of a power-split hybrid electric vehicle. IEEE Trans Control Syst Technol. 2012;20:593–603.
28. Hernandez EM. A natural observer for optimal state estimation in second order linear structural systems. Mech Syst Signal Proc. 2011;2938–47.
29. Zhang H, Shi Y, Wang J. On energy-to-peak filtering for nonuniformly sampled nonlinear systems: a markovian jump system approach, fuzzy systems. IEEE Trans. 2014;212–22.
30. Zhang H, Wang J. State estimation of discrete-time takagi-sugeno fuzzy systems in a network environment. IEEE Trans Cybern. 2015;45(8):1525–1536.
31. Erazo K, Hernandez EM. A model-based observer for state and stress estimation in structural and mechanical systems: experimental validation. Mech Syst Sig Process. 2014;141–52.
32. Togun N, Baysec S, Kara T. Nonlinear modeling and identification of a spark ignition engine torque. Mech Syst Signal Process. 2012;26:294–304.
33. Rakotomamonjy A, Le Riche R, Gualandris D, Harchaoui Z. A comparison of statistical learning approaches for engine torque estimation. Control Eng Pract. 2008;16:43–55.
34. Vong C, Wong P, Li Y. Prediction of automotive engine power and torque using least squares support vector machines and Bayesian inference. Eng Appl Artif Intell. 2006;19:277–87.
35. Wang Y, Krishnaswami V, Rizzoni G. Event-based estimation of indicated torque for IC engines using sliding-mode observers. Control Eng Pract. 1997;5:1123–9.
36. Doostdar P, Keighobadi J. Design and implementation of SMO for a nonlinear MIMO AHRS. Mech Syst Signal Process. 2012;32:94–115.
37. Gao B, Chen H, Ma Y, Sanada K. Design of nonlinear shaft torque observer for trucks with Automated manual transmission. Mechatronics. 2011;21:1034–42.
38. Francoa J, Grigoriadisb MAFA. Real-time brake torque estimation for internal combustion engines. Mech Syst Signal Process. 2008;338–61.

Chapter 6
Design of a Differential-Based PSD

PSD is the core component of series-parallel hybrid electric vehicle (SPHEV). Current research on PSD can be divided into two main aspects: one is that tycoons in global automobile manufacturing, such as Toyota and General Motors (GM), put forward different PSDs successively. Toyota's representative is Toyota Hybrid System (THS) [1], which is equipped on the Prius. GM also proposed a series of configurations, including dual-planetary [2–4] and triple-planetary [5–9] configurations. The other research aspect is that numerous studies on the motion characteristics, modeling, and control strategy of PSDs have been conducted [10–13]. Although considerable studies have been devoted to the configuration characteristics and control strategy of the existing several classic PSDs, the research on the design of a new, simple, and practical PSD is quite rare.

In this chapter, considering that the traditional differential and THS are both based on planetary gear mechanism and have similar kinematics characteristics, the difference is considered as a prototype, and its retrofit is used as a differential-based PSD (DPSD). Retrofit design, rigid-flexible virtual prototype simulation verification and bench test are carried out to analyze the performance of DPSD.

6.1 Analysis for Using the Traditional Differential as a DPSD

According to the simulation and analysis of DPSD in Fig. 3.7, during the whole NEDC, the speed difference between the left and right side gear shafts of the differential is continuously large, as shown in Fig. 6.1.

Table 6.1 shows the simple statistics of the speed difference.

According to Fig. 6.1 and Table 6.1, when the differential is used as a PSD in HEVs, the speed difference between the right and left side gear shafts is very big. In the traditional car drive axle, the planetary gears of the differential start rotating

X. Zeng and J. Wang, *Analysis and Design of the Power-Split Device for Hybrid Systems*, https://doi.org/10.1007/978-981-10-4272-0_6

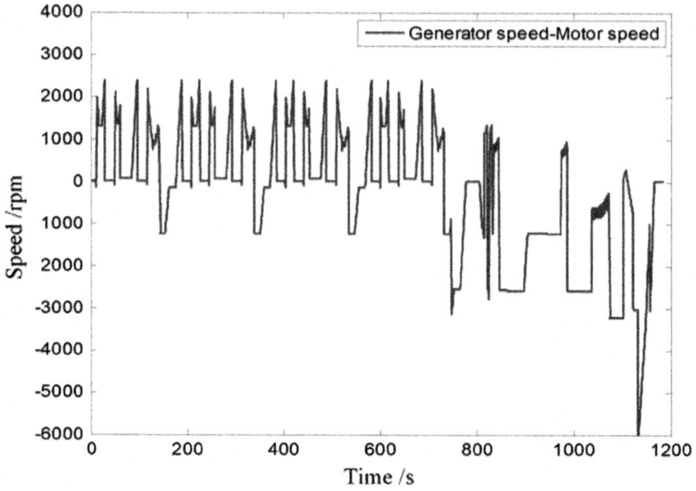

Fig. 6.1 Speed difference between two side gear shafts under NEDC

Table 6.1 Statistics of the speed difference

Item	Average	Maximum
Forward speed difference/rpm	581	2401
Reverse speed difference/rpm	1684	6210

only when the speeds of the left and right drive wheels are different (like steering or one side of the wheel slipping). At this point, the planetary gear rotation speed is relatively small (generally less than 500 rpm) and the duration is short. Therefore, in traditional vehicles, when differential planet gear and planet gear shaft have direct contact with each other, slipping occurred can be reduced by putting lubricating oil between the planet gear and the planet gear shaft. However, when the differential is used as a PSD in HEVs, a severe sliding friction between planet gear and planet gear shaft is observed, which can lead to the failure of the differential. This failure will clearly affect the normal power transmission of the vehicle.

6.2 Retrofit Design of Traditional Differential

To reduce the wear caused by the speed difference between two-side gear shafts when traditional differential is used as PSD, the contact relationship of the planet gear and the planet gear shaft must be retrofitted. A retrofitted design scheme that can meet the demand of allowing a sustained large speed difference between the right and left side gear shafts is proposed as follows.

Needle bearing is installed between planet gear and planet gear shaft to turn the contact relationship from sliding friction to rolling friction. To install the needle bearing, the structure of the planet gear shaft is redesigned, and heat treatment is implemented on its contact surface with the needle roller and the planet gear bore surface to improve surface hardness and avoid indentation.

6.2.1 Bearing Selection

To ensure the strength of the planet gear shaft, the shaft diameter and needle bearing should be considered before installation. The space between planet gear and planet gear shaft should be limited but could not be too small. The needles of K class needle bearing are independent, which shows that the bearing does not have inner or outer rings and can thus be installed directly. In addition, the bearing needs a small radial space. Therefore, selection of the K class needle bearing is essential and the needle bearing is shown in Fig. 6.2.

6.2.2 Design of Planet Gear Shaft

To provide axial positioning for the needle bearing, the original planet gear shaft is redesigned by adding a shaft shoulder. To meet the design requirement, the material of 20MnTi is applied. Meanwhile, partial heat treatment is used on the interface between needle bearing and shaft to increase surface hardness. As the integral differential is used as the prototype of the DPSD, for the purpose of convenient assembly, two sleeves should be installed on both sides of the planet gear shaft respectively. A pinhole should also be processed on one side of the planet gear shaft and the sleeve mounted on it. Finally, a pin is used to fix the planet gear shaft and the sleeve to the differential housing. Figures 6.3 and 6.4 show the redesigned planet gear shaft and the sleeves respectively. Figure 6.5 illustrates the fitting components.

Fig. 6.2 K class needle bearing

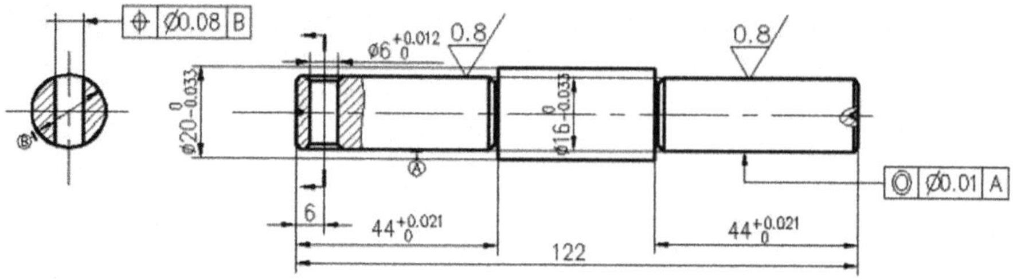

Fig. 6.3 Redesigned planet gear shaft

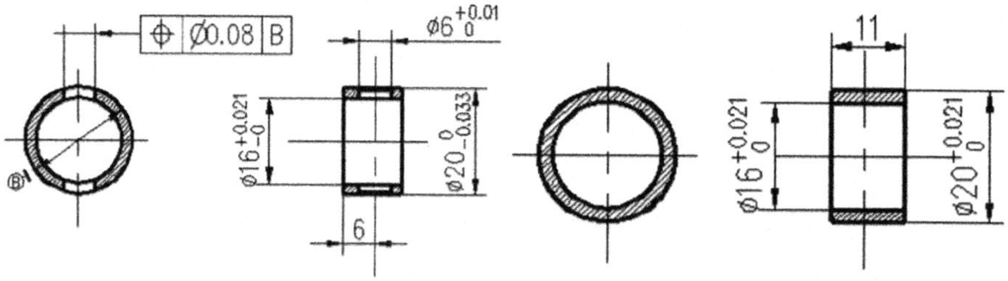

(a) Left side sleeve (b) Right side sleeve

Fig. 6.4 Sleeves

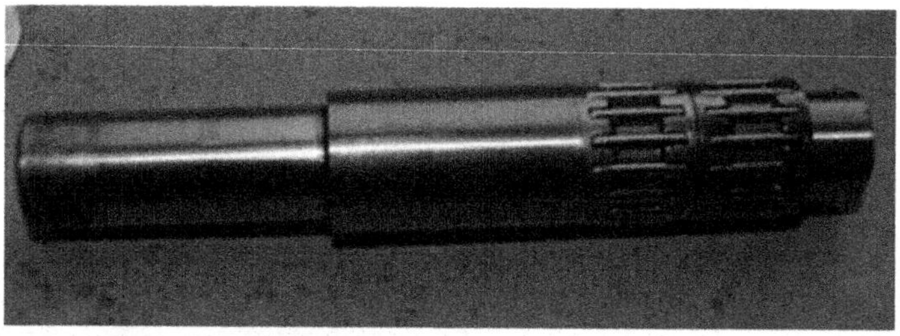

Fig. 6.5 Fitting components of the redesigned planet gear shaft and K class needle bearing

6.2.3 *Structure of DPSD*

To reduce the cost of the retrofit design, the DPSD uses the original parts and structure of traditional differential except for the two modified parts mentioned earlier. Figure 6.6 shows the tradition differential used for retrofitting. Figure 6.7 shows the DPSD's structure after the retrofit design [14].

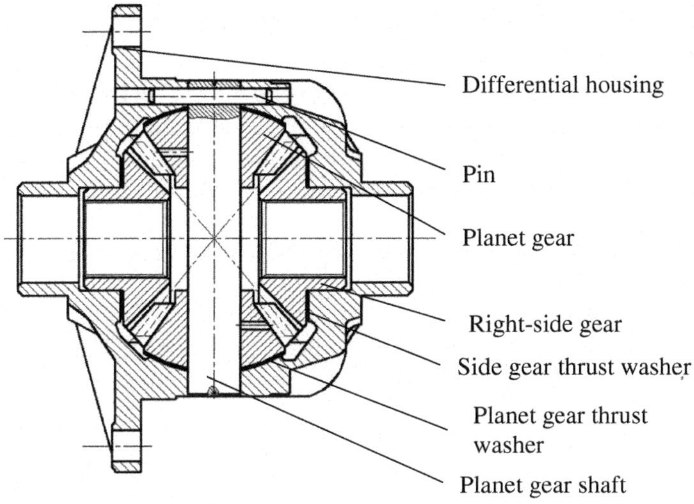

Fig. 6.6 Traditional differential

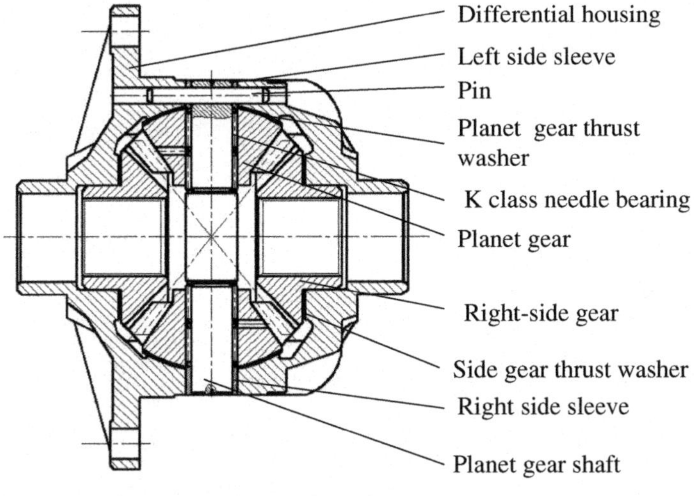

Fig. 6.7 Structure of DPSD

6.3 Verification of DPSD

On the verification of DPSD, first, the finite element analysis is tested. Second, software simulation is used to verify whether DPSD will become invalid under NEDC. At last, bench test is conducted for the further verification of correctness.

6.3.1 Simulation Verification of DPSD

A rigid-flexible virtual prototype of the DPSD is built in CATIA and ADAMS. As shown in Fig. 6.8 finite element analysis is conducted on the planet gear shaft strength and deformation through rigid-flexible co-simulation. It is verified by software simulation whether the verification of DPSD will fail under the whole NEDC.

Table 6.2 shows the maximum load of the planet gear shaft under NEDC according to the simulation results.

The parameters mentioned in Table 6.2 are applied to the virtual prototype in the simulation calculations. The simulation results are as follows: Fig. 6.9 shows the stress on the planet gear shaft; Fig. 6.10 shows the strain on the planet gear shaft;

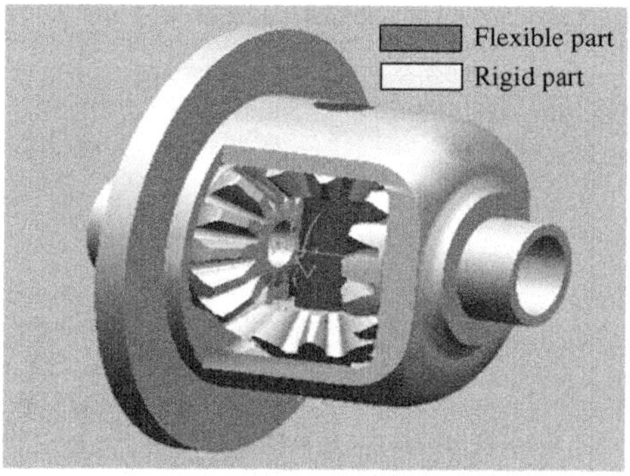

Fig. 6.8 Rigid-flexible virtual prototype of the DPSD

Table 6.2 Maximum load of the planet gear shaft under NEDC

Left side gear shaft torque/Nm	Differential housing speed/rpm	Right side gear shaft speed/rpm
−51	1016	4335

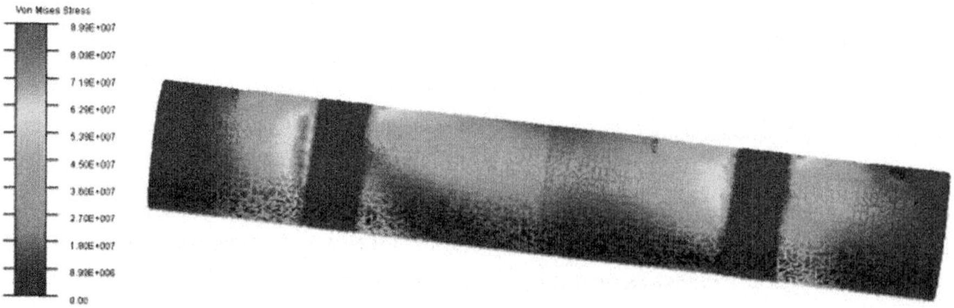

Fig. 6.9 Stress on the planet gear shaft

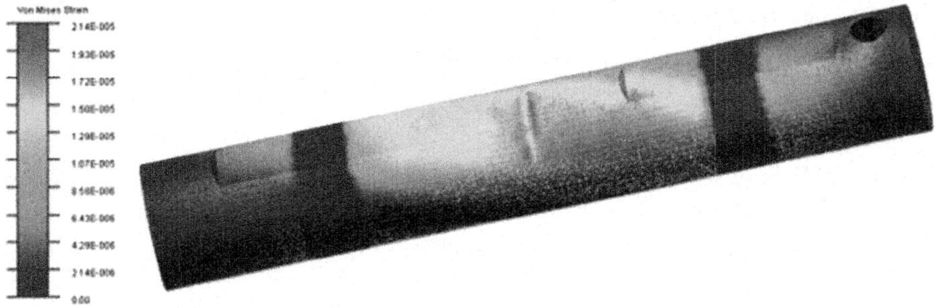

Fig. 6.10 Strain on the planet gear shaft

According to the results, the maximum stress is 89.9 MPa and the maximum deformation is 0.0058 mm, which are both within the allowable strength and deformation.

Based on the movement characteristics of the DPSD, the speed relationship can be obtained as follows:

$$n_4 = (n_0 - n_3)\frac{z_1}{z_4} \tag{6.1}$$

where n_3 is the rotation speed of the side gear shaft, n_0 is the rotation speed of the differential housing rotating around the axle of the side gear shaft, n_4 is the rotation speed of the planet gear shaft, z_1 is the teeth number of side gear, z_4 is the teeth number of planet gear.

When n_0 = 1016 rpm, n_3 = 4335 rpm, z_4 = 14, z_1 = 10 are substituted into Eq. (6.1), the result is n_4 = − 4647 rpm.

Based on the simulation results under NEDC, the axial force (F_{a1}) of the planet gear during gear transmission is 517.7 N. Meanwhile, the centrifugal force (F_{a2}) of the planet gear can be calculated by the following equations:

$$r_1 = \frac{mz_r}{2} = 37.8mm \tag{6.2}$$

$$F_{a2} = m_1\left(\frac{\pi z_r}{30}\right)^2 r_1 = 64.5N \tag{6.3}$$

where m_1 is the mass of the planet gear (i.e., 0.151 kg), and r_1 is the reference radius of side gear, $r_1 = m \cdot z_1/2 = 37.8mm$, m = 5.4.

Total axial force of the spherical washer caused by the planet gear is as follows:

$$F_a = F_{a1} + F_{a2} = 582.2N. \tag{6.4}$$

The movement and loading conditions for stress and temperature co-simulation can be obtained based on the maximum load of planet gear shaft under NEDC (see Table 6.2). The result is shown in Table 6.3.

Table 6.3 Simulation conditions for ABAQUS

Rotation speed of the planet gear/rpm	Total axial force of the spherical washer/N
−4647	582.2

With the help of stress and temperature fields coupled finite element analysis, the stress, strain, and temperature distribution of the planet gear thrust washer can be obtained quickly and accurately [15]. Figures 6.11 to 6.14 show the simulation results.

Figure 6.11 shows that the stress on most parts of the surface is about from 2.45 to 3 MPa; the stress on the inner ring is greater than that in the outer ring; and the maximum stress is 4.377 MPa (less than the allowable stress). As shown in Fig. 6.12, the deformation on most parts of the surface is approximately from 0.098 to 0.19 mm. The relatively large deformation concentrated in the inner ring, which

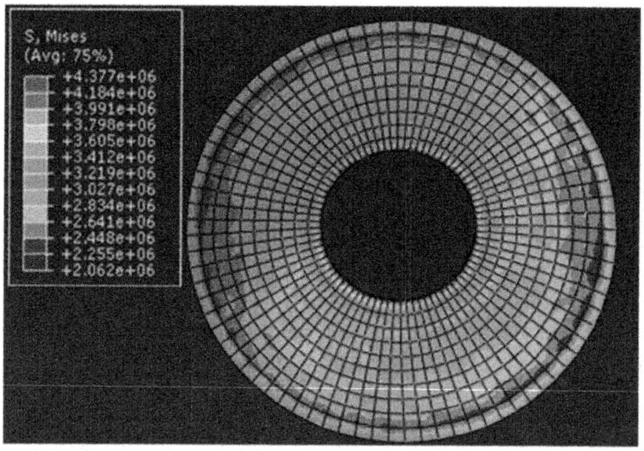

Fig. 6.11 Stress on the planet gear thrust washer

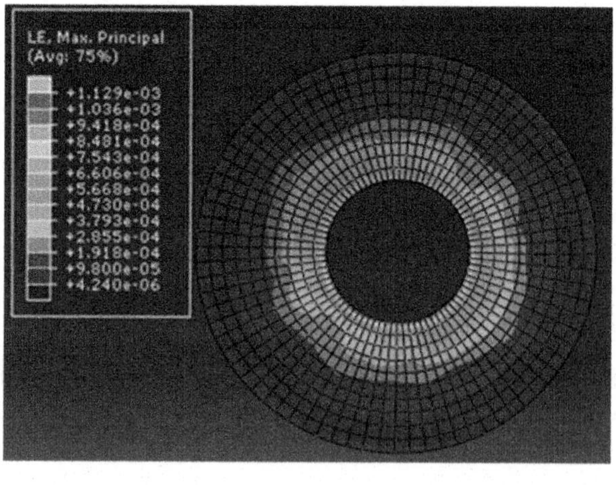

Fig. 6.12 Deformation on the planet gear thrust washer

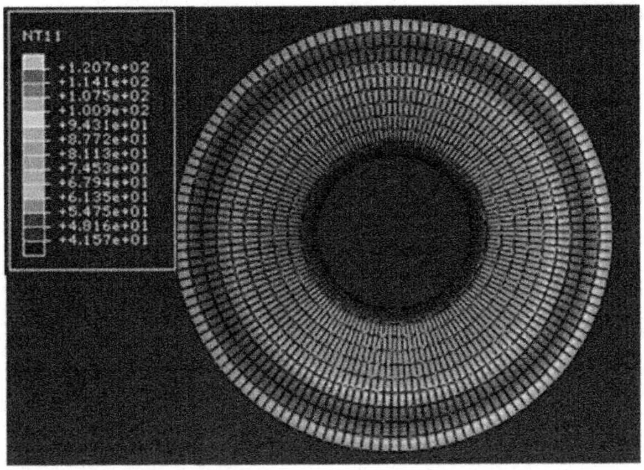

Fig. 6.13 Temperature on the planet gear thrust washer

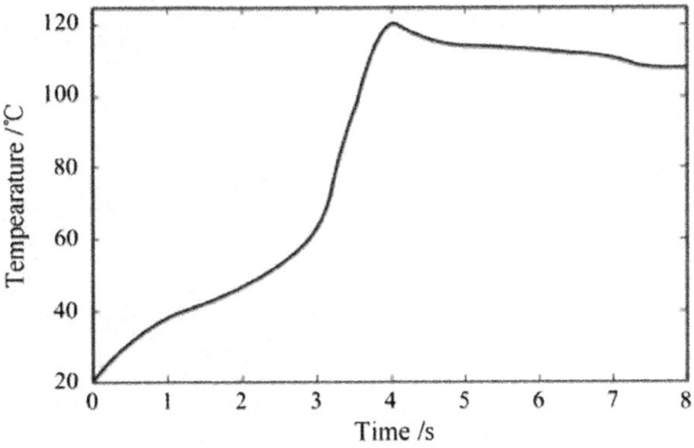

Fig. 6.14 Temperature curve of a node in the highest temperature region

led to an average inner-ring deformation of about 0.75 mm, but a maximum deformation of below 1.129 mm. Figure 6.13 shows that maximum temperature, which is approximately 120 °C, concentrates near the edge of the thrust washer, and that relatively high temperature, which is approximately 100 °C, concentrates on both sides of the part with maximum temperature and the inner part of the thrust washer. In Fig. 6.14, the temperature of the planet gear thrust washer is shown to reach its maximum, which is 120.7 °C, at 5 s, and then stabilizes at around 108 °C.

Simulation results show that maximum stress, deformation, and temperature are all within the allowed values, which means that DPSD will not be abraded to burn eclipse and meet the design requirements.

6.3.2 Bench Test of DPSD

To verify further the feasibility of the retrofit design, several typical operating points of the NEDC are selected for bench test. The Baoke electronic power train

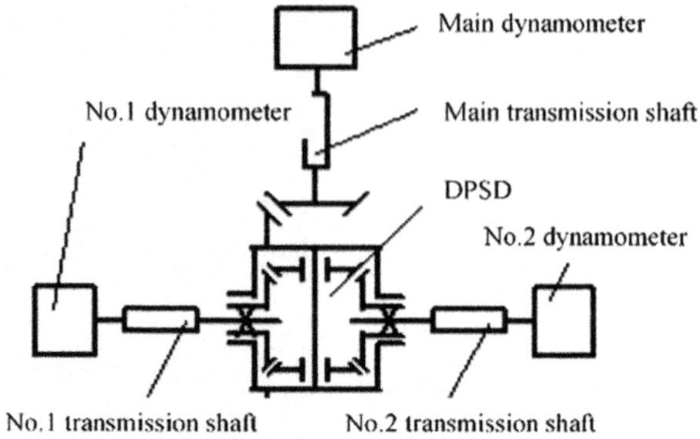

Fig. 6.15 Layout of the bench test

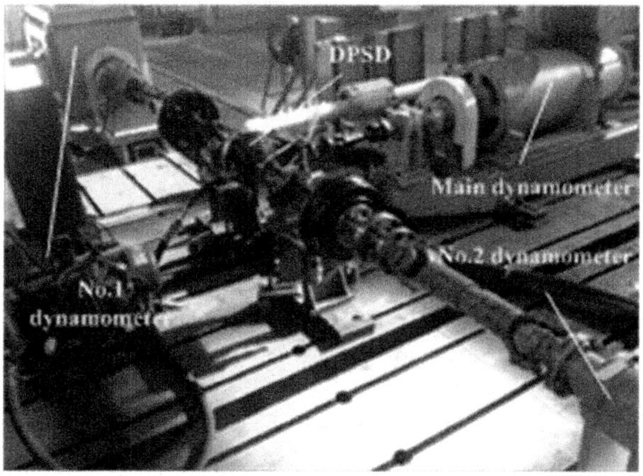

Fig. 6.16 DPSD testing bench

test bench is used. The bench is arranged based on the connection relationship shown in Fig. 6.15. The engine is substituted by the main dynamometer; the generator and motor are substituted by the No.1 dynamometer and the No.2 dynamometer respectively, as shown in Fig. 6.16.

As the speed of the main dynamometer is limited, the bench can be used to test the working condition of the DPSD when the HEV is driving in the low-velocity segment (velocity less than 50 km/h) of the NEDC. The program for bench test is shown in Table 6.4.

The torque of the No.1 dynamometer and the speeds of the main dynamometer and the No.2 dynamometer are adjusted according to Table 6.4. Speed, torque, and temperature of each test point are then recorded. During the bench test, each power source works in accordance with the established testing program shown in Table 6.4. The results of the bench test at 15 km/h are shown in Figs. 6.17, 6.18 and 6.19, including the speed, torque, and power of the input and output shaft of DPSD. As shown in Fig. 6.19, the internal friction power of the DPSD decreases to approximately 500 W after retrofitting.

Table 6.4 Bench test program of the DPSD

Test velocity/kmh^{-1}	Main dynamometer speed/rpm	No. 2 dynamometer speed/rpm	No. 1 dynamometer speed/Nm					
5	0	−182	\	10	0	−10	\	\
10	0	−363	20	10	0	−10	−20	\
15	0	−545	20	10	0	−10	−20	−30
20	0	−726	20	10	0	−10	−20	−30
25	3000	627	20	10	0	−10	−20	−30
30	3000	445	20	10	0	−10	−20	−30
35	3000	264	20	10	0	−10	−30	\
40	3000	82	20	10	0	−30	\	\
45	4000	412	20	10	−30	\	\	\
50	4000	230	20	10	0	−10	20	−30

Note Positive torque data represent driving; negative torque data represent driven; positive speed data represent forward; and negative speed data represent reverse.

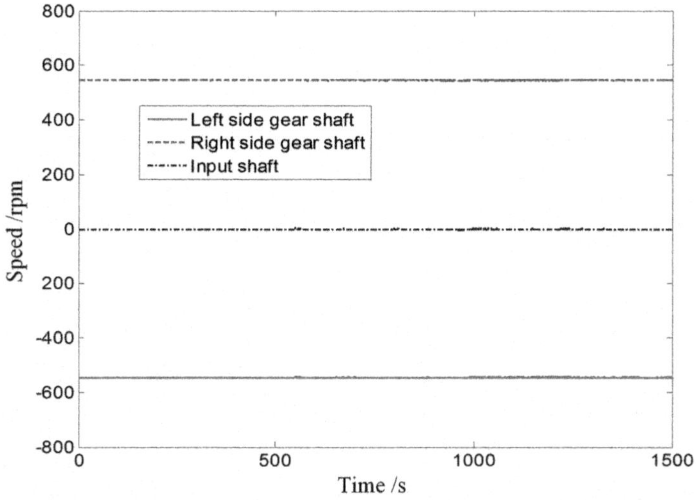

Fig. 6.17 Speed of DPSD shafts

During the test, the planet gear, planet gear shaft, thrust washers, and other parts of the DPSD all work smoothly without any acute noise. The conditions of the main parts (Fig. 6.20) are as follows: planet gear shaft is fine without any wear scars or indentation cracks, needle bearing is not curved or broken, side gear shaft and gear thrust washers are undamaged, location pinhole of the left-side sleeve only has a slight deformation, planet gear has a few scratches on the back, and planet gear thrust washers have some abrasions without any failure.

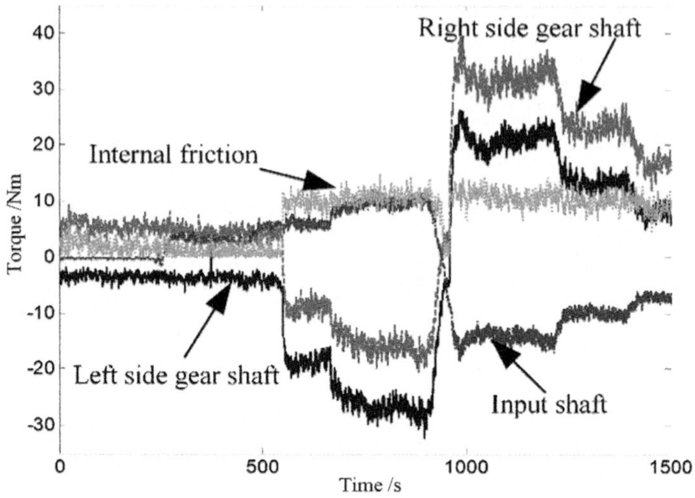

Fig. 6.18 Torque of DPSD shafts

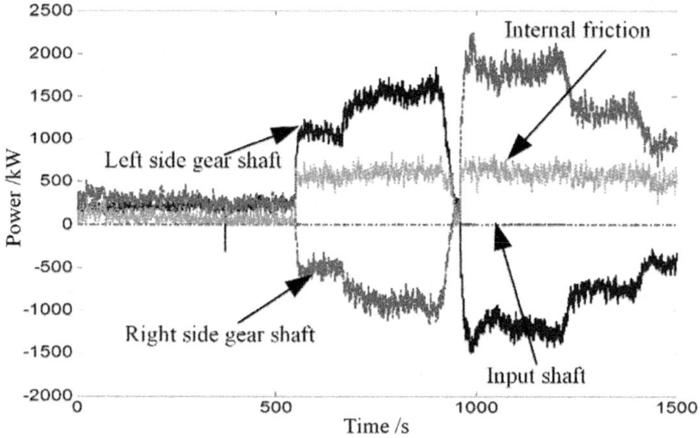

Fig. 6.19 Power of DPSD shafts

To verify the necessity and advantage of retrofitting further, the same bench test using the original differential was conducted. When the test reached the 15 km/h point, an unusual noise was heard from the differential. Subsequently, the test was terminated and the differential was disassembled. Figure 6.21 shows that the housing and the planet gear shaft became attached to each other, and the planet gear shaft was broken.

1)Planet gear, 2)Planet gear thrust washer, 3)Left side sleeve, 4)Needle bearing, 5)Planet gear shaft, 6)Side gear thrust washer

Fig. 6.20 Main parts of DPSD after bench test

Fig. 6.21 Failure of differential after bench test

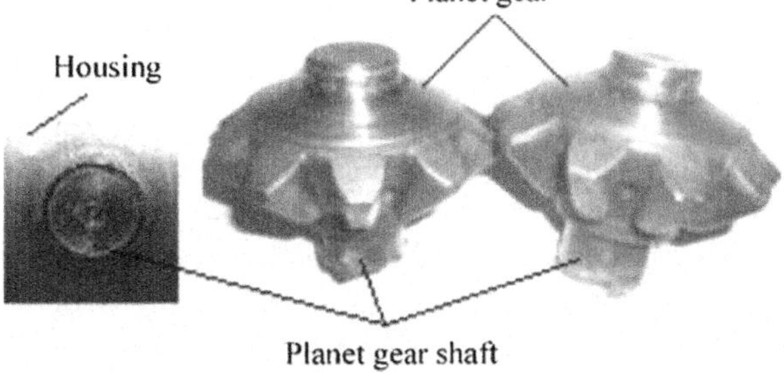

Table 6.5 Decreasing internal friction of the differential and the DPSD

Velocity/kmh^{-1}	Internal friction power/W		Degree of improvement/%
	Differential	DPSD	
5	246.8	183.7	25.6
10	630.3	388.7	38.3
15	1247.1	518.4	58.4

The friction powers of the differential and the DPSD at the test velocity of 5–15 km/h were compared. After retrofitting, the friction power decreased. This degree of improvement shows an increasing trend as the velocity increased, as shown in Table 6.5 and Fig. 6.22.

The result of the bench test shows that after retrofitting, the planet gear, planet gear shaft, needle bearing, and thrust washers do not have any major failure. Moreover, retrofitting effectively reduced the friction power of the differential.

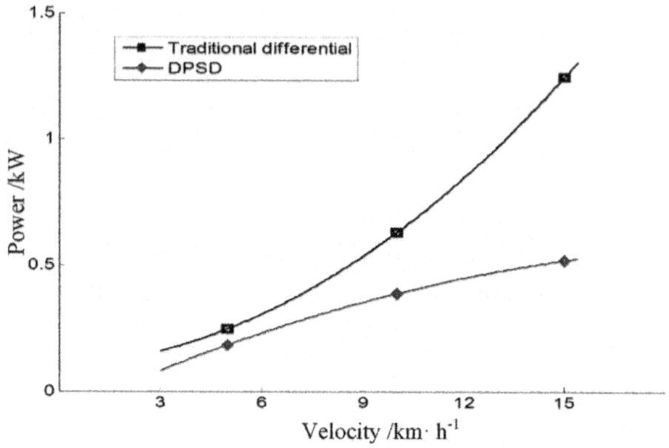

Fig. 6.22 Internal friction power

6.4 Summary

This chapter proposes a new PSD based on the traditional differential for HEV. A simple retrofit design scheme, including adding a needle bearing and redesigning the planet gear shaft, is proposed. Then, the rationality of the retrofit is verified by finite element analysis and bench test. Results show that after the retrofit design, the DPSD can adapt to the torque and speed of every test point of the experimental scheme. At the same time, no obvious signs of failure are observed when the original lubricating oil system is used. In addition, friction power is effectively reduced. Consequently, the feasibility of applying the retrofitted differential as the PSD is verified. However, according to the experiment results, we can see that the DPSD is also wear after the bench test. Therefore, it is necessary to optimize the design of the structure in the follow-up work.

References

1. Koide T, Matsui H, Nada M. Hybrid vehicle drive system having two motor/generator units and engine starting means. U.S. patent 5934395. 10 Aug 1999.
2. Schmidt MR. Two-mode, input-split, parallel, hybrid transmission. U.S. patent 5558588. 24 Sep 1996.
3. Holmes AG. Two range electrically variable power transmission. U.S. patent 6945894. 20 Sep 2005.
4. Holmes AG, Klemen D, Schmidt MR. Electrically variable transmission with selective input split, compound split, neutral and reverse modes. U.S. patent 6527658. 4 Mar 2003.
5. Holmes AG, Schmidt MR. Hybrid electric power train including a two-mode electrically variable transmission. U.S. patent 6478705. 12 Nov 2002.
6. Schmidt MR. Two-mode, compound-split electro-mechanical vehicular transmission. U.S. patent 5931757. 3 Aug 1999.
7. Grewe TH. Defining the general motors 2-mode hybrid transmission. 2007.
8. Schmidt MR. Electro-mechanical power train. U.S. patent 5935035. 10 Aug 1999.

9. Raghavan MN, Bucknor K, Hendrickson JD. Electrically variable transmission having three planetary gear sets and three fixed interconnections. U.S. patent 7238131. 3 Jul. 2007.
10. Sasaki S. Toyota's newly developed hybrid power train. In: Power Semiconductor Devices and ICs, 1998. ISPSD 98. Proceedings of the 10th International Symposium on 1998 p. 17–22. IEEE.
11. Miller JM. Hybrid electric vehicle propulsion system architectures of the e-CVT type. Power Electron IEEE Trans. 2006;21(3):756–67.
12. Liu J, Peng H. Control optimization for a power-split hybrid vehicle, In: Proceedings of American control conference (ACC.2006);2006 June; Minneapolis, America. p. 6.
13. Liu J, Peng H. Modeling and control of a power-split hybrid vehicle. Control Syst Technol IEEE Trans. 2008;16(6):1242–51.
14. Zeng XZ, Wang QN, Song DF, He L, Yu YT, Jin LJ, Yu YB, Wang PY, Wang W, Wang JX. A differential used for hybrid electric vehicle power split device. China patent 201339701, 2009-11-04.
15. Söderberg A, Andersson S. Simulation of wear and contact pressure distribution at the pad-to-rotor interface in a disc brake using general purpose finite element analysis software. Wear. 2009;267(12):2243–51.

Chapter 7
Thermal Analysis of Differential-Based PSD

To figure out the changing rules of temperature field and provide rational allocation of lubrication, the thermal analysis is indispensable. A method based on thermal network model is proposed, which is used to predict the study-state temperature distribution under typical driving condition. Then thermal elastohydrodynamic lubrication (TEHL) line contact of bevel gears in DPSD is approximately computed based on a theory in which a bevel gear is equalized to an equivalent spur gear. Finally, finite element method is used to analyze the temperature distribution of the tooth of gear.

7.1 Thermal Analysis of DPSD Based on Thermal Network Method

Power loss not only reduces the efficiency, but also generates large amount of heat, which will directly affect the mechanical transmission performance [1].

Currently, the thermal analysis methods include the numerical method and lumped parameter method (LPM). For LPM, although the accuracy is not as good as the numerical method, the temperature node selection is flexible, convenient and time-saving. The most common method of LPM is the thermal network method. It is suitable for the heat transfer analysis on simple geometrical shape or the model which can be simplified appropriately under certain conditions. In this section, the analysis process of thermal network method on the overall temperature field of DPSD is introduced.

X. Zeng and J. Wang, *Analysis and Design of the Power-Split Device for Hybrid Systems*, https://doi.org/10.1007/978-981-10-4272-0_7

7.1.1 Introduction of Thermal Network Method

Thermal network method is used to analyze the steady-state and transient temperature field [2]. The distribution of node location and the density of thermal network can be flexibly created according to practical needs. A thermal node corresponds to one temperature of a certain point of the fluid medium (lubricating oil, ambient air, etc.). Then, according to the actual situation, the different thermal resistance is linked to form an integrated thermal network system between the relevant nodes [3].

To solve the transient temperature response, the thermal flux q flowing into the node should be expressed as:

$$\rho_i C_{pi} V_i \frac{dT_i}{d_t} = q_i (i = 1, 2, \ldots, n) \tag{7.1}$$

where i is the sequence number of nodes. ρ, C_p and V are the node density, specific heat and volume, respectively. dTi/dt is the temperature derivative of time. N nodes can establish n transient equations. The equations can be solved to obtain the transient temperature response of the working system.

7.1.2 Thermal Network Model of DPSD

Based on the principle of LPM, 25 nodes are chosen to describe DPSD. Each node represents an isothermal body or surface. The distribution of nodes is shown in Fig. 7.1. The location of each node of the thermal network is listed in Table 7.1.

The thermal network of DPSD is shown in Fig. 7.2. According to Kirchhoff current law, heat balance equation can be expressed as:

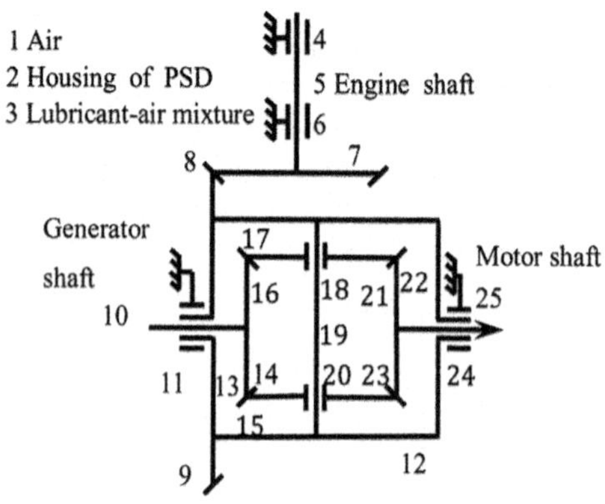

Fig. 7.1 Distribution of temperature nodes

Table 7.1 Location of each node of thermal network

No.	Location
1	Air
2	Housing of DPSD
3	Lubricant-air mixture
4	Taper roller bearing
5	Engine shaft I
6	Taper roller bearing
7	Spiral bevel gear
8	Meshing point of spiral bevel gear
9	Spiral bevel gear
10	Generator shaft II
11	Taper roller bearing
12	Differential housing
13	Drive shaft gear
14	Contact surface between drive shaft gear and planet gear
15	Planet gear
16	Contact surface between drive shaft gear and planet gear
17	Planet gear
18	Needle roller bearing
19	Needle roller bearing shaft
20	Needle roller bearing
21	Contact surface between drive shaft gear and planet gear
22	Drive shaft gear
23	Contact surface between drive shaft gear and planet gear
24	Taper roller bearing
25	Generator shaft III

$$p_i + \sum \frac{T_j - T_i}{R(i,j)} - \sum \frac{T_i - T_k}{R(i,k)} = 0 \tag{7.2}$$

where P_i is the power loss of node i, T_i, T_j and T_k are the temperature of node i, j and k, respectively. $R(i,j)$ is the thermal resistance between node i and node j. $R(i,k)$ is the thermal resistance between node i and node k.

The temperature is affected by power loss and thermal resistance. The thermal radiation can be ignored due to the small temperature difference [4].

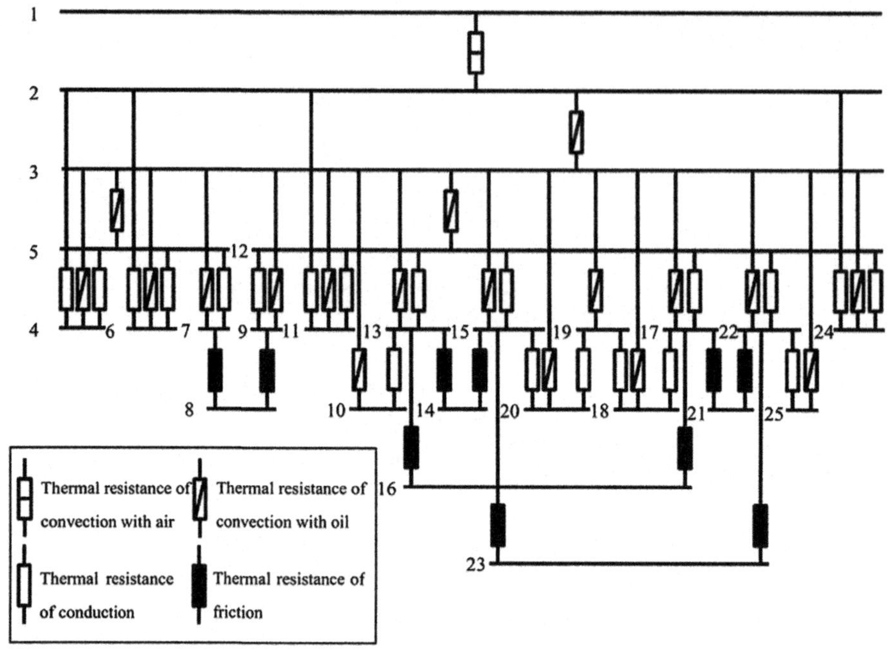

Fig. 7.2 Thermal network of DPSD

7.1.3 *Power Loss*

By analyzing the transmission principle and structural characteristics of DPSD, the power losses include the power loss of the gear meshing, rolling bearing loss, churning loss, windage loss and friction loss. These power losses are majorly converted into heat, which will cause the high temperature. Therefore, it is important to build a reasonable computing model of power losses of DPSD.

7.1.3.1 Power Loss of Gear Meshing

Power loss of the gear meshing includes sliding friction loss and rolling friction loss. In this section, the Anderson method is used to calculate the power loss. The bevel gear model is equivalent to the spur gear model [5]. The total loss is:

$$N = N_s + N_t \tag{7.3}$$

The average sliding friction loss N_s and the average rolling friction loss N_t are expressed as:

$$N_s = \frac{fF_n V_s}{1000} \tag{7.4}$$

$$N_t = 9000 \frac{bV_t h\varepsilon}{\cos \beta_b} \tag{7.5}$$

where f is the friction coefficient. F_n is the average tooth surface normal load of bevel gear. b is the tooth width. h is the average film thickness. ε is the contact ratio. β_b is the helical angle of the base circle. V_t is the relative sliding speed and V_s is the rolling speed. V_s and V_t can be calculated by Eqs. (7.6) and (7.7).

$$V_s = 0.02618\, n_{v1}\, g_z \frac{z_{v1} + z_{v2}}{z_{v2}} \tag{7.6}$$

$$V_t = 0.2094\, n_{v1} \left[r_1 \sin \alpha - 0.125\, g_z \frac{z_{v2} - z_{v1}}{z_{v2}} \right] \tag{7.7}$$

where n_{v1} is the equivalent speed of driving gear. g_z is the length of meshing line. z_{v1} and z_{v2} are the equivalent teeth number of the driving and driven gears. r_1 is the pitch radius of driving gear, and α is the pressure angle.

7.1.3.2 Rolling Bearing Loss

The rolling bearing loss N_f is the product of the friction torque of the rolling bearing M and angular velocity of the rolling bearing inner ring ω.

$$N_f = M\omega \tag{7.8}$$

The total power loss includes load friction torque, viscous friction torque and roller end-flange friction torque [6]. The friction torque of rolling bearing M can be expressed as:

$$M = M_1 + M_v + M_f \tag{7.9}$$

When the rolling bearing exists the friction loss of the elastic hysteresis and local differential slip, the calculation formula of M_1 can be expressed as:

$$M_1 = f_1 F_\beta d_m \tag{7.10}$$

where M_1 is the coefficient related to the bearing type and the load. f_1 is the coefficient related to the bearing type and the load. F_β is the load generated by the friction torque, and d_m is the average diameter of the bearing.

M_v is the viscous friction torque, which can be expressed as:

$$M_v = \begin{cases} 160 \times 10^{-7} f_0 d_m^3 & v_0 n \leq 2000 \\ 10^{-7} f_0 (v_0 n)^{2/3} d_m^3 & v_0 n \geq 2000 \end{cases} \tag{7.11}$$

where f_0 is the coefficient related to the bearing type and lubrication mode. v_0 is the kinematic viscosity of lubricant, and n is the rotational speed of the bearing inner ring. M_f should be considered when the rolling bearing supports the radial and axial loads simultaneously, which can be expressed as:

$$M_f = f_f F_a d_m \tag{7.12}$$

where f_f is the coefficient related to the structure type of bearing and the lubrication mode, and F_a is the axial load of bearing.

7.1.3.3 Friction Loss

Based on the analysis of DPSD structure, the internal friction torque mainly includes T_{rf}, T_{rb}, T_{rsw} and T_{rsn}. T_{rf} is generated by relative motion between the back sphere of planet gear and the differential housing, whose direction is opposite to the rotation of the planet gear. T_{rb} is produced by the normal force between needle roller bearing and planet gear, whose direction is opposite to the rotation of the planet gear. T_{rsw} and T_{rsn} are caused by the friction between the back sphere of drive shaft gears and differential housing [7].

$$T_{rf} = \mu_{rf} \cdot T_0 \frac{\sin\alpha \cdot \sin\theta \cdot \left[cos\left(ar\ \tan\left(\frac{d+d_k}{4R_s}\right)\right)\right]}{4r\cos\alpha} \cdot (d + d_k) \tag{7.13}$$

$$T_{rb} = \frac{\mu_{rb} \cdot M_0 \cdot \sin\alpha \cdot \cos\theta}{r \cdot \cos\alpha} \cdot d_k \tag{7.14}$$

$$T_{rsw} = \frac{\mu_{rs} \cdot \left(\frac{T_0}{2r} - \frac{T_{rf} + T_{rb}}{2r}\right) \cdot \sin\alpha \cdot \sin\gamma}{\cos\alpha} \cdot r \tag{7.15}$$

$$T_{rsn} = \frac{\mu_{rs} \cdot \left(\frac{T_0}{2r} + \frac{T_{rf} + T_{rb}}{2r}\right) \cdot \sin\alpha \cdot \sin\gamma}{\cos\alpha} \cdot r \tag{7.16}$$

where T_0 is the torque of the differential housing. θ is the pitch cone angle. d_k is the mounting-hole diameter of the planet gear. d is the sphere diameter. R_s is the sphere radius. γ and r are the pitch cone angle and pitch radius of sun gear. μ_{rf} is the friction coefficient between the back sphere of planet gear and the differential housing. μ_{rb} is the friction coefficient between the planet gear shaft and the inner hole of the planet gear, and μ_{rs} is the friction coefficient between the sun gear and the differential housing.

7.1.3.4 Churning Loss

The churning loss is the entrainment loss, which is affected by many factors, including the gear modulus, the viscosity of lubricant, the operating temperature, etc. The rotating parts, which contact with the lubricant, lead to churning loss. The churning loss is greater when the lubricant infiltrates is deeper. Hence, it is difficult to determine the precise model of the churning loss. Referring to the first part of British Standard BSISO/TR14179, the empirical methods [8] for solving the churning loss are given as follows.

The churning loss related to the outside diameter of shaft can be expressed as:

$$N_{WG} = \frac{7.37 \times f_g v n^3 D^{4.7} L}{A_g \cdot 10^{26}} \tag{7.17}$$

The churning loss related to the end face of gear can be expressed as:

$$N_{WG} = \frac{1.474 \times f_g v n^3 D^{5.7} L}{A_g \times 10^{26}} \tag{7.18}$$

The churning loss related to the gear can be expressed as:

$$N_{WG} = \frac{7.37 \times f_g v n^3 D^{4.7} B \left(\frac{R_f}{\sqrt{\tan \beta}} \right)}{A_g \times 10^{26}} \tag{7.19}$$

where D is the outside diameter of gears. f_g is the infiltration factor of gear, which ranged from 0 to 1. A_g is the configuration constant. L is the component length. B is the face width. β is the helical angle. v is the kinematic viscosity of lubricant. R_f is the roughness factor, and n is the rotation speed of gear.

7.1.3.5 Windage Loss

The windage loss is rotational energy loss in the DPSD, which is affected by many factors, including the rotational speed of gear, the concentration of oil mist, the diameter of gear, etc. Under the condition of low speed, the windage loss of gear is small in total power loss, which usually can be ignored. The windage loss can be expressed as:

$$N = C\left(1 + 2.3\frac{t}{R}\right)\rho^{0.8}n^{2.8}r^{4.6}\mu^{0.2} \tag{7.20}$$

where C is the constant 2.04×10^{-8}. t is the gear width. n is the rotation speed of gear. ρ and μ are the density and viscosity of the lubricant-air mixture respectively.

The power loss of DPSD can be obtained under the different driving conditions based on the established computing model, which provides the basis for building the heat balance equations.

7.1.4 Thermal Resistance Calculation

Thermal resistance is a key parameter in thermal analysis, which includes conduction, convection and radiation [9–12]. The impact of radiation can be ignored because of the small temperature difference between components. The calculation equations of the thermal resistance are summed up from a large number of experiments.

The thermal resistance of conduction can be divided into two conditions.

For flat plate,

$$R = \frac{L}{\mathrm{k}A} \tag{7.21}$$

where L is the thickness of the flat plate. k is the thermal conductivity. A is the area of transfer heat.

For cylinder,

$$R = \frac{\ln\left(\frac{r_2}{r_1}\right)}{2\pi Lk} \tag{7.22}$$

where r_1 and r_2 are the inner and outer radius of cylinder. L is the axial length in the heat flow direction.

The thermal resistance of convective can be also divided into two conditions.

For flat plate,

$$R\frac{1}{hA} \tag{7.23}$$

where h is the coefficient of the convective heat transfer, and A is the area of convective.

For cylinder,

$$R = \frac{1}{2\pi rhL} \tag{7.24}$$

where r is the cylindrical radius, and h is the coefficient of convective heat transfer.

The thermal balance equation can be established for each node in the thermal network based on Eq. (7.2). Then, the heat balance equations of the whole thermal network system are established, and the temperature value at each node of the thermal network can be calculated.

7.1.5 Results

The lubricating oil properties are shown in Table 7.2. The operating conditions of the combined drive and engine drive mode are shown in Table 7.3.

Under the combined driving mode, the distribution of temperature field of DPSD is shown in Fig. 7.3. It can be found that the temperatures of planet gear and sun gear are higher, but the differences are smaller. There are four meshing points between the planet gear and sun gear, whose power loss is large, and eventually this will lead to higher temperature. In addition, the larger thermal conductivity coefficient of the gear material makes that the temperature difference of gears is smaller. Under combined driving mode, the temperature curves of the lubricant and planet gear against air temperature are shown in Fig. 7.4.

Under the engine driving mode, the distribution of the temperature field of DPSD is shown in Fig. 7.5. Compared to the combined driving mode, the transfer torque and the power loss are smaller. Hence, the temperature of node is lower and the temperature difference between planet gear and sun gear is also smaller. Under this mode, the temperature curves of lubricant and planet gear against the air temperature are shown in Fig. 7.6.

Table 7.2 Lubricating oil properties

Temperature (°C)	Density (kg/m^3)	Thermal conductivity W/(m·K)	Kinematic viscosity 10^{-5} (m^2/s)	Prandtl number
40	880.6	0.1414	9.07	1121
100	846.2	0.1361	1.15	160

Table 7.3 Typical operating conditions for the two modes

Operating mode	No.	Torque (Nm)	Rotating speed (rpm)
Combined driving mode	I	17.39	4002.04
	II	−32.45	2547.2
Engine driving mode	I	15.58	1579.39
	II	29.07	686.94

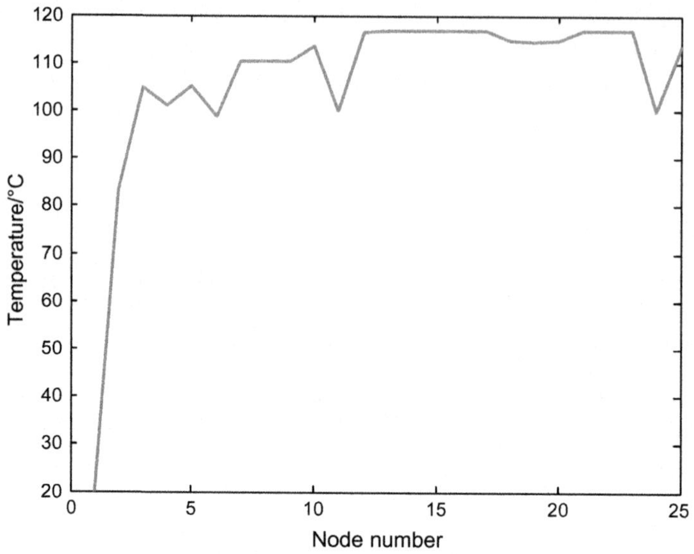

Fig. 7.3 Temperature distribution under combined driving mode

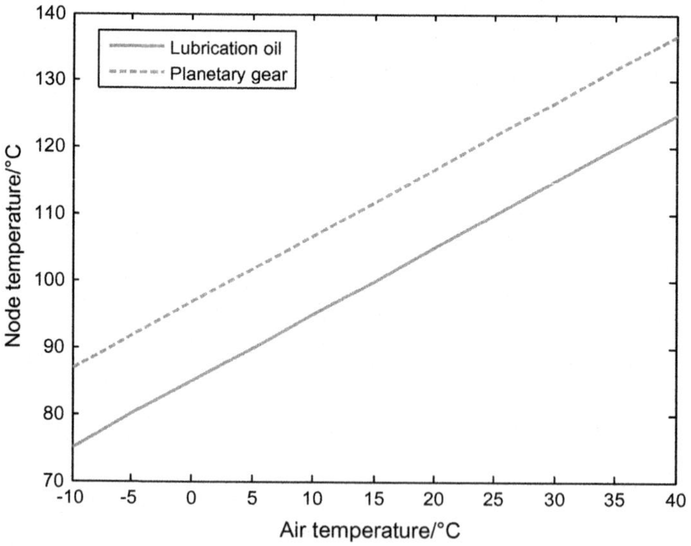

Fig. 7.4 Temperature curves of lubricant and planet gear against air temperature under combined driving mode

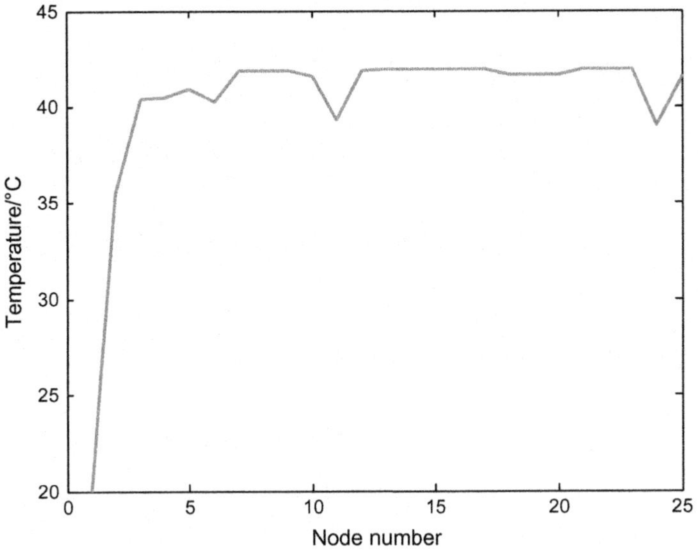

Fig. 7.5 Temperature distribution under engine driving mode

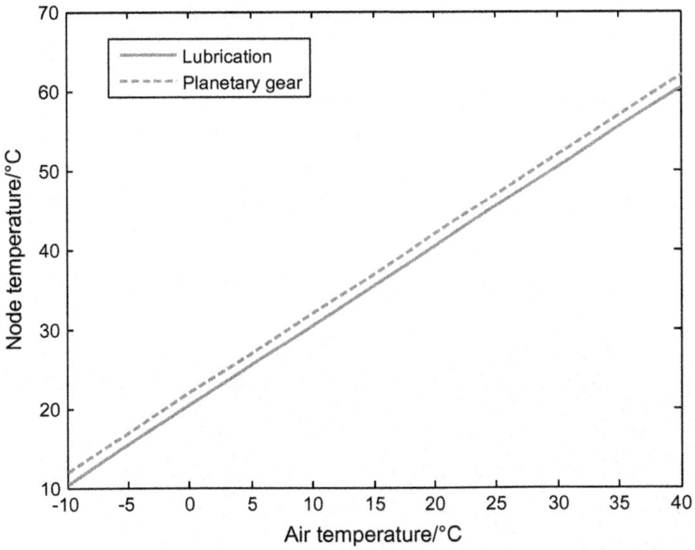

Fig. 7.6 Temperature curves of lubricant and planet gear against air temperature under engine driving mode

7.1.6 Summary

By establishing the thermal network model and calculating the temperature field of DPSD under the typical driving conditions, the DPSD temperature is analyzed. The temperatures of lubricant and the planet gear are close to linear variation with the air temperature in engine driving mode and combined driving mode. The above results show that the thermal network method can be effectively used to predict the steady-state temperature distribution and the changing law of temperature in DPSD.

7.2 Temperature Analysis of Oil Film

During the test of DPSD, there is fracture failure in the planet gear shaft and burn in the tooth surface, as shown in Fig. 7.7. The cause includes lubrication failure, bearing failure, planet gear shaft failure, etc. Though DPSD has the same operational principal with traditional differential, the speed difference of DPSD is larger and the working time is longer. And this causes the complete difference in lubrication system design. So it is necessary to analyze their internal lubrication phenomenon. The analysis content refers to thermal elastohydrodynamic lubrication (TEHL) [13], which includes the pressure, thickness and temperature rise of oil film.

Radial sliding, rolling speed of the tooth surface and tooth surface load change with the gear rotation. In order to simplify the calculation, the TEHL problem of bevel gears is equivalent to the TEHL problem of equivalent spur gear, and the mathematical model is established. Then, the mathematical model is used to solve the lubrication characteristics between planet gear and sun gear when the planet gear shaft is under the condition of maximum load [14].

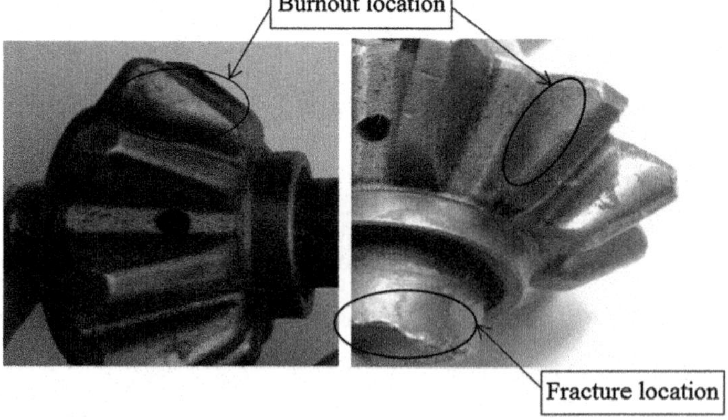

Fig. 7.7 Gear tooth failure

7.2.1 Mathematical Model of Bevel Gear Lubrication

Firstly, a mathematical model is established based on TEHL model of linear gear contact of spur gear, including generalized Reynolds equation, oil film thickness equation, energy equation, heat section equation, load balance equation and bevel gear correlation equations [15]. In the spur gear TEHL model, the requirements of the gear parameters include the equivalent radius, the normal force of tooth surface and the integrated speed [16].

7.2.1.1 Geometrical Model of Bevel Gear

As shown in Fig. 7.8, equivalent straight gear tooth width is equal to the bevel gear tooth width. The tooth thickness is equal to the mid-point tooth thickness of the bevel gear. The spherical involute surface of the bevel gear is also approximately simplify to a plane involute surface. Therefore, the geometric simplification model of the bevel gear mainly includes the equivalent radius of the tooth contact surface, the unit length of the contact line, normal force of tooth surface and integrated speed of contact surface.

1. **Equivalent radius R of contact surface**

   The meshing of the equivalent spur gear pair is shown in Fig. 7.9. The intersection point between N_1N_2 (line of action) and O_1O_2 (line of centers) is chosen as the research point.

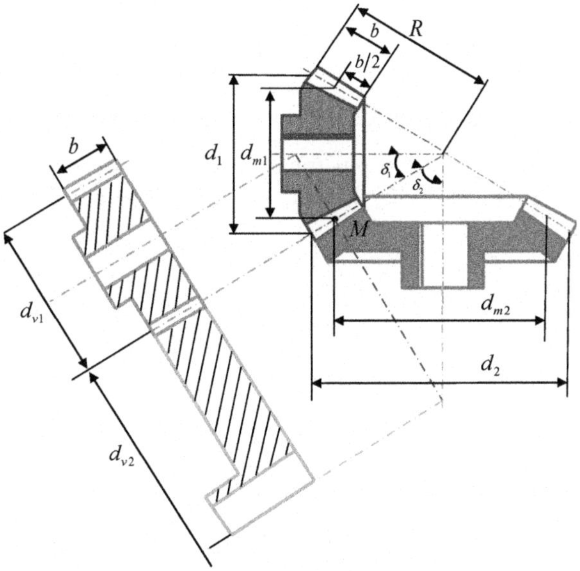

Fig. 7.8 Equivalent spur gear model of bevel gear

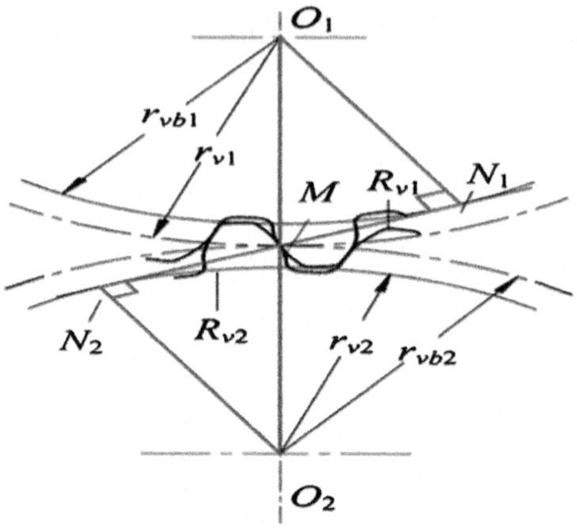

Fig. 7.9 Meshing in the equivalent spur gear pair

Considering the geometrical relationship of bevel gears and the equivalent radius equations of TEHL, the equivalent radius at point M can be expressed as:

$$R = \frac{R_{v1} R_{v2}}{R_{v1} + R_{v2}} = \frac{r_{m1} r_{m2} \sin \alpha}{r_{m1} \cos \delta_2 + r_{m2} \cos \delta_1} \tag{7.25}$$

2. **Normal force W_n on unit length**

According to the force analysis of bevel gears, the normal force on unit length of the contact line at point M is expressed as:

$$W_n = \frac{KFZ^2}{b \cos \alpha} \tag{7.26}$$

where K is a coefficient related to equivalent gears. F represents peripheral force of bevel gears, and Z is a coefficient expressing the contact ratio of equivalent gears.

3. **Equivalent speed at tooth contact point**

Based on the geometrical relationship of bevel gears and the equivalent speed equation of TEHL, the equivalent speed at point M can be expressed as:

$$U_e = \frac{u_{v1} + u_{v2}}{2} = \frac{\pi}{120} \left(\frac{n_1 d_{m1}}{\cos \delta_1} + \frac{n_2 d_{m2}}{\cos \delta_2} \right) \tag{7.27}$$

7.2.2 *TEHL Model*

According to TEHL line contact model of spur gears, the mathematical model of bevel gears is built, which includes Reynolds equation, film thickness equation, energy equation, thermal interface equations and load balance equation.

(1) Reynolds equation

The generalized Reynolds equation, which allows the viscosity and the density of the oil to vary along the film thickness direction, can be expressed as:

$$\frac{\partial}{\partial x}\left[\left(\frac{\rho}{\eta}\right)_e h^3 \frac{\partial p}{\partial x}\right] = 12u_e \frac{\partial(\rho^* h)}{\partial x} \tag{7.28}$$

where $\left(\frac{\rho}{\eta}\right)_e = 12(\eta_e\rho'_e/\eta'_e - \rho''_e)$, $\rho^* = [\rho'_e\eta_e(u_b - u_a) + \rho_e u_a]/u_e$, $\rho_e = \frac{1}{h}\int_0^h \rho d\ell$, $\rho'_e = \frac{1}{h^2}\int_0^h \int_0^\ell \frac{d\ell'}{\eta^*} d\ell$, $\rho'_e = \frac{1}{h^3}\int_0^h \rho \int_0^\ell \frac{\ell' d\ell'}{\eta^*} d\ell$, $\frac{1}{\eta_e} = \frac{1}{h}\int_0^h \frac{d\ell}{\eta^*}$, $\frac{1}{\eta'_e} = \frac{1}{h^2}\int_0^h \frac{\ell d\ell}{\eta^*}$, $\eta^* = \eta\left(\frac{\tau}{\tau_0}\right)/\sinh\left(\frac{\tau}{\tau_0}\right)$.

Boundary conditions for Eq. (7.28) are as follows:

$$\begin{cases} p(x_{in}) = p(x_{out}) = 0 \\ p(x) \geq 0 \quad (x_{in} < x < x_{out}) \end{cases} \tag{7.29}$$

(2) Film thickness equation

The film thickness is composed of rigid displacement, geometric film thickness and elastic deformation, whose equation can be expressed as:

$$h(x) = h_{00} + \frac{x^2}{2R} - \frac{2}{\pi E'}\int_{x_{in}}^{x_{out}} p(x')\ln(x - x')dx' \tag{7.30}$$

(3) Viscosity equation

The equation about viscosity-pressure-temperature relationship proposed by Roelands is chosen in this paper, which can be expressed as:

$$\eta = \eta_0 \exp\left\{(9.67 + \ln\eta_0)\left[-1 + \left(1 + 5.1\times10^{-9}\overline{P}p_H\right)^{z_0} \times \left(\frac{T - 138}{T_0 - 138}\right)^{-S_0}\right]\right\} \tag{7.31}$$

(4) Energy Equation

Ignoring influence of heat radiation, thermal conductivity and gravity, the energy equation can be expressed as:

$$C_p\left(\rho u\frac{\partial T}{\partial x} - q\frac{\partial T}{\partial \ell}\right) = K\frac{\partial^2 T}{\partial \ell^2} - \frac{T}{\rho}\frac{\partial \rho}{\partial T}u\frac{\partial p}{\partial x} + \eta^*\left(\frac{\partial u}{\partial \ell}\right)^2 \tag{7.32}$$

where $q = \frac{\partial}{\partial x}\int_0^\ell \rho u d\ell'$.

(5) Thermal interface equation

To determine the temperature on contact surfaces, the gear is considered as a half-space body with a moving heat source. Therefore, the heat conduction equations of gears are built, as shown in Eq. (7.33).

$$\begin{cases} T(x,0) = \frac{K}{\sqrt{\pi \rho_1 c_1 u_1 K_1}} \int\limits_{-\infty}^{x} \frac{\partial T}{\partial \ell_a}\Big|_{x,0} \frac{ds}{\sqrt{x-s}} + T_0 \\ T(x,h) = \frac{K}{\sqrt{\pi \rho_2 c_2 u_2 K_2}} \int\limits_{-\infty}^{x} \frac{\partial T}{\partial \ell_b}\Big|_{x,h} \frac{ds}{\sqrt{x-s}} + T_0 \end{cases} \tag{7.33}$$

where $T(x,0)$ is the temperature on the contact surface between planet gear and oil film. $T(x,h)$ is the temperature on the contact surface between drive shaft gear and oil film.

The continuous condition of the heat flow of the contact surface is given as:

$$\begin{cases} K\frac{\partial T}{\partial \ell}\Big|_{\ell=0} = K_1 \frac{\partial T}{\partial \ell_a}\Big|_{\ell_a=0} \\ K\frac{\partial T}{\partial \ell}\Big|_{\ell=h} = K_2 \frac{\partial T}{\partial \ell_b}\Big|_{\ell_b=0} \end{cases} \tag{7.34}$$

(6) Load balance equation

The load equation is given as:

$$W = \int\limits_{x_{in}}^{x_{out}} p dx. \tag{7.35}$$

7.2.3 Numerical Solution

A complete numerical solution to TEHL line contact of a spur bevel gear is obtained by combining multi-grid method, Jacobi iteration method and scanning method [17–19]. The pressure loop and the temperature loop are included in the numerical analysis of TEHL, which form a whole loop. The flow chart is shown in Fig. 7.10. Jacobi iteration method is chosen for the pressure loop. Scanning method is chosen for the temperature loop and multi-grid method is chosen for the whole loop. In order to simplify the calculation, all the related variables are dimensionless.

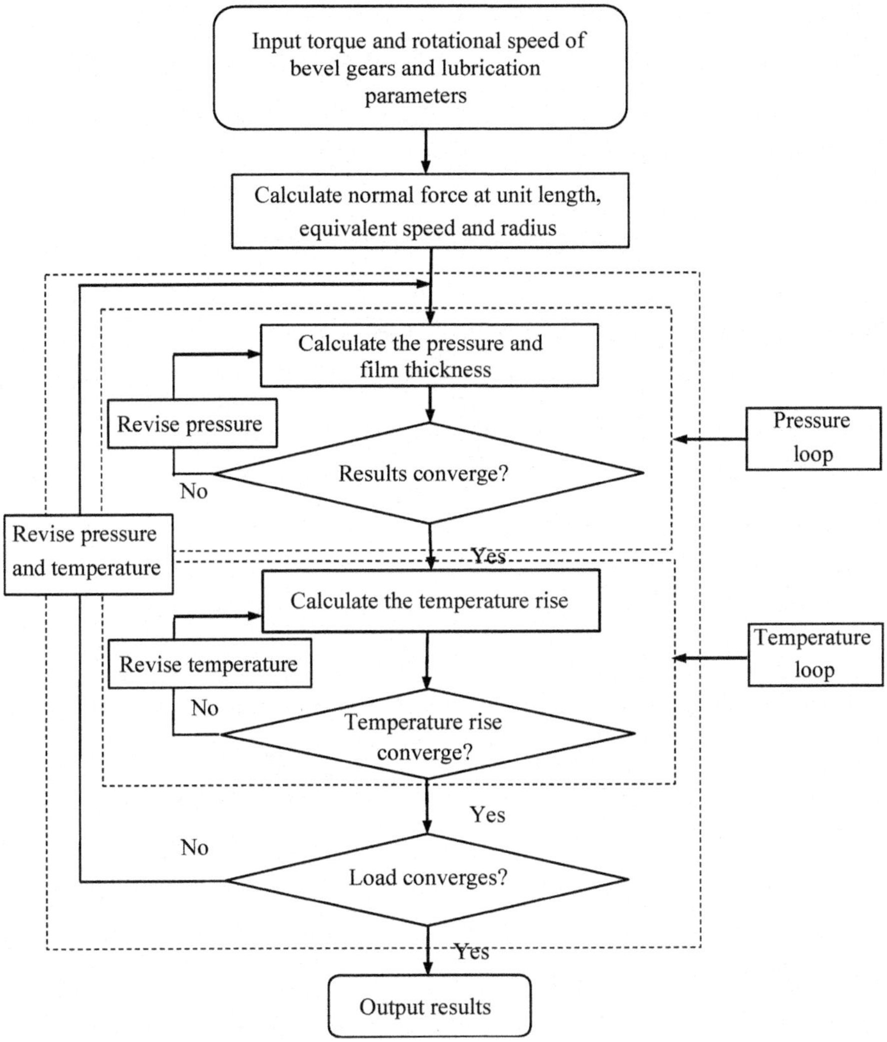

Fig. 7.10 Flow chart of TEHL of bevel gears

7.2.4 Results

As the housing is considered as the reference system and the influence of the rotation of housing on the lubrication is ignored, the lubrication of meshing gears in DPSD is simplified as a steady-state issue about a pair of bevel gears. In such case, when THEL of gears in DPSD is analyzed, the autorotation speed of planet gears and the rotational speed relative to the housing of drive shaft gears are adopted during calculating.

Table 7.4 Parameters of bevel gears in DPSD

Modulus (mm)	Pressure angle (°)	Number of teeth	Density (kg/m^3)
$m = 5.4$	$\alpha = 22.5$	$z_1 = 10$ $z_2 = 14$	$\rho = 7800$
Equivalent elastic modulus (Pa)	Coefficient of tooth width	Tooth width at pitch circle (mm)	Reference cone angle (°)
$E = 2.06e^{11}$	$\psi_R = 0.2846$	$b_1 = 15.4$ $b_2 = 15.4$	$\delta_1 = 35.53$ $\delta_2 = 54.47$

Notice Subscript 1 and 2 represent planet gear and drive shaft gear respectively

Table 7.5 Rotation speed and torque of bevel gears and housing

	Output torque (Nm)	Absolute rotational speed (rpm)	Rotational speed relative to housing (rpm)
Planet gear	–	1110.2	1110.2
Left drive shaft gear (A)	−30	1816	793
Right drive shaft gear (B)	−35	230	−793
Housing	65	1023	0

The parameters of bevel gears in DPSD are shown in Table 7.4.

The rotational speed and torque of bevel gears and the housing under maximum load condition are shown in Table 7.5.

Figures 7.11, 7.12 and 7.13 show the distributions of film pressure, film thickness and temperature rise at point A and B when DPSD is under maximum load condition.

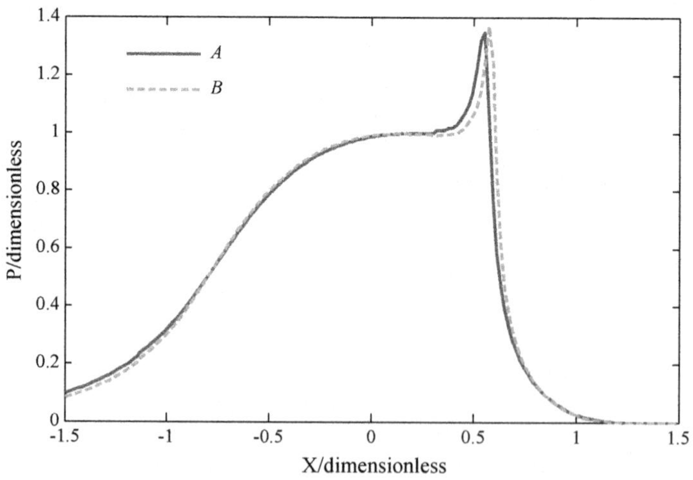

Fig. 7.11 Distributions of film pressure

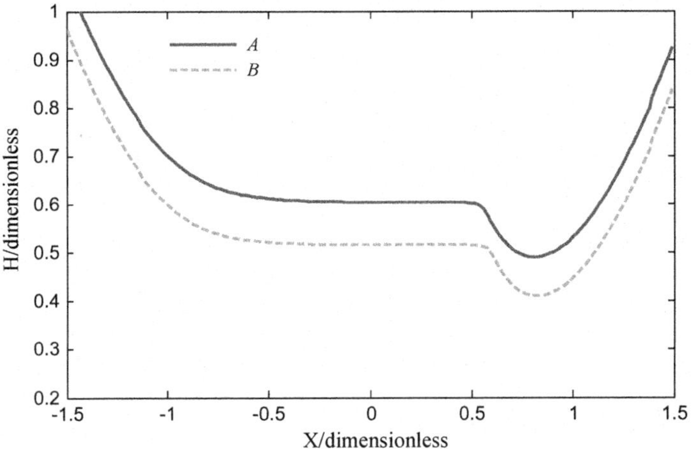

Fig. 7.12 Distributions of film thickness

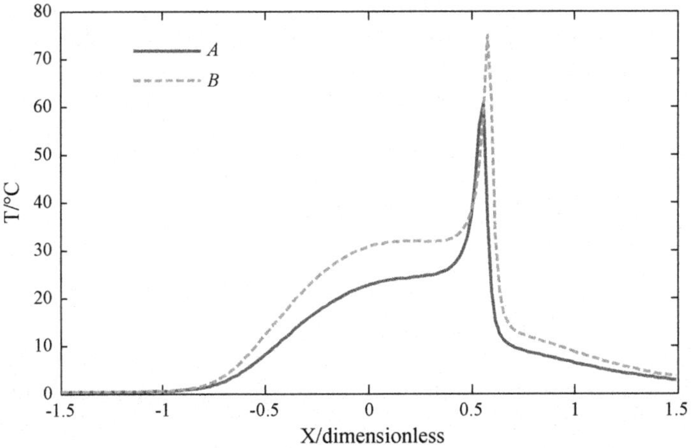

Fig. 7.13 Distributions of temperature rise

The pressure distribution of point A agrees with that of point B. Film thickness at point A is bigger than that at point B evidently and temperature rise at point A is lower than that at point B. This mainly results from the difference of output torques from left and right drive shaft. According to the indoor experiment, the reason for the difference is frictions of two parts. One is that between planet gears and their shafts, and the other is that between back cone of gears and the housing.

The research provides some approximate reference data for the design of lubrication and cooling system of DPSD. However, as the solution of TEHL line contact of bevel gears in DPSD is approximate, the equivalent model needs to be improved in the future.

7.3 Temperature Analysis of Gear Tooth

At present, the methods of thermal analysis of gears include analytic method, numerical method and experimental method. In this section, the thermal analysis of the gear tooth is analyzed by finite element method (FEM).

7.3.1 *Finite Element Model*

One tooth as an example is analyzed as the temperature distribution of each tooth is assumed the same. Different tooth surfaces have different thermal boundary conditions. Generally, the overall surface is divided into end-face, meshing face and top-face, as shown in Fig. 7.14. Where A_4 and A_7 are end-face. A_2 and A_6 are meshing face, and A_5 is the top-face [20]. By setting the material properties, the heat transfer coefficient and the convective heat transfer coefficient of the tooth surface, the finite element model are shown in Fig. 7.14.

7.3.2 *Results*

The tooth surface friction coefficient is set to 0.1. The temperature distribution of the tooth is shown in Fig. 7.15.

It is obvious that temperature distribution of the tooth surface is not uniform. And two crests appear in the tooth root and tooth top. The maximum value appears near the tooth root. The gluing of gears usually occurs there. The distribution of the large temperature gradient seriously reduces the load carrying capacity and operation performance of the gear.

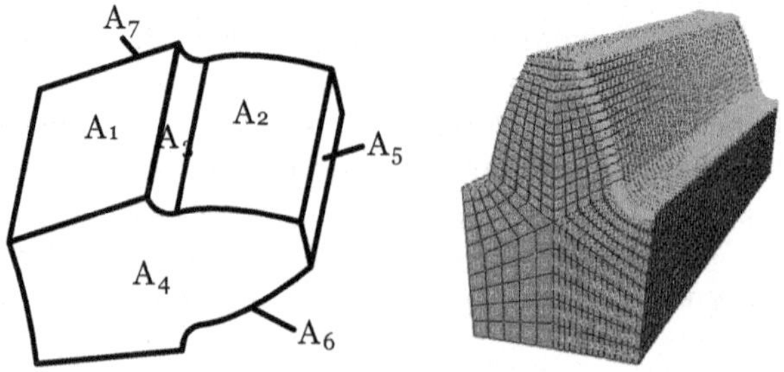

Fig. 7.14 Finite element model of one tooth

Fig. 7.15 **a** Temperature distribution of *right view*. **b** Temperature distribution of *front view*. **c** Temperature distribution of *left view*

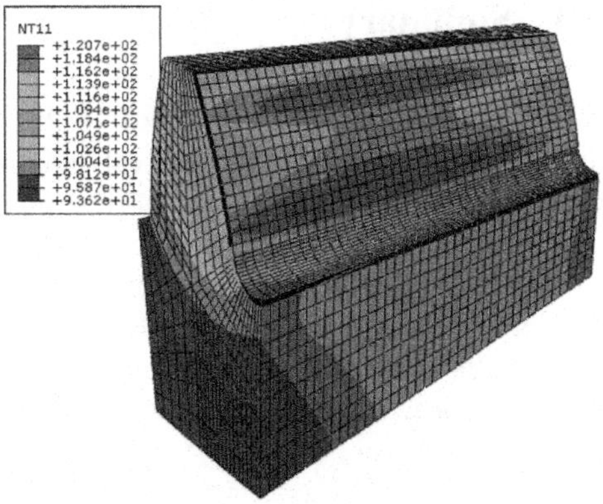

(a) Temperature distribution of right view

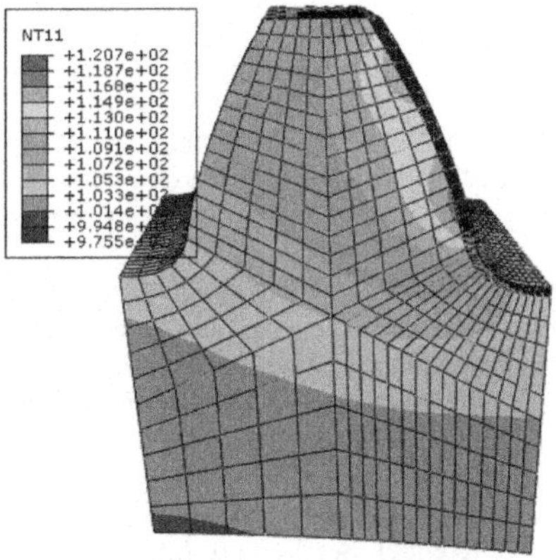

(b) Temperature distribution of front view

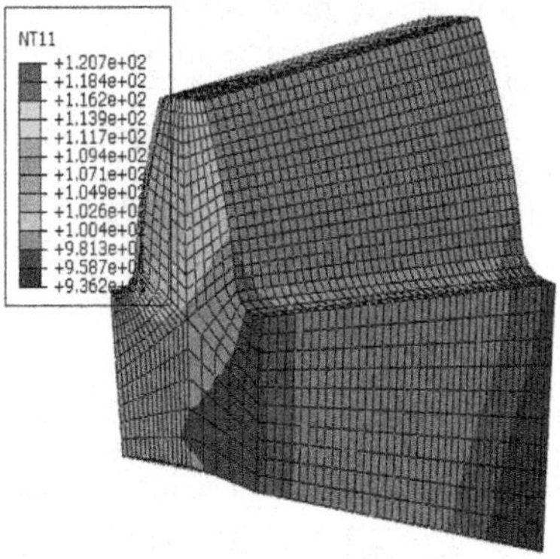

(c) Temperature distribution of left view

7.4 Summary

In this chapter, the changing rules of DPSD temperature is analyzed by establishing the thermal network model and calculating the temperature field of DPSD under typical driving conditions. The lubrication properties of planet gears and drive shaft gears are obtained by the approximate solution of TEHL line contact of bevel gears in DPSD. However, the equivalent model needs to be improved in the future. The temperature distribution of the tooth of gear is analyzed by finite element method. The results show that the maximum temperature of the tooth surface appears near the tooth root, which provides the references for anti-gluing design of the gears.

References

1. Zheng S. Failure analysis and test validation on differential coupling device in HEV. Jilin University; 2010.
2. Wang JX, Wang XP, Yang ZY, et al. Analysis of heat transfer in power split device for hybrid electric vehicle using thermal network method. Adv Mech Eng. 2014;2014(1):1–9.
3. Liu N. Analysis of thermal characteristics on power-split device for HEV. Jilin University; 2014.
4. Piacentino A. Thermal analysis and new insights to support decision making in retrofit and relaxation of heat exchanger networks. Appl Therm Eng. 2011;31(16):3479–99.
5. Anderson NE, Loewenthal SH. Spur-gear-system efficiency at part and full load. National Aeronautics and Space Administration, Scientific and Technical Information Office, 1980.
6. Harris T, Michael NK. Rolling bearing analysis (2 volume Set). Boca Raton: CRC Press; 2006.
7. Xiao WY, Wang SH. Mechanical analysis for planetary gear of common bevel gear differential. Sci Technol Inf. 2007;14:35–6.
8. Moldovean G, Butuc BR, Bozan CA. On the power losses of cylindrical and bevel gears used in wind turbines and tracking systems for photovoltaic platform. In: SYROM 2009. Springer Netherlands; 2009. p. 300–350.
9. Incropera FP, Lavine AS, DeWitt DP. Fundamentals of heat and mass transfer. New York: Wiley; 2011.
10. K&K Associates, Developers of Thermal Analysis Kit. Thermal Network Modeling Handbook, 1999–2000.
11. Kaviany M, Kanury A. Principles of heat transfer. New York: Wiley; 2002.
12. de Gevigney JD, Changenet C, Ville F, Velex P. Thermal modelling of a back-to-back gearbox test machine: application to the FZG test rig. Proc Inst Mech Eng Part J J Eng Tribol. 2012;266(6):501–15.
13. Wen SZ. Principles of tribology. 4th ed. Beijing: Tsinghua University Press; 2012.
14. Zhang YH, Li ZX, Feng SS, et al. Approximation solution for TEHL of bevel gears in PSD. J Appl Math Phys. 2013;1:93–7.
15. Larsson R. Transient non-Newtonian analysis of an involute spur gear. Wear. 1997;207: 67–73.
16. Tang L. TEHL numerical analysis of power coupling device based on multigrid method. Jilin University; 2013.
17. Wang YQ, Yi XJ. Non-Newtonian transient thermoelastohydrodynamic lubrication analysis of an involute spur gear. Lubr Sci. 2010;22(10):465–78.

18. Wang J, Qu SY, Yang PR. Simplified multigrid technique for the numerical solution to the steady-state and transient EHL line contacts and the arbitrary entrainment EHL point contacts. Tribol Int. 2001;34(3):191–202.
19. Qiu LH, Xin YH, Wang T, Jiang S. A calculation of bulk temperature and thermal deflection of gear tooth about profile modification. J Shanghai Jiaotong Univ. 1995;02:79–86.
20. Chen L. The temperature analysis of planet gear reducer based on ANSYS. Nanjing University of Aeronautics and Astronautics; 2009.

Chapter 8
Engineering Analysis of Differential-Based PSD

In this chapter, two aspects of engineering analysis is introduced to operate the engineering analysis for DPSD. One is to introduce gear generation method and contact stress analysis of spur bevel gears in DPSD. A detailed finite element model is created. Then contact analysis of gear pairs is conducted. The other is to introduce a rigid-flexible coupled simulation method for DPSD. The rotational speed of planet gear and output torque of planet gear shaft is analyzed by the rigid-flexible coupled model of DPSD established.

8.1 Stress Analysis of Spur Bevel Gears in DPSD

This section introduces the method of gear tooth profile generation. A detailed finite element model of mating gear pairs is provided. In ABAQUS, contact formulation is based on a surface-to-surface discretization, a small-sliding tracking approach and a slave-master contact algorithm. Pinion is set slave surface and gear is set as master surface.

8.1.1 Nonlinear Analysis of Gear Tooth Contact

Mesh type and size in the finite element model are important. Tooth contact analysis of spur bevel gear is taken as an example to introduce the method of mesh. The tooth of spur bevel gear consists of tip cone and back cone.

X. Zeng and J. Wang, *Analysis and Design of the Power-Split Device for Hybrid Systems*, https://doi.org/10.1007/978-981-10-4272-0_8

8.1.1.1 Gear Tooth Division

Both tip cone and back cone of the spur bevel gear are divided into six auxiliary intermediate surface. Then a tooth is divided into six subvolumes [1], as shown in Fig. 8.1.

8.1.1.2 Meshing of Tooth

In order to achieve a reliable contact stress, the finite element model of whole gear is divided into two regions, the contact region and non-contact region. The contact region is usually not larger than four times the Hertzian's contact half-width [2]. To obtain reasonable results, correspondingly the mesh size of the tooth contact region is one-tenth or less of the Hertzian half-width, as shown in Fig. 8.2. To reduce the computational consumption, the mesh of non-contact region can be sparse [3], as shown in Fig. 8.3. The two conical surfaces are meshed in the same proportion.

$$a^2 = \frac{4PR}{\pi E^*} \tag{8.1}$$

The Hertzian's contact half-width can be obtained from the above formula, where $\frac{1}{E^*} = \frac{1-v_1^2}{E_1} + \frac{1-v_2^2}{E_2}$, $\frac{1}{R} = \frac{1}{R_1} + \frac{1}{R_2}$. E_1, E_2 are the elastic modulus of the two gear materials. V_1, V_2 are the Poisson's ratio of two gear materials. R_1, R_2 are the contact radius of the two gears.

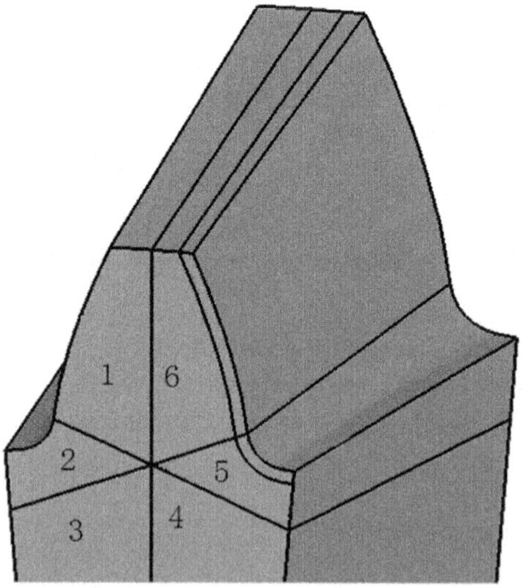

Fig. 8.1 Gear tooth division

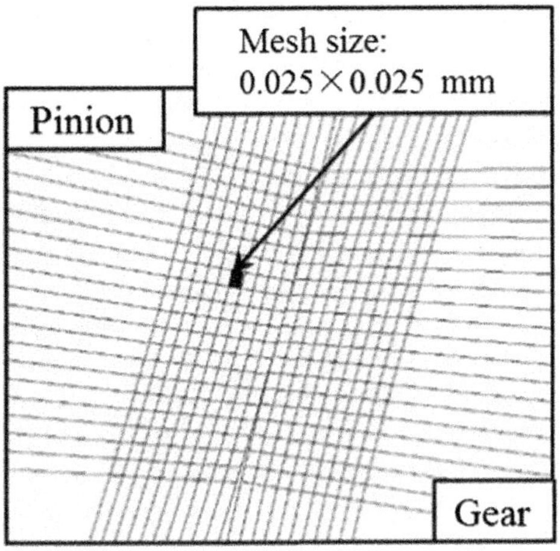

Fig. 8.2 Meshes at contact region

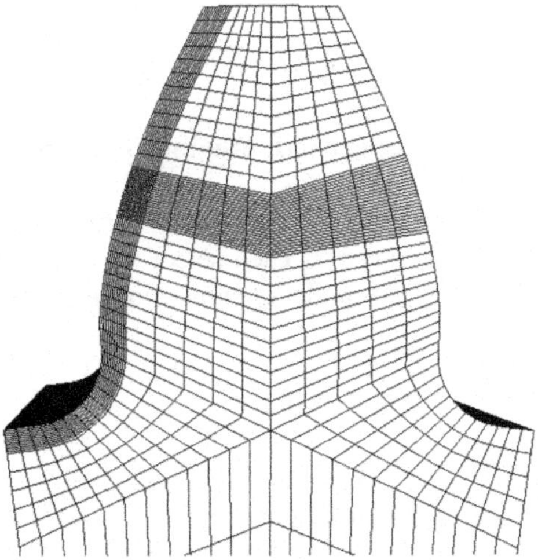

Fig. 8.3 Elements of the back cone

8.1.1.3 Generation of Finite Element Model

3D finite element model of spur bevel gear is obtained by dragging 2D elements from back cone surface to tip cone surface, as shown in Fig. 8.4.

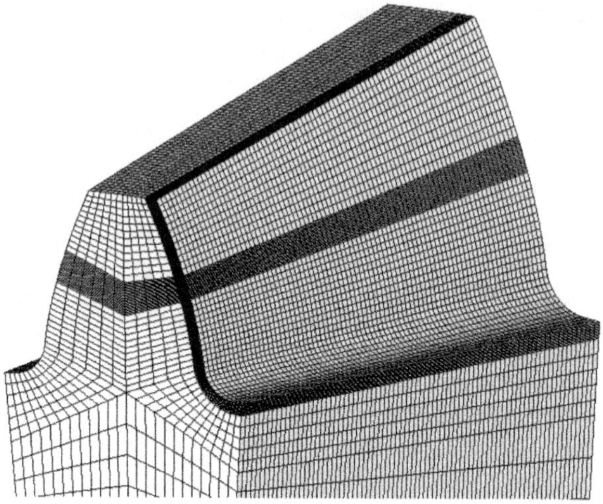

Fig. 8.4 Mesh whole tooth

8.1.2 Definition of Properties in Finite Element Model

8.1.2.1 Import INP File into ABAQUS

Click on the main menu, **File** → **Import** → **Model**, and set **File Filter** to ***.cae**, as shown in Fig. 8.5. Then select the created **INP** file and import it into ABAQUS. The imported model is shown in Fig. 8.6.

8.1.2.2 Repair Model

Because the imported model is default to only one part, it needs to be repaired in order to do the contact analysis [4]. First, remove the default **Model-1**. Find **Model-1** in the model tree. Right click and select **Delete** option as shown in Fig. 8.7. Then make the parts be separated from each other by copy. Each of them is created as a new part. In the model tree, click to open **bsg model** → **Parts**. Right click **Part-1**. Select **Copy** option. In the pop-up **Part Copy** dialog, check **Separate disconnected regions into parts** option. Click **OK**. The new parts are created. The specific process is shown in Figs. 8.8 and 8.9.

8.1.2.3 Define Material Properties

In the module, set the function module to **Property**. Click the icon to pop up the material property settings dialog. Change **Name** to **Steel**. Click **General** option. Set the material density to 7.9e−9, and switch to **Mechanical** option to set the elastic properties of the material. Then select **Elasticity** → **Elastic** in turn and set Young's

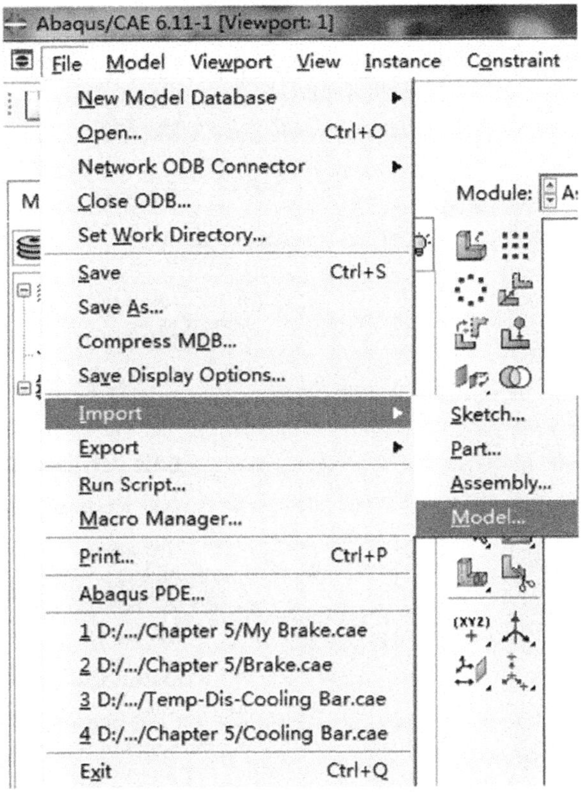

Fig. 8.5 Model import option

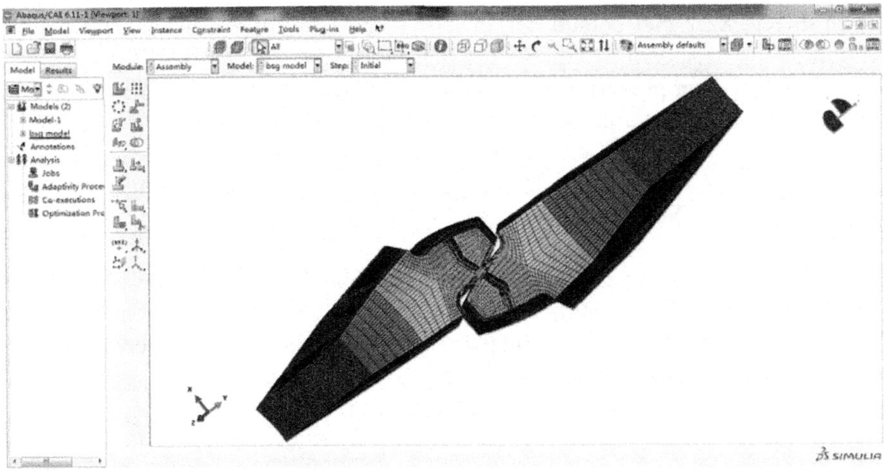

Fig. 8.6 Imported model

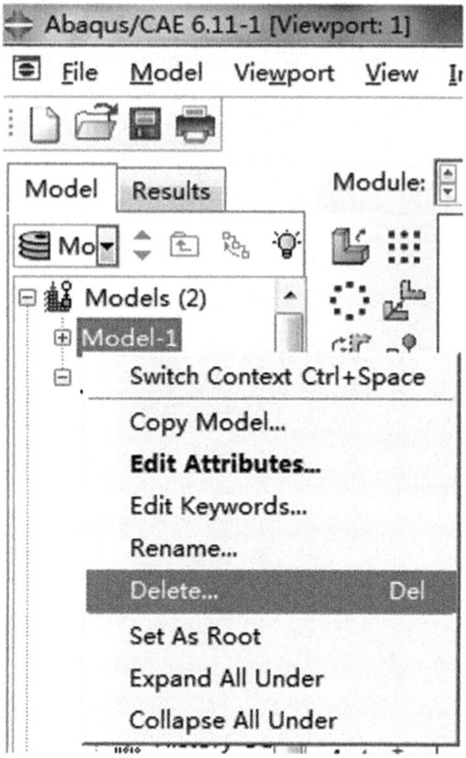

Fig. 8.7 Delete the default model

Modulus to 210000 and set Poission's Ratio to 0.3, as shown in Fig. 8.10. Click **OK**.

Create section properties. Click the icon to pop up **Create Section** dialog as shown in Fig. 8.11. Set default parameters. Click **Continue** icon to pop up **Edit Section** dialog as shown in Fig. 8.12. Click **OK**.

To assign section properties, click on the icon to pop up dialog as shown in Fig. 8.13, and select two parts in turn to define section properties.

8.1.2.4 Create a New Assembly

Set module of the current situation to assembly. In the model tree, click **Model** → **Assembly** → **Instance** option in turn and remove the current **Instance**, as shown in Fig. 8.14. The destructed assembly is deleted. Then create a new assembly. Select **Instance** and right click to select **Create** option, as shown in Fig. 8.15, and select two parts (Ctrl-click). Click **OK**.

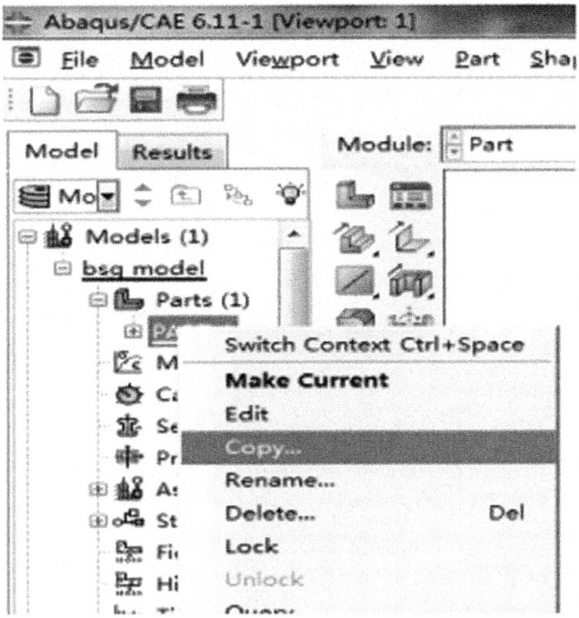

Fig. 8.8 Copy a part

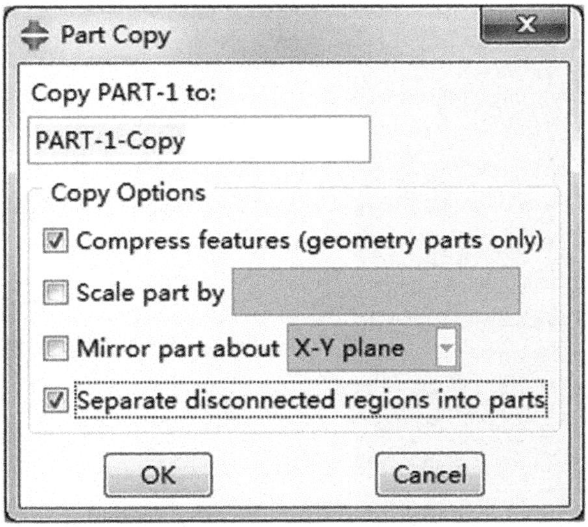

Fig. 8.9 Copy part options dialog

8.1.2.5 Create Contact Relation

The first step is to create the local coordinate system.

Create the shaft of the gear inner circle. Switch the module to **Interaction**. Left click the icon in tools bar and hold down the left button to select **Create Datum Axis: 3 Points on circle** mode. And then select any three points on the circle shown

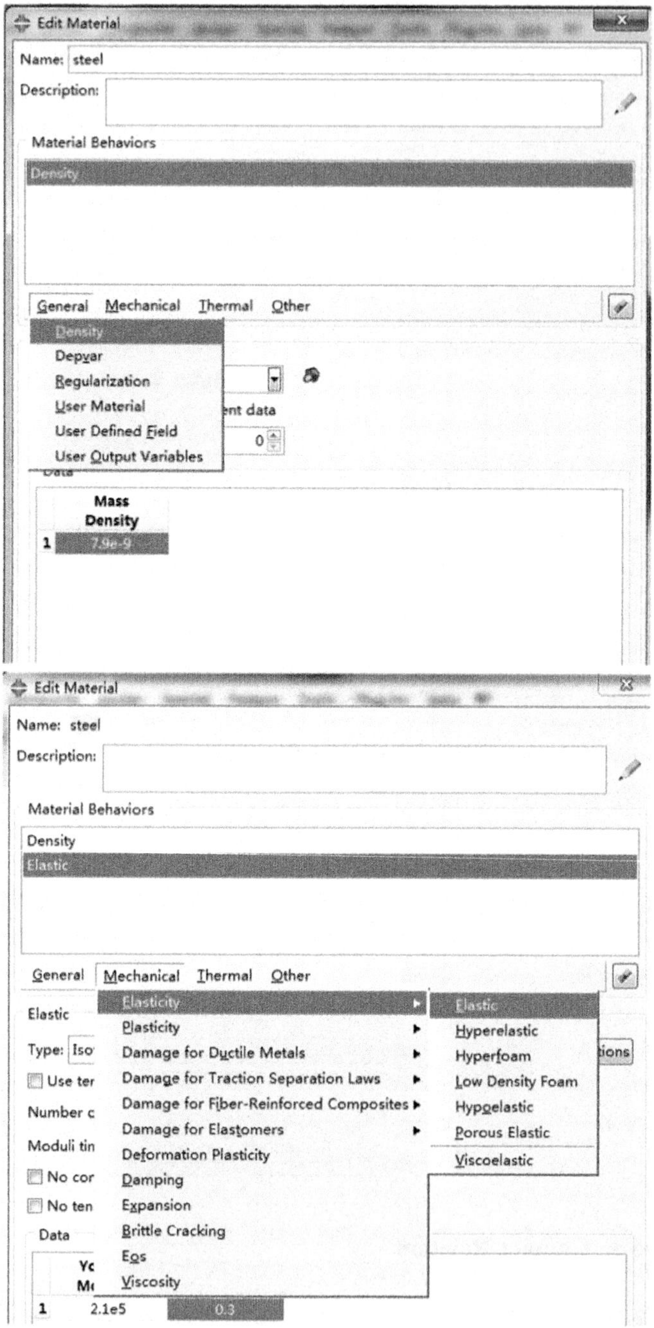

Fig. 8.10 Define material properties

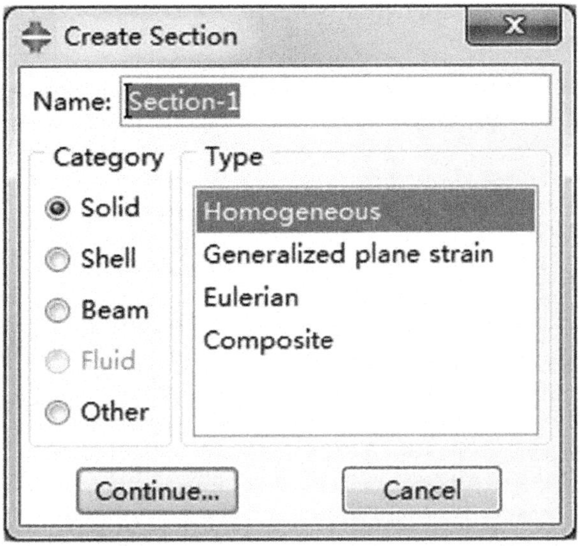

Fig. 8.11 Define section properties

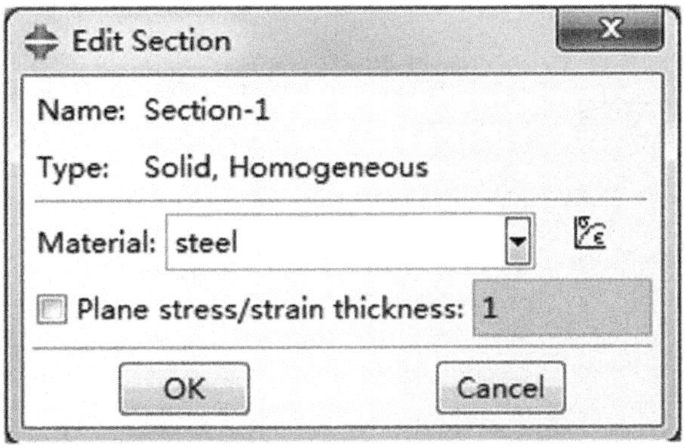

Fig. 8.12 Section properties dialog

in Fig. 8.16. Similarly create another axis of the gear. The created axis is the x-axis of the local coordinate system, and then create the origin of the coordinate system and another point on the axis to determine the direction of the x-axis. Left click on the icon in tools bar. Hold down the left button and select **Create Datum Point: Project Point** mode. Another two points on the axis are built by selecting any two nodes on the part and project them onto the axis. Select one of the points to create the reference point. Click the icon in tools bar. Select the point on the axis, and create the reference points. Finally, create a local coordinate system. Left click the icon in tools bar and hold down **Create Datum CSYS: 3 Points**. The pop-up

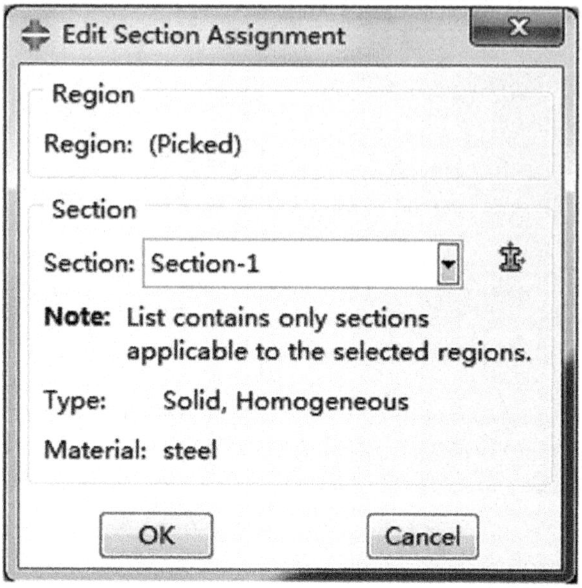

Fig. 8.13 Assign section properties

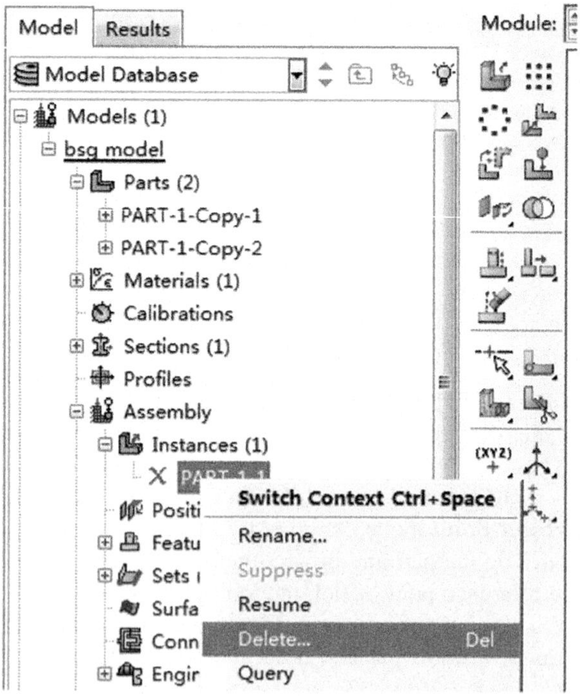

Fig. 8.14 Delete destructed assembly

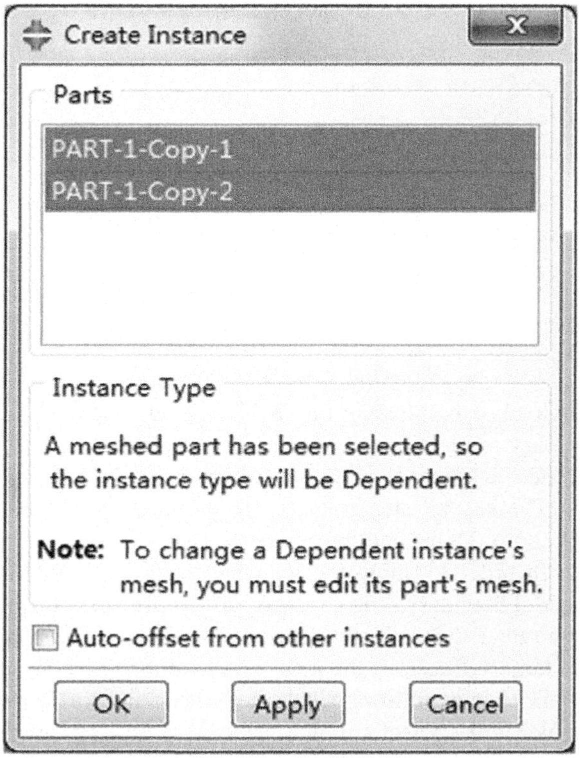

Fig. 8.15 Create a new assembly

Fig. 8.16 Three optional points on the circle

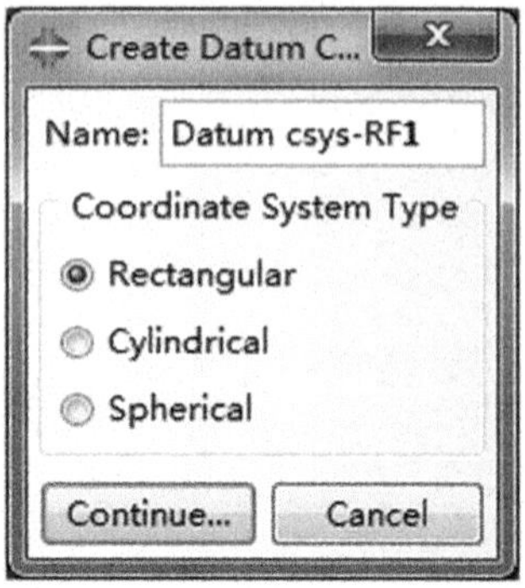

Fig. 8.17 Coordinate system properties

dialog box is shown in Fig. 8.17. The name is modified to **Datum csys-RF1**, which is easy to identify. And select **Rectangular**. Click **Continue**. The local coordinate system is created, as shown in Fig. 8.20.

The second step is to create the coupling of the tooth boundary's degree of freedom (DOF) and the reference point.

In order to apply the load and determine the motion conditions of the tooth model boundary, it is necessary to couple the nodes from partial boundary of tooth with the DOF of the reference points. Because this part of the nodes is in the region of large stiffness, so the DOF is zero treated as a "rigid node". The coupling process of DOF is as follows. Click the icon in tools bar to pop up **Create Constraint** dialog box. Select coupling type to **Coupling**, as shown in Fig. 8.18. Then corresponding to the prompt area Choose type of region: Geometry Mesh, select **Geometry**. Corresponding to Select the constraint control points Done, select the reference created point. Click **Done** icon. When Select the constraint region type: Surface Node Region appears in the prompt area, select **Node Region**, and then select the reference point corresponding to each tooth boundary shown in Fig. 8.19. The coupling relations are completed, as shown in Fig. 8.20.

The third step is to create contact relation of the tooth surface [5].

Unlike the usual fixed boundary conditions, the contact boundary conditions are constantly changing with the elastic deformation of the model during the analysis. The process is as follows: Click the icon . **Create Interaction** dialog box appears, as shown in Fig. 8.21a. Select **Surface-to-Surface** option. **Edit Interaction** dialog box is popped up, as shown in Fig. 8.21b. Click on **Contact Interaction Property** to pop up **Create Interaction Property** dialog box. Select **Contact** option, and pop up **Edit Contact Property** dialog box, as shown in Fig. 8.21c. Click **Mechanical** → **Tangential Behavior**. Select **Penalty** in **Friction Formulation**, and set **Friction Coefficient** to 0.15. Then set the normal properties on the contact face. Click **Mechanical** → **Normal Behavior**. The settings of the parameters are shown in Fig. 8.21c.

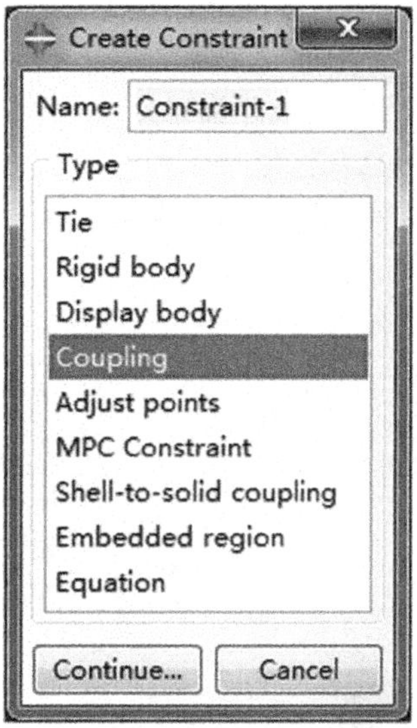

Fig. 8.18 Select interaction type

8.1.3 Load Definition and Solution

Switch the module to **Load** and define the boundary conditions. The DOF of the reference nodes and the DOF of the geometric boundaries of the tooth are coupled to each other in the above definition. The boundary conditions and loads only need to be defined at the reference points [6]. Define the boundary conditions for reference point1. Click the icon in tools bar to pop up **Create Boundary Condition** dialog box, as shown in Fig. 8.22a. Set the boundary type to **Symmetry/Antisymmetry/Encastre**. The following appears in the prompt area at the bottom left of the dialog and click **Geometry** icon. Select reference point 1. **Edit Boundary Condition** dialog box is popped up, as shown in Fig. 8.22b. Select **Encastre.** The boundary of reference point 1 is defined.

The boundary definition for reference point 2 is similar to 1. But the boundary type is **Displacement/Rotation** in **Create Boundary Condition** dialog box. The prompt area shows . Click **Geometry** icon. Select reference point 2, and then pop up the corresponding **Edit Boundary Condition** dialog box, as shown in Fig. 8.22c. Click the arrow button in the rectangle box. Select the local coordinate system created at reference point 2 as the reference coordinate system. Check the option dialog shown in Fig. 8.22c. The DOF is zero except for axial rotation.

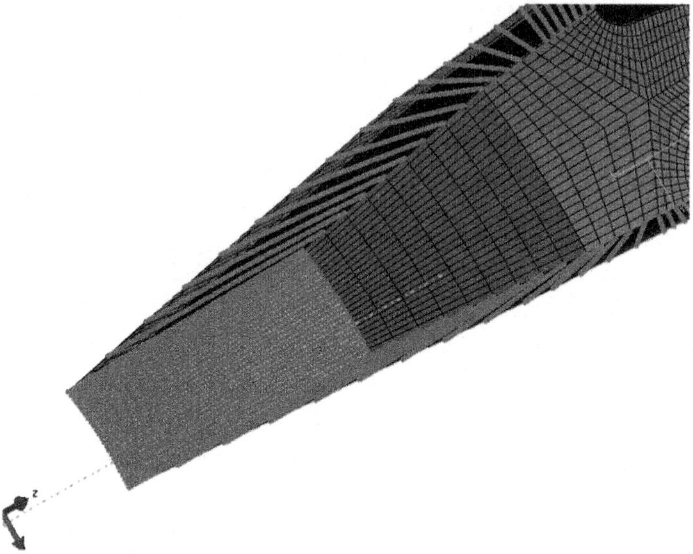

Fig. 8.19 Select nodes to be coupled

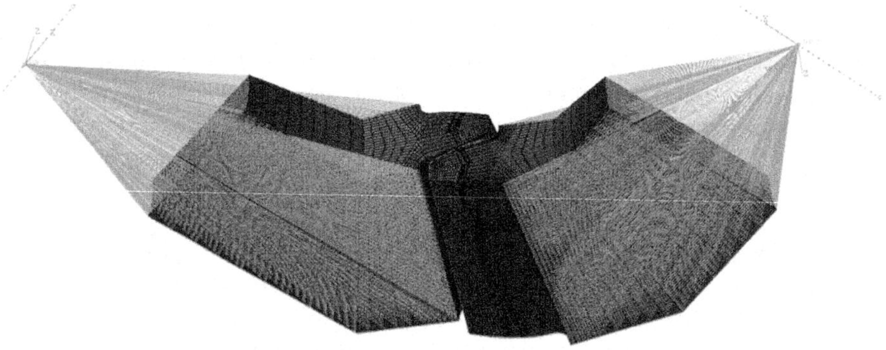

Fig. 8.20 Model after coupling DOF

Click the icon in tools bar to pop up **Create Load** dialog box as shown in Fig. 8.23a. Select **Moment** type and the prompt area will show . Click **Geometry** icon to select reference point 2. The pop-up **Edit Load** dialog box is shown in Fig. 8.23b. Click the arrow button in the rectangular box. Select the local coordinate system on the reference point 2 as the reference coordinate system. Other definitions are shown in Fig. 8.23b. So far, all boundary conditions and loads are defined.

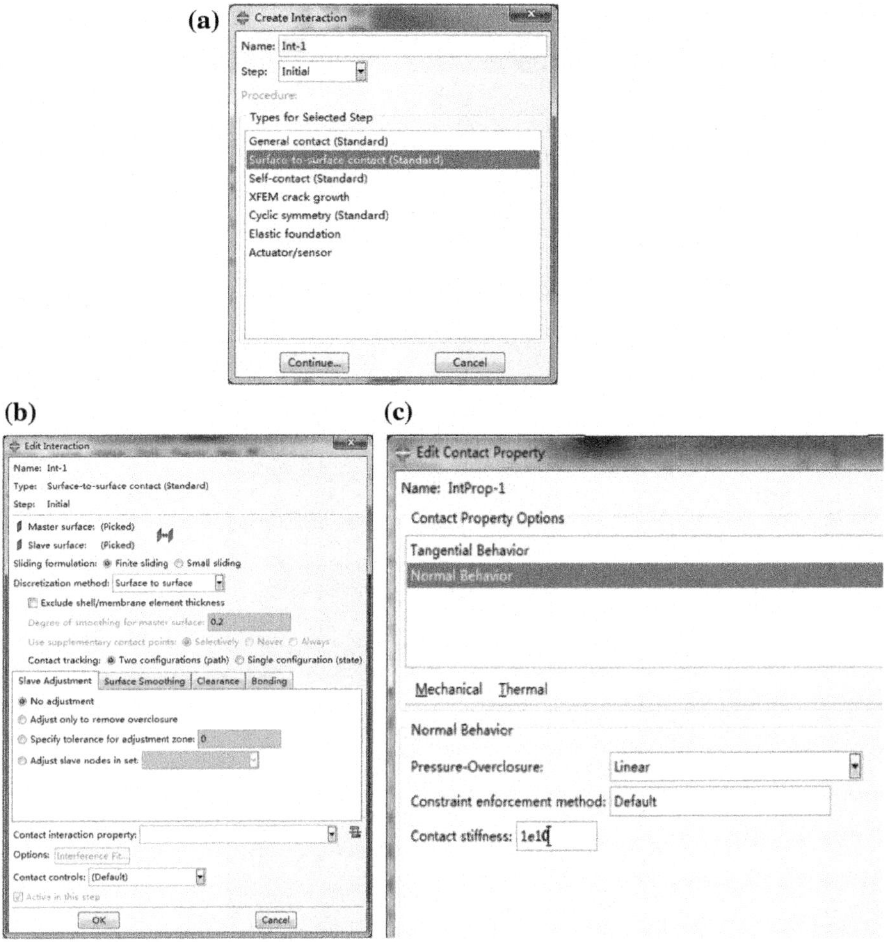

Fig. 8.21 **a** Select contact type. **b** Define contact properties. **c** Contact properties dialog

Switch the state of the module to **Job**. Click the icon in tools bar, then **Create Job dialog** box is popped up as shown in Fig. 8.24a. Rename **Job** to **Job-bsg**. Click **Continue**. Keep the default settings. Click **OK**. Then **Job Manager** dialog box shown in Fig. 8.24b will be popped up. Click **Submit** icon and the **Job** is in the analysis state.

8.1.4 Results

Click **Result** icon to **Visualization** module. The calculation result is shown in Fig. 8.25. There is large stress when two teeth are in the contact region. In addition,

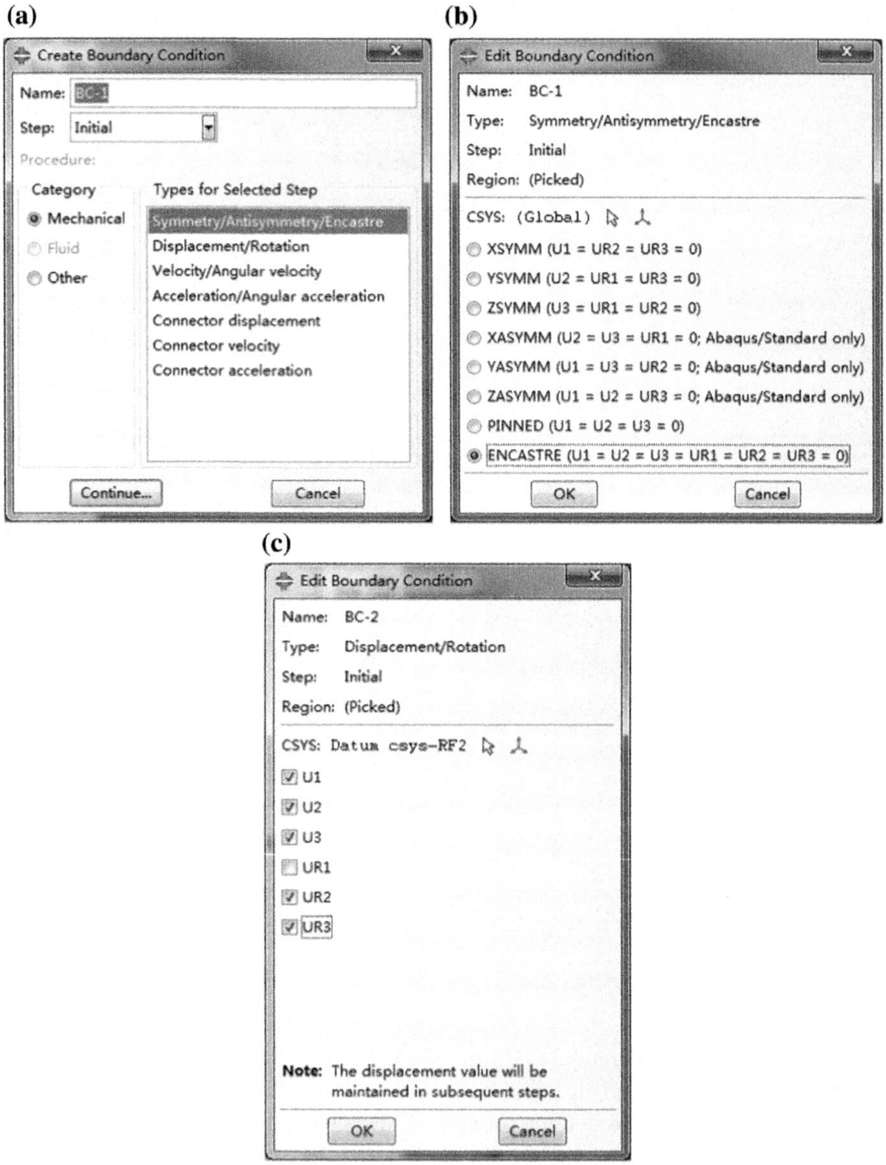

Fig. 8.22 **a** Boundary type selection. **b** Fixed boundary definition. **c** Definition of DOF constraint

bending stress is also large at the tooth root. Figure 8.26a, b show the tooth contact stress and the tooth root bending stress respectively. It can be seen that both contact stress and bending stress are distributed approximately uniformly along the tooth-width direction.

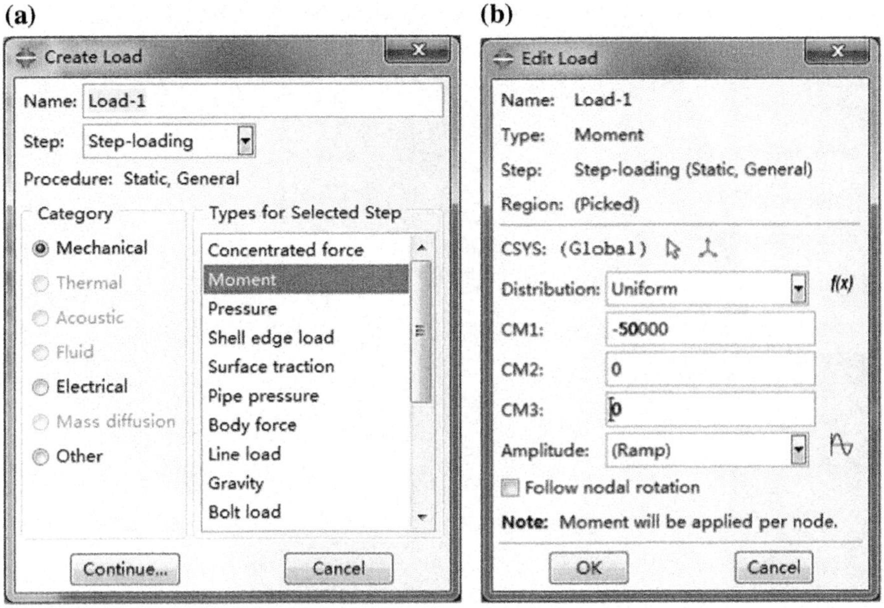

Fig. 8.23 **a** Select the load type. **b** Load definition

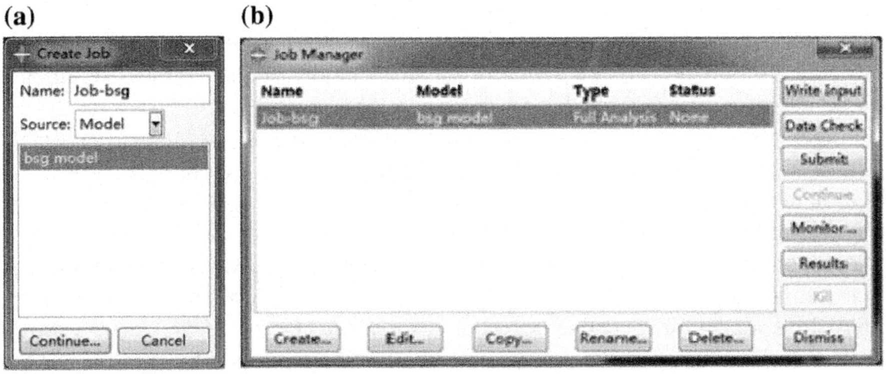

Fig. 8.24 **a** Job created dialog. **b** Job manager

The deformation of each part of the tooth is shown in Fig. 8.27, which shows that the stiffness of the back conical surface is small when the tooth is in contact. As can be seen in Fig. 8.28a, contact stress in the direction of tooth-width is distributed almost uniformly. It can be concluded that the load is distributed uniformly in

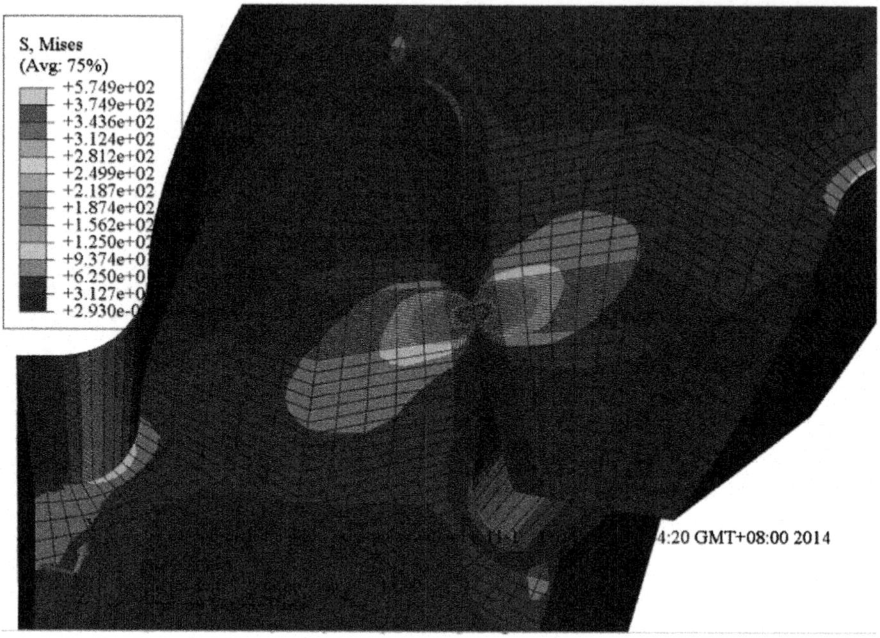

Fig. 8.25 Stress of contact region

tooth-width direction. The displacement of the contact region shows a different distribution trend due to the nonuniform distribution of the stiffness along the contact region.

In real gear, a small chamfer is processed on the boundary of the gear end-face. This case is discussed about sudden change of interface. Figure 8.28 shows the result in the case of right angle. The curve in Fig. 8.28 expresses the variation of the stress along the direction of tooth-height to the top. The nodes are on the boundary of a section in the direction of the tooth-width. Curve 1 shows the change trend of stress on boundary of back conical surface as the position of the nodes changes. Curve 2 shows the same change trend on the small end-face of teeth. Curve 3 shows the boundary points on the intermediate section of the tooth-width direction and the meanings are the same as those of the curves 1 and 2. The values of curves 1 and 2 are significantly larger than those of curve 3 at the nodes of the contact region, but the stress on the nodes of the non-contact region is almost the same, which illustrates that the geometric mutation of the boundary causes stress concentration in the contact stress.

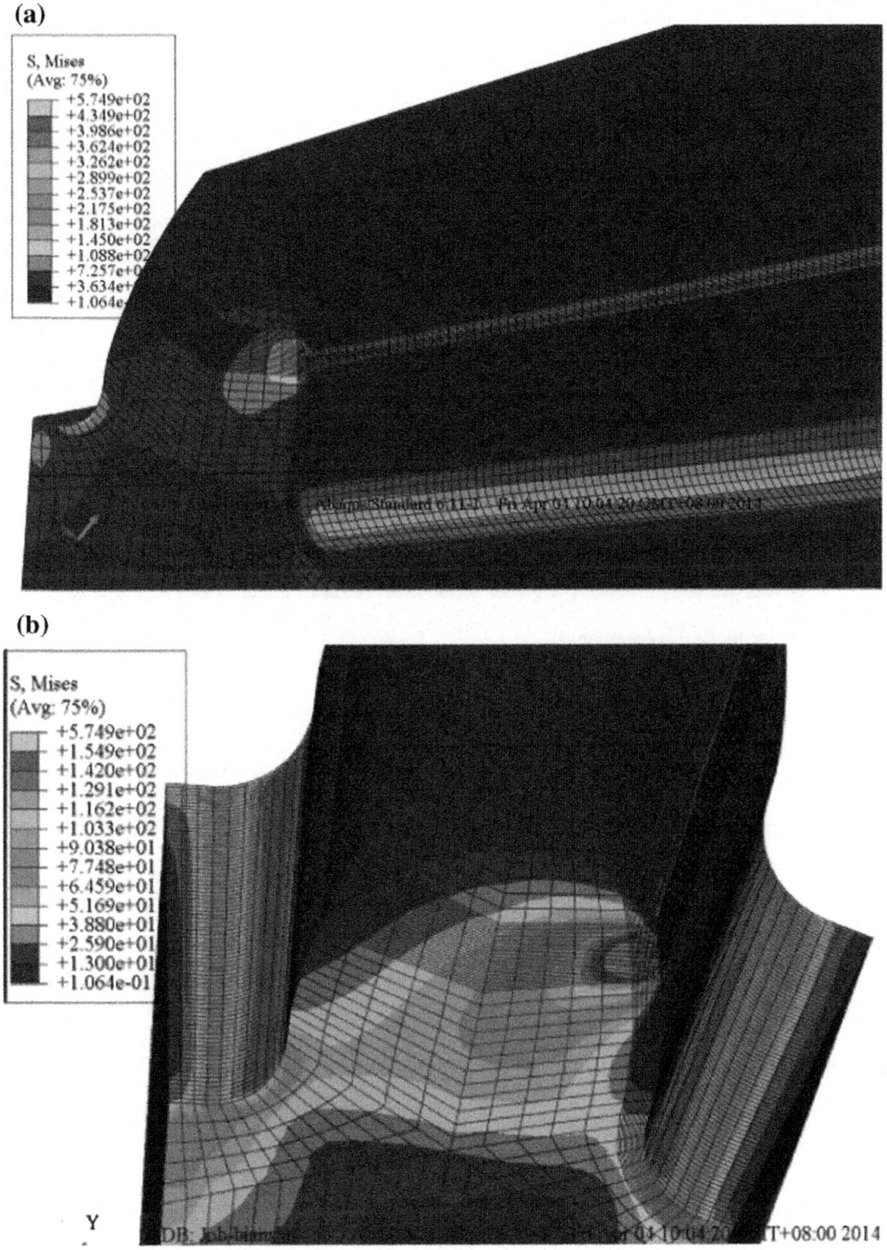

Fig. 8.26 **a** Contact stress along the tooth-width direction. **b** Bending stress along the tooth-width direction

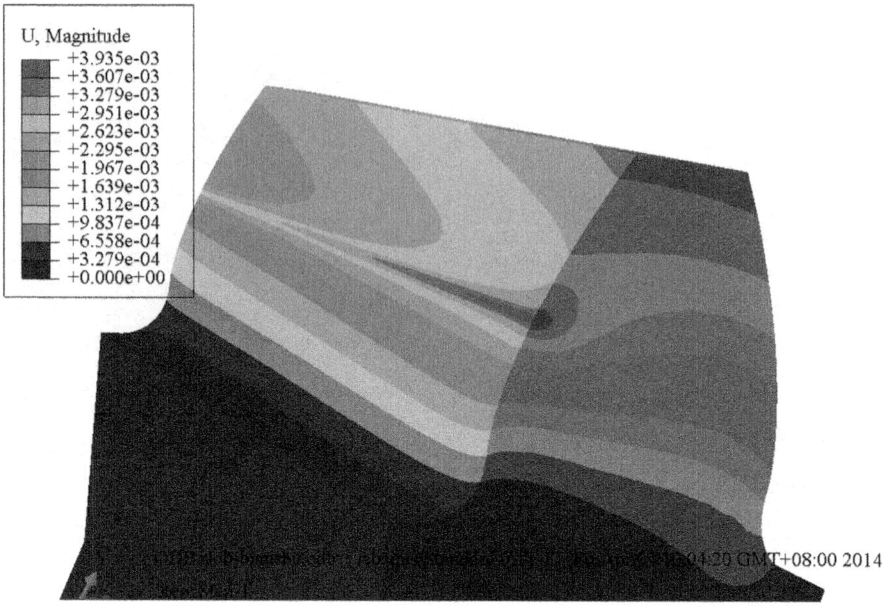

Fig. 8.27 Gear tooth displacement

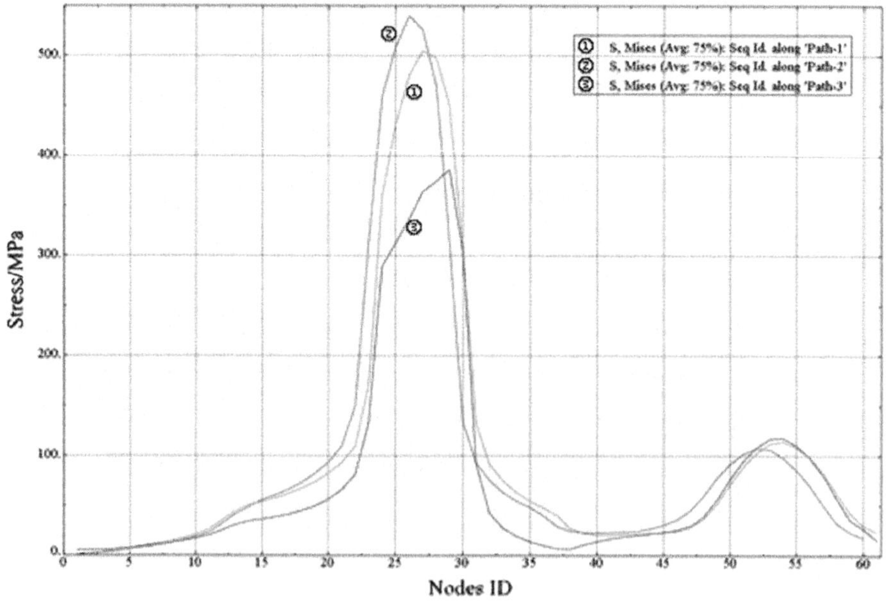

Fig. 8.28 Distribution of contact stress at tooth-height directions

8.2 Dynamic Contact Analysis of DPSD

Firstly, a rigid multibody dynamics model is created. Then, the planet gear shaft is processed flexibly and the rigid shaft is replaced. The rigid-flexible coupling model is generated. The correctness of the model is verified by the experimental data. Finally, the angular velocity analysis of the planet gear and the stress analysis of the planet gear shaft are carried out under the normal condition (including the case of maximum load) and the failure condition.

8.2.1 Create a Rigid Multibody Model of DPSD

Before the rigid-flexible coupling analysis, the rigid model of DPSD is firstly created to prepare for the flexible body to be imported.

8.2.1.1 Define the Modeling Environment [7, 8]

(1) Run **MSC Adams**. In welcome section, select **Create a new model**. **Model Name** is named DPSD (Fig. 8.29).
(2) Confirm **Earth Normal** (-global Y) in Gravity dialog and **Units** is MKS—m, kg, N, s, deg, as shown in Fig. 8.30. Click **OK** after confirmed.
(3) In **Settings** menu, select **Working Grid**. In parameter settings dialog, set **Size** to 20 mm in both *X* and *Y* directions and set **Spacing** in *X* and *Y* directions to 1 m. Click **OK** after finished.

8.2.1.2 Import the Model

Select **Import** in **File** menu. In the settings dialog, select the format of the imported file in menu of **File Type**. Double click space of **File To Read** to select the path of the imported file and modify name of the model in **Model Name** field just as shown in Figs. 8.31 and 8.32.

Click **OK** to display the 3D model of DPSD in **Adams** dialog as shown in Fig. 8.33.

8.2.1.3 Add Constraints and Loads

After the model is successfully imported, the next step is to add constraints and loads to the model.

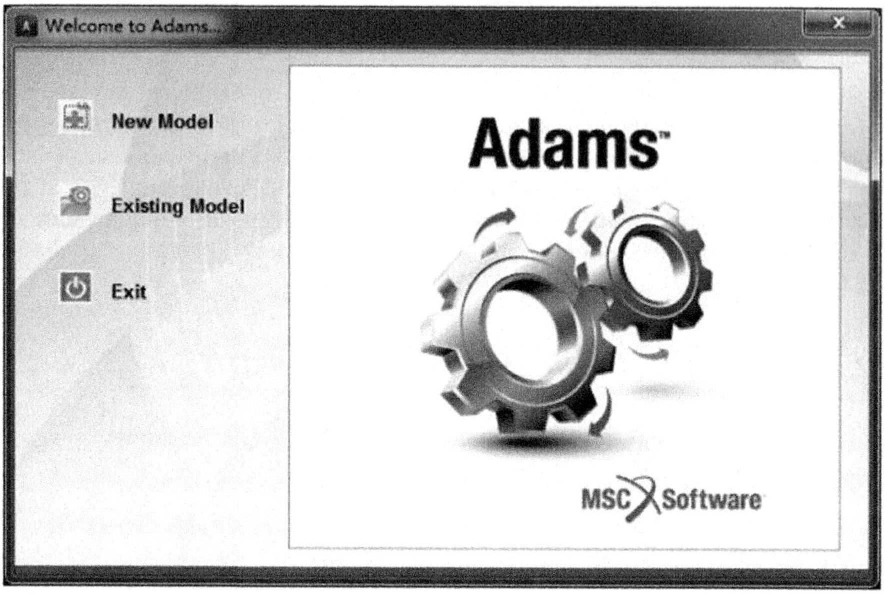

Fig. 8.29 Adams welcome screen

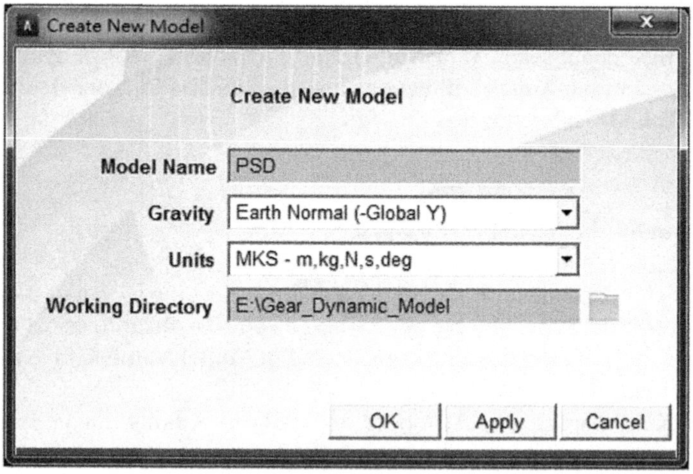

Fig. 8.30 Setting section

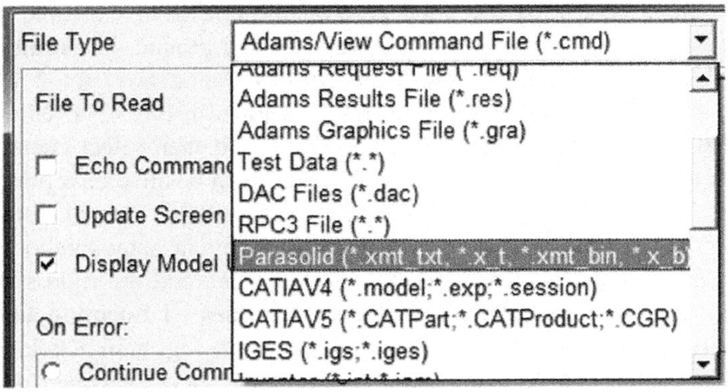

Fig. 8.31 Neutral file type

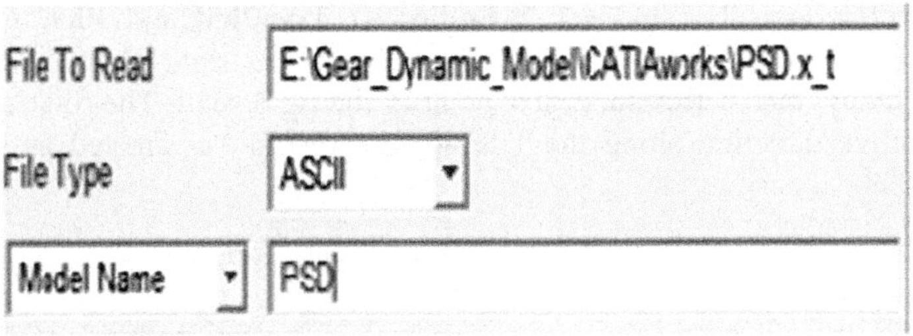

Fig. 8.32 Neutral file path and naming

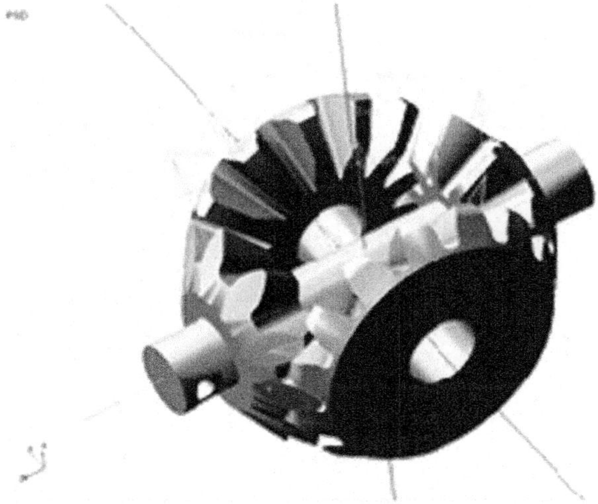

Fig. 8.33 DPSD 3D model

(1) Firstly, create a revolute between gear on the left drive shaft and ground, a revolute between gear on the right drive shaft and ground. In **Connector** tools bar, select revolute . In parameter settings dialog box, set **2 Bodies—1 Location** and **Pick Geometry Feature**, as shown in Fig. 8.34. Click gear on the right drive shaft and Ground respectively and then select center point of gear on the right drive shaft. The rotation shaft is in positive direction along the *Y* shaft. The revolute is created as shown in Fig. 8.35. The revolute between gear on the left drive shaft and ground is created by the same method as above.
(2) To create a revolute between the two planet gears and the rigid shaft, select revolute in **Connector** tools bar and set **2 Bodies—1 Location** and **Normal To Grid** in parameter settings dialog, as shown in Fig. 8.36. Click one of the planet gears and the rigid shaft respectively, then select the center point of the planet gear to create a revolute as shown in Fig. 8.37. The revolute of another planet gear and rigid shaft is created by the same method as above.
(3) To create a revolute between a rigid shaft and the ground, select revolute in **Connector** tools bar and set **2 Bodies—1 Location** and **Pick Geometry Feature** in parameter settings dialog. Click the rigid shaft and Ground respectively, then select the center point of the rigid shaft. The rotation shaft is in positive direction along the *Y* shaft, the revolute is created as shown in Fig. 8.38.

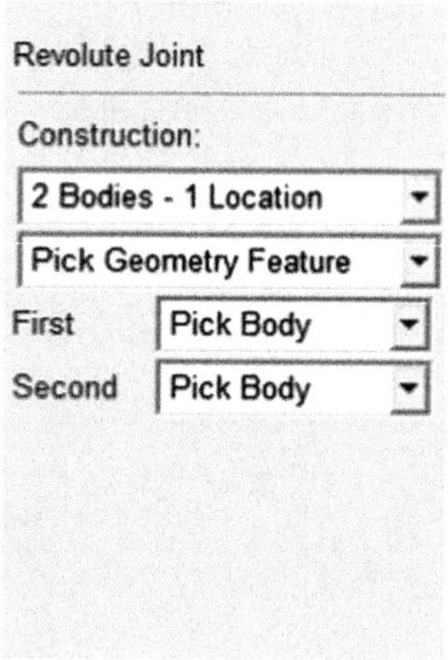

Fig. 8.34 Parameter settings

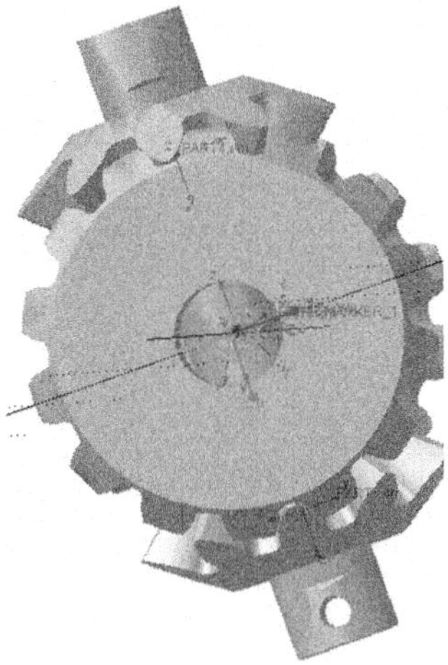

Fig. 8.35 Revolute between gear on the right side of shaft and ground

(4) The contact force is added between the each gear. In **Force** tools bar, select the contact force. In parameter settings dialog, set **Contact Type** to **Solid To Solid**. Right click space of **I Solid (s),** then select **Contact** → **Pick** and click gear on the right drive shaft. Similarly, right click space of **J Solid (s)**. Select **Contact** → **Pick** and click the planet gear, so two contact entities are selected as shown in Fig. 8.39. In contact force setting, set **Impact** to **Normal Force**. Stiffness is set to 1e5. Force Exponent is set to 2.2. Damping is set to 10. Penetration Depth is set to 0.1 and click **OK** to complete the contact settings, as shown in Fig. 8.40. Contact between gear on the right drive shaft and the other planet gear, contact between gear on the left drive shaft and the two planet both can be created by the same method.

(5) Apply the load torque to the gear on left drive shaft. Select torque  in **Force** tools bar. **Run—time Direction** is set to **Space Fixed**. **Construction** is set to **Pick Feature**, and **Characteristics** is set to **Custom** in parameter settings dialog, as shown in Fig. 8.41. Click the gear on left drive shaft, then select the center point of it. The rotation shaft creates the torque in positive direction along the *Y* shaft. In **Modify Torque** dialog box, import the user-defined input torque function and click **OK** to complete the creation of the load torque, as shown in Fig. 8.42.

(6) To apply drive to the gear on right drive shaft, select the drive of revolute icon  in **Motion** tools bar. Click **Joint 3** in the workspace to complete the drive application as shown in Fig. 8.43. The DPSD housing drive application is the same as above.

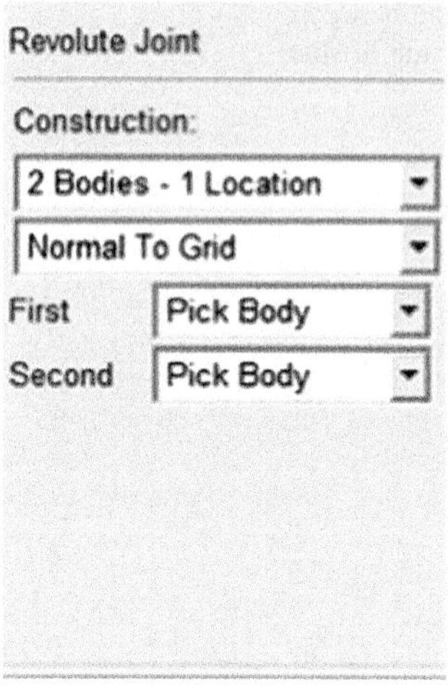

Fig. 8.36 Parameter settings

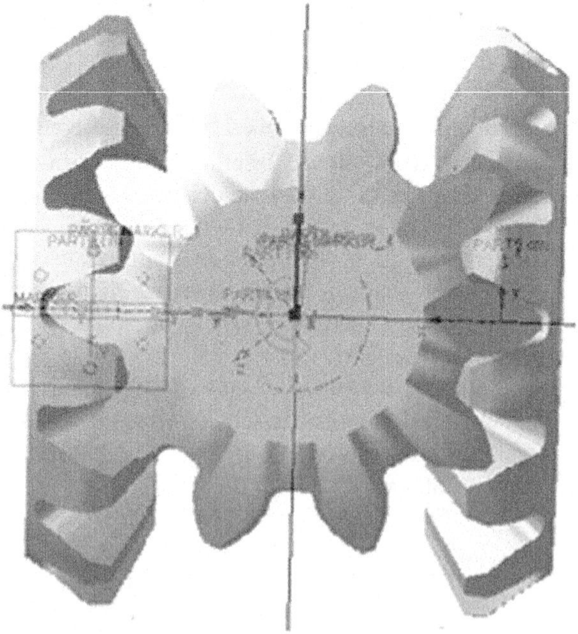

Fig. 8.37 Revolute between small gear and rigid shaft

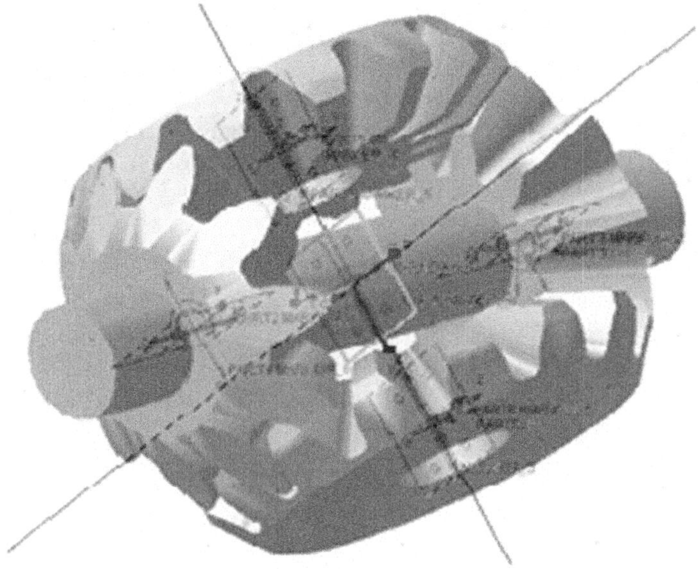

Fig. 8.38 Revolute between rigid shaft and ground

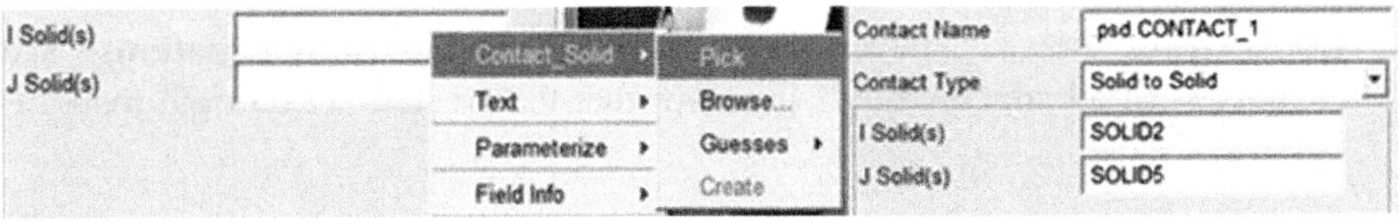

Fig. 8.39 Contact entity selection

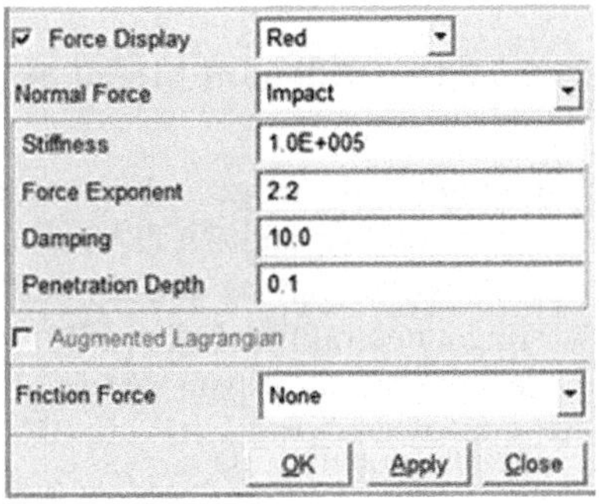

Fig. 8.40 Contact parameter settings

(7) Verify correctness of DPSD rigid model
Right click . Click the alternative icon . Perform model redundant constraint check for the rigid model. The dialog box is shown in Fig. 8.44. When

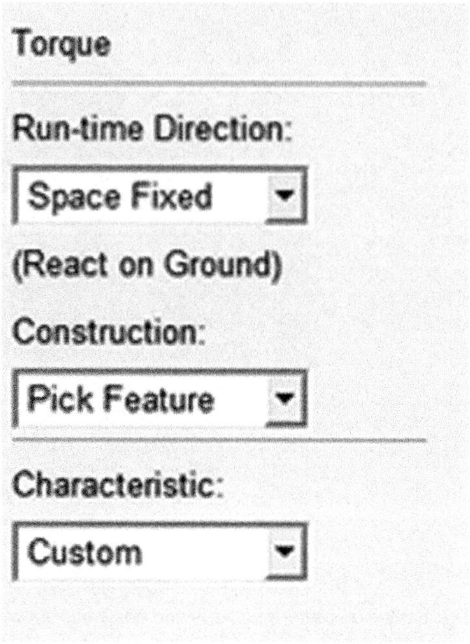

Fig. 8.41 Torque settings

Information shows "**There are no redundant constraint equations**" and "**Model verified successfully**", it proves that the created DPSD rigid model is correct.

8.2.2 Rigid-Flexible Model of DPSD [7, 9–12]

8.2.2.1 Generation of Modal Neutral File

1. Import the planet gear shaft model created in the 3D software into ANSYS 14.0 Since ANSYS and CATIA have a dedicated data section, files of the planet gear shaft model can be directly imported from CATIA to ANSYS without generating a neutral file. Select **File** → **Import** → **CATIA V5** in ANSYS menu bar. Select the import path and import the planet gear shaft as shown in Fig. 8.45.
2. Parameter definition
   (1) Element type
   Two types of elements are defined for planet gear shafts: structural elements and mass elements. Planet gear shafts are the three-dimensional solid structure and are not compressible elastic material, so select **SOLID 185** for its structural element. The purpose of making the planet gear shaft flexible is

Fig. 8.42 Load torque

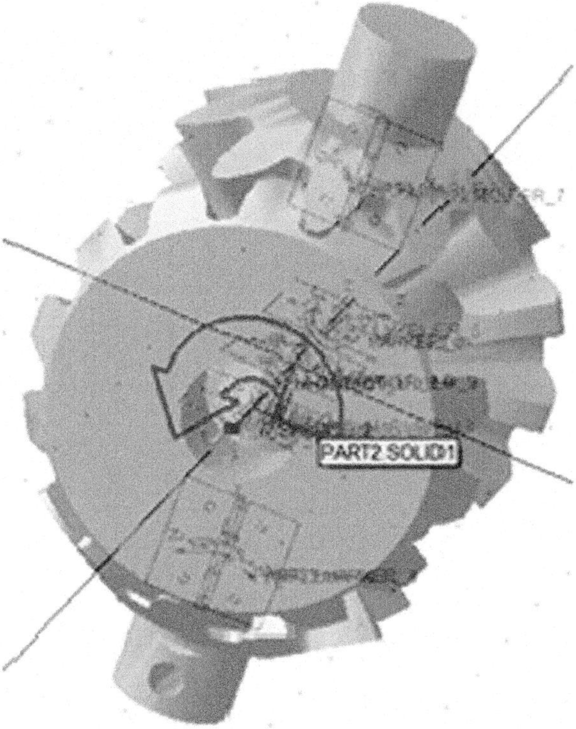

Fig. 8.43 Define the driver

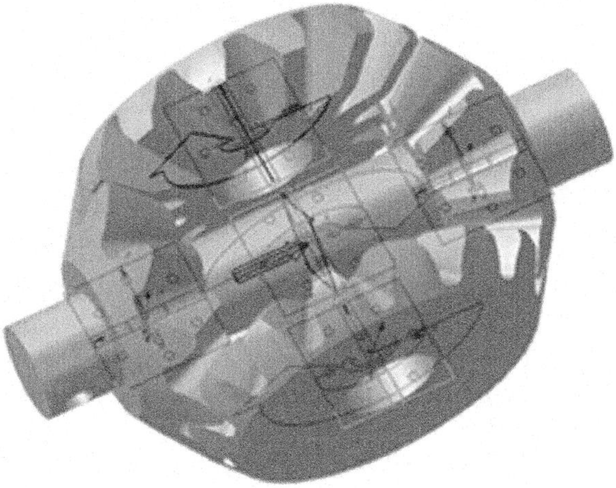

to make dynamic rigid-flexible couple analysis of the entire DPSD. Its mass need to be defined, so select MASS 21 for its mass element, as shown in Fig. 8.46. The menu path for defining the element type is: **Main Menu → Preprocessor → Element Type → Add/Edit/Delete**.

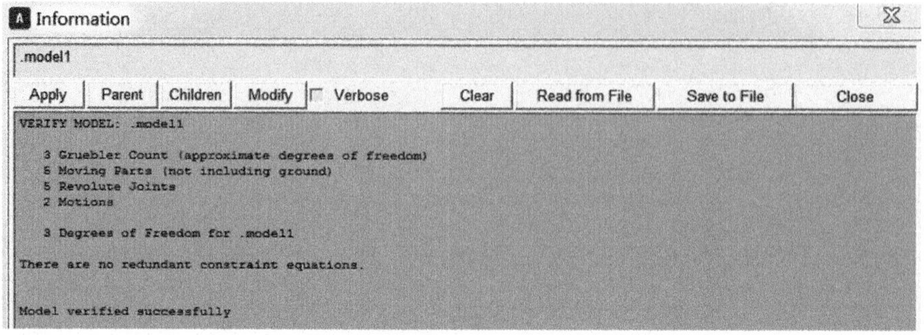

Fig. 8.44 Model redundant constraint check

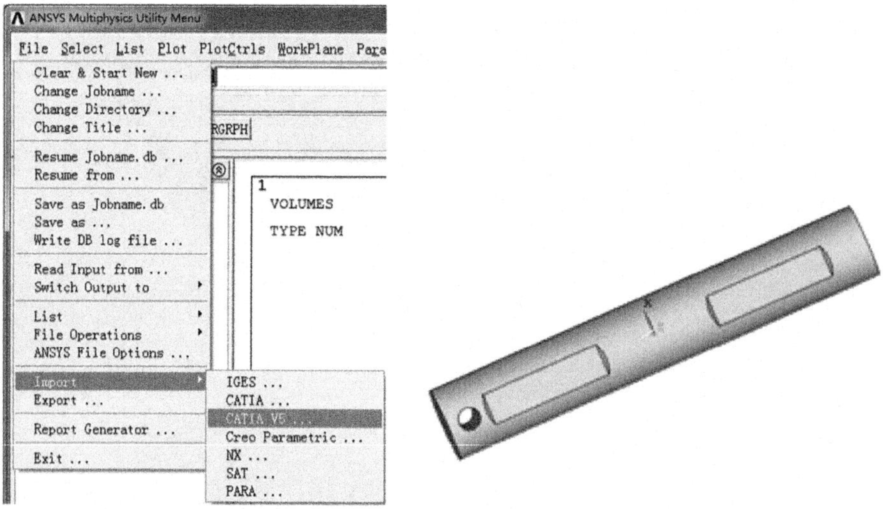

Fig. 8.45 Import models into ANSYS

(2) Edit MASS 21 mass element
Edit MASS 21. Fill the properties in the dialog box. It is a very small value, such as 1e−9 etc.
The menu path is: **Main Menu → Preprocessor → Real Constants → Meshing → Add/Edit/Delete.**

(3) Material parameters
The planet gear shafts mentioned in this book are steel. Set the properties of the material. Set Elastic modulus to 2e+11. Set Poisson ratio to 0.3. Set Density to 7850. As shown in Fig. 8.47. The menu path is: **Main Menu → Preprocessor → Material Props → Material Models**.

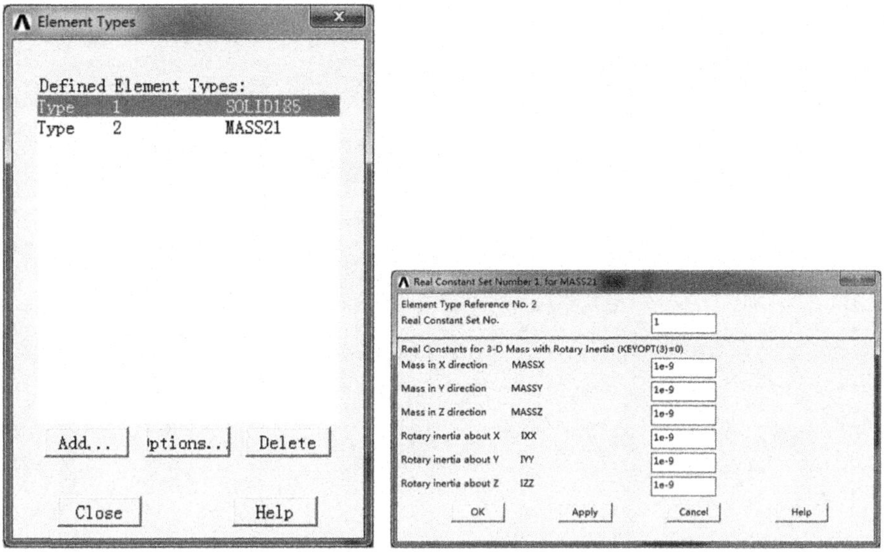

Fig. 8.46 Adding a mass element

3. Meshing and generating rigid regions

   (1) Meshing
   In **Main Menu**, select **Preprocessor**, then click **Meshing** and select **MeshTool** to mesh the planet gear shaft, as shown in Fig. 8.48.

   (2) Create **Keypoints**
   Select **Preprocessor** in **Main Menu**. Click **Create** in **Modeling**. Select **Keypoint** and click **In Active CS.**
   The dialog of creating a Keypoint is showed in Fig. 8.49. Create Keypoint at (0, 0, 0), (0, 123123, 0), (0, −123123, 0). Number them 80001, 80002, 80003 respectively. Note that the number of Keypoints created can not coincide with the node number of the model element here. Otherwise, it will cause deformation of the original model. The created Keypoint is showed in Fig. 8.50.

   (3) Keypoints Meshing
   Select **Mass 21** element. Mesh the keypoints (80001, 80002, 80003) created in the previous step to create **Interface Nodes**, which will automatically generate **Mark points** after being imported into ADAMS. The connection is created by these points and other rigid or soft bodies, as shown in Fig. 8.51. The menu path is:

Main Menu → Preprocessor → Meshing → Mesh Attributes → Default Attribs.

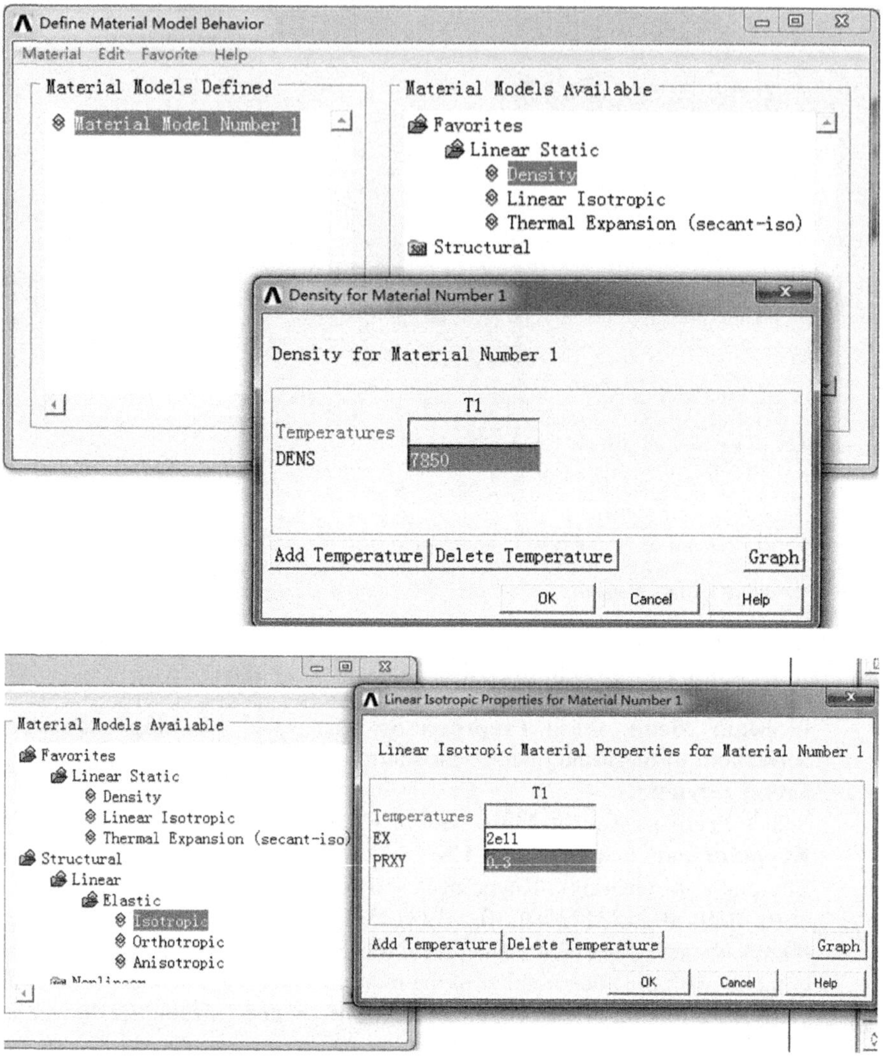

Fig. 8.47 Material properties settings

(4) Create a rigid zone
To create a rigid region (In ADAMS, as a non-deformation region connected to the outside, it is essential), select **Nodes** near **Interface Nodes** and connect to them. The number of connection points must be greater than or equal to 2, so there are at least two rigid regions. Select **Interface Node**. Click **Apply** and select the surrounding **Nodes**, the created **Interface Nodes** as shown in Fig. 8.52.

Fig. 8.48 Meshing freely

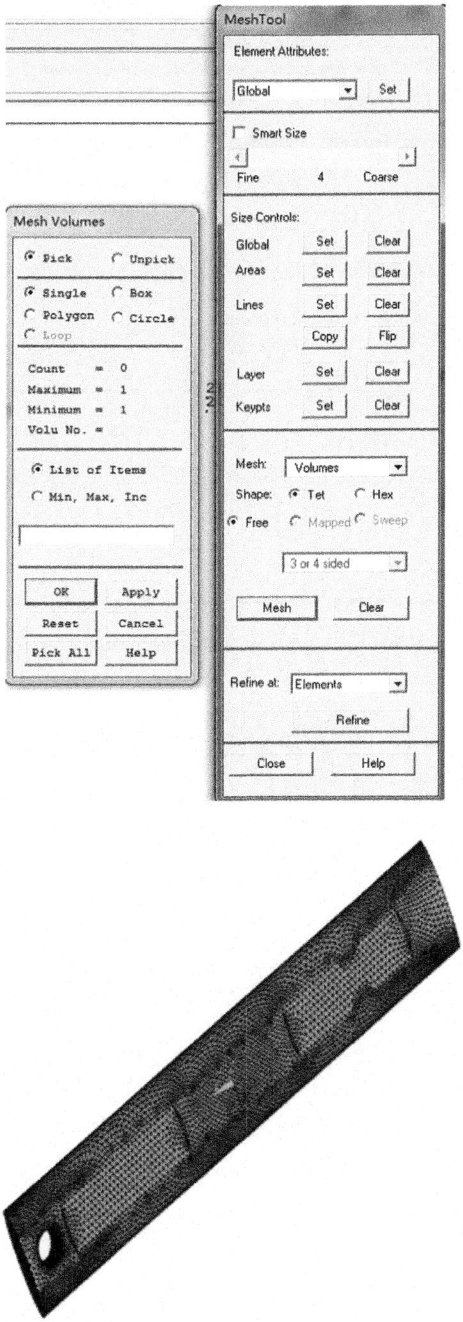

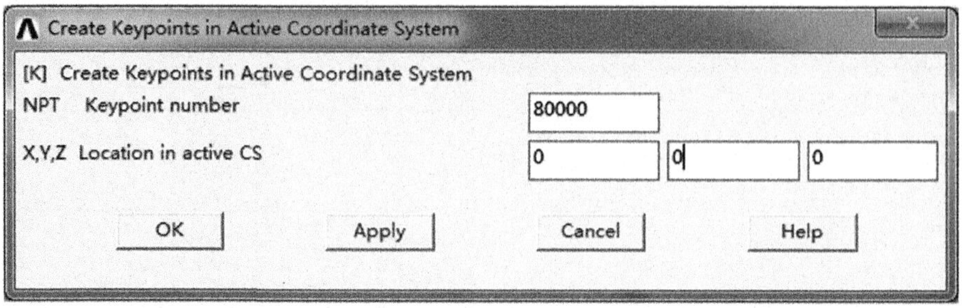

Fig. 8.49 Add a keypoint

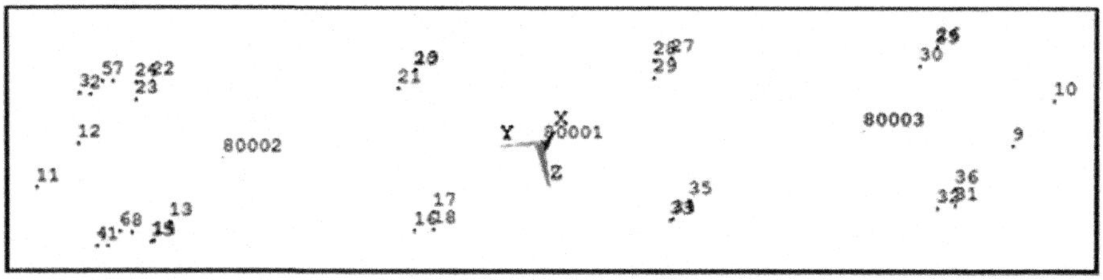

Fig. 8.50 Each node of planet gear shaft

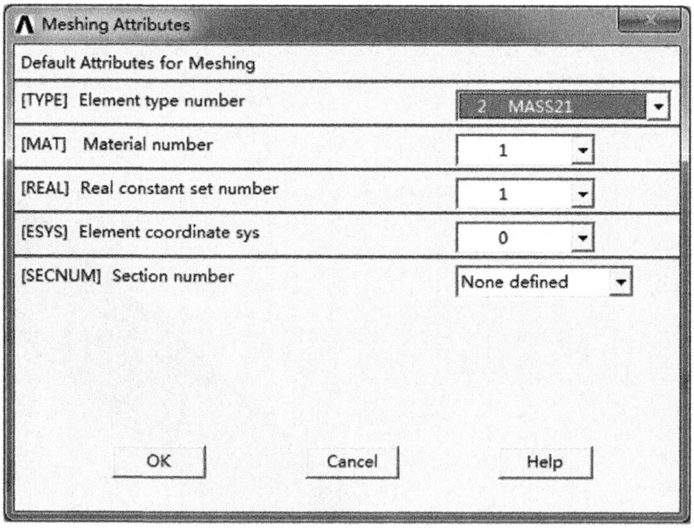

Fig. 8.51 Meshing of keypoints

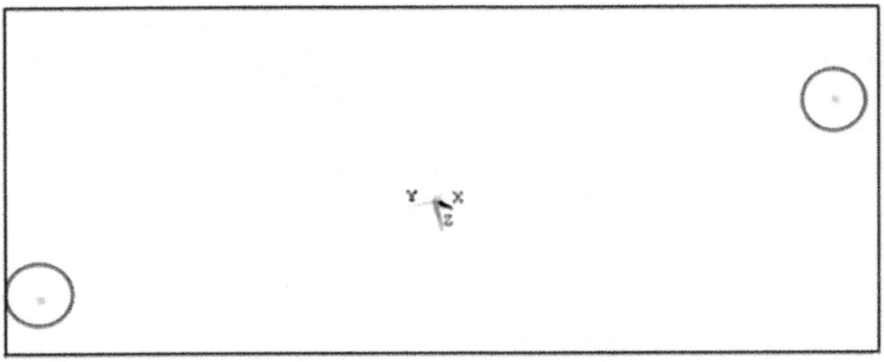

Fig. 8.52 Generation of section nodes

The menu path is: **Main Menu → Preprocessor → Coupling Ceqn → Rigid Region** and use the same method to create three rigid regions as shown in Fig. 8.53.

(5) Generate the mnf file
Perform the command **Solution-&gt> Adams Connection-&gt> Export to Adams**. The nodes to be selected is the ones that create rigid regions as

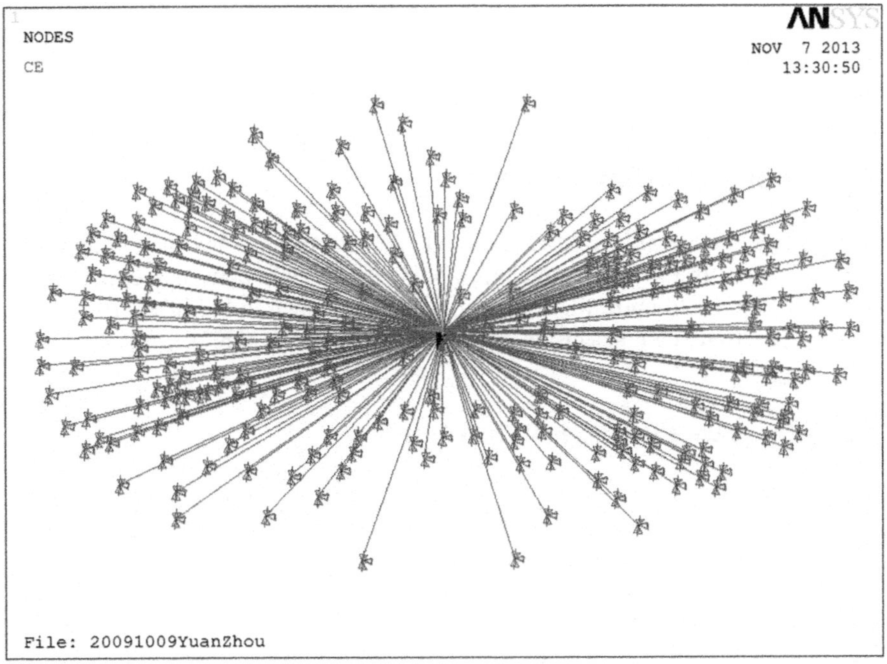

Fig. 8.53 Creation of a rigid region

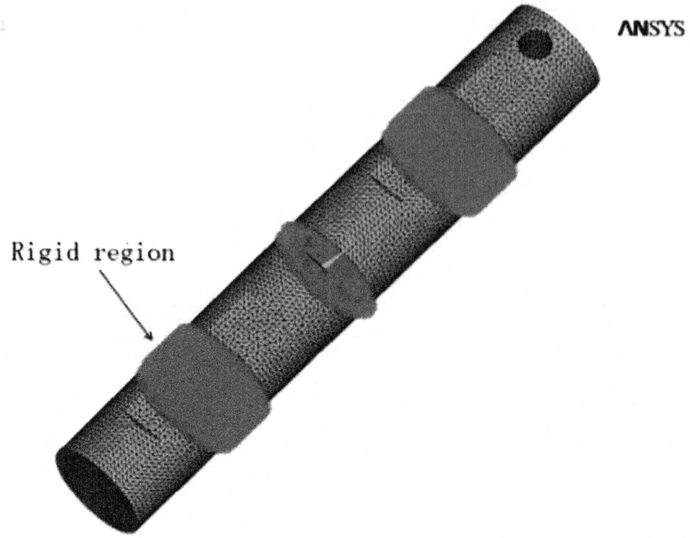

Fig. 8.54 Planet gear shaft with rigid region

shown in Fig. 8.54 (only interface nodes). The unit of output is SI, which can generate * .mnf file. **Element Result** is set to **Include Stress and Strain**. Select the storage path and click on the icon Solve and create export file to ADAMS to complete the mnf neutral file generation (Fig. 8.55).

8.2.2.2 Import Modal Neutral File

Click **Bodies**, in **Flexible Bodies** tools bar, select **Adams/Flex** icon. Import the created mnf file. In the modify dialog, modify the flexible body name in **Flexible Body Name** and set import file format to **MNF**, as shown in Fig. 8.56. Right click and select **Browse**, then select the created mnf file, click "**OK**".

8.2.2.3 Input the Experimental Data [13]

Click **File Import** to open the dialog and set **File Type** to **Test Data** (*. *). Select **Create Splines**, then double click **File To Read** and select**.txt** file to record the experimental data. Set **Independent Column Index** to 1 (In the text file, make the first column of data as arguments) and click the **OK**, as shown in Fig. 8.57. The experimental data includes load torque of left drive shaft, input rotational speed of right drive shaft and input rotation speed of DPSD housing.Make them to curves. The curves in Adams/PostProcessor are shown in Fig. 8.58.

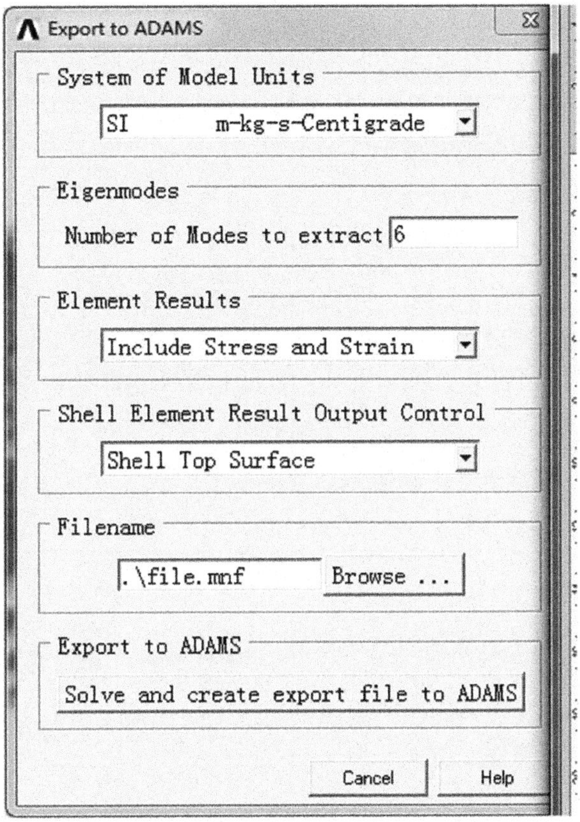

Fig. 8.55 Generation of mnf file

8.2.2.4 Definition of Drive

In the process of replacing a rigid shaft with a flexible shaft, the constraint and drive associated with the rigid shaft are removed. The revolute pair between flexible shaft and two planet gears needs to be applied. The revolute pair and drive between flexible shaft and the ground also need to be applied. The application procedure is described in 8.2.2.

8.2.3 Dynamic Simulation and Validation of DPSD

8.2.3.1 Validation of DPSD

In Simulation tools bar, click the icon  then pop up a **Simulation Control** dialog. The simulation time is set to 0.344 s. The simulation step is set to 1/1000.

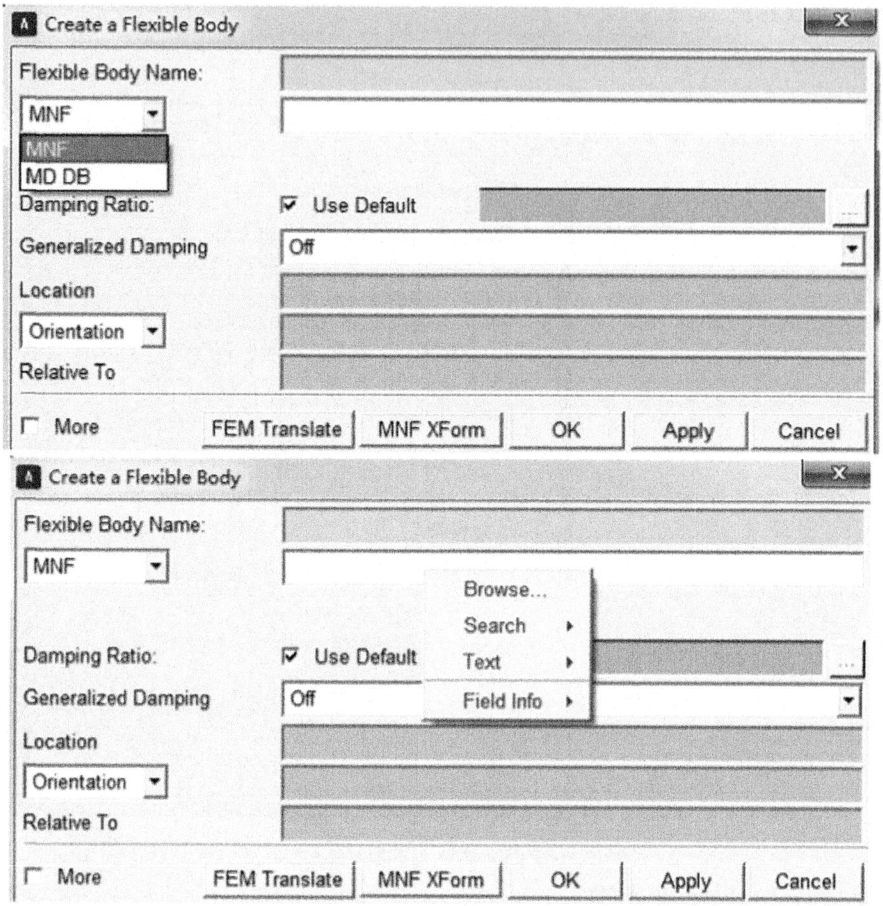

Fig. 8.56 Import modal neutral file to Adams

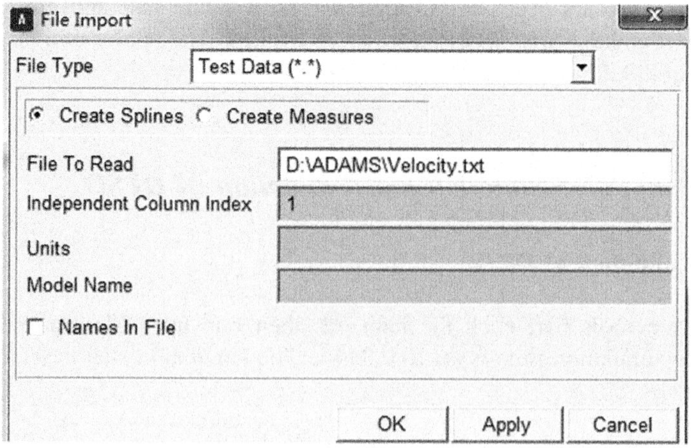

Fig. 8.57 Import of experimental data

(a)
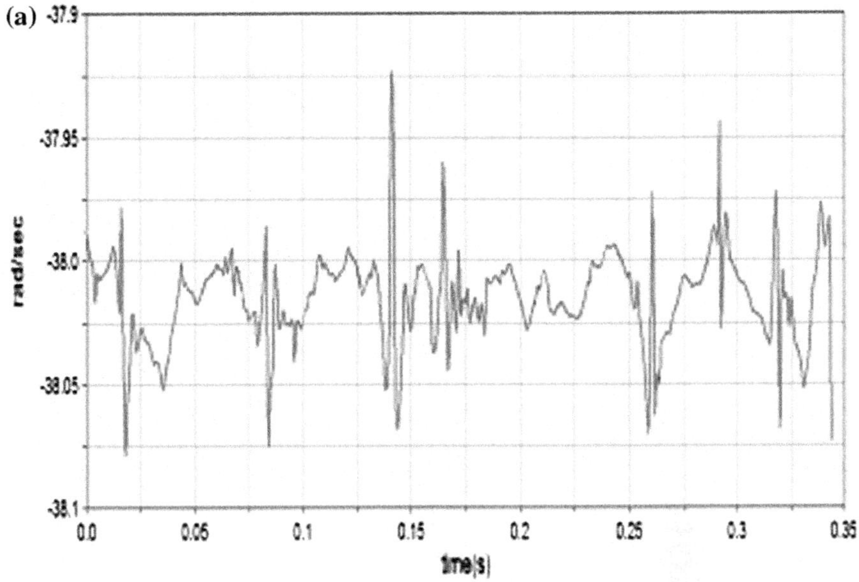

(b)
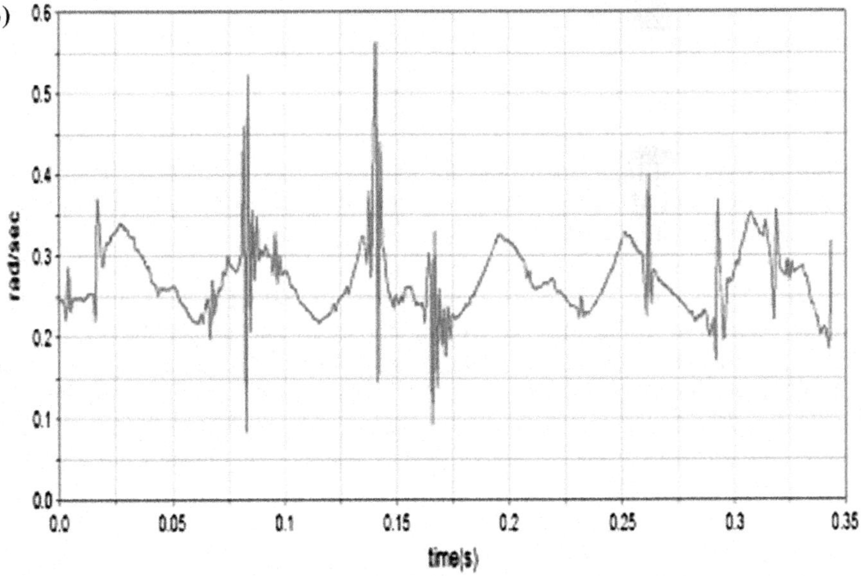

Fig. 8.58 **a** Torque of *left* drive shaft. b Torque of DPSD housing. c Torque of right drive shaft

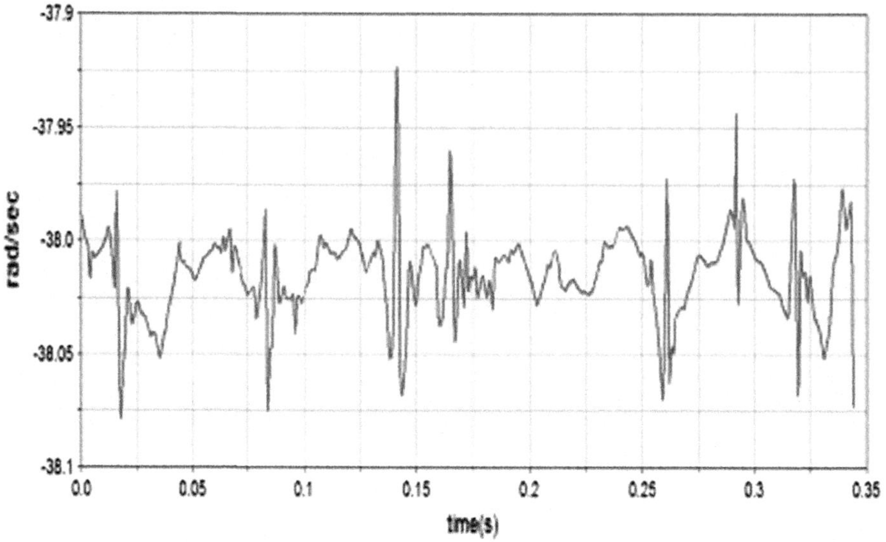

Fig. 8.58 (continued)

The simulation type is set to **Dynamic**, as shown in Fig. 8.59. When finished, click the icon ▶|.

After the simulation is complete, in **Design Exploration** tools bar, click the icon to create a new measurement and the pop-up dialog is shown in Fig. 8.60. Take measuring torque of right drive shaft for example, select **MOTION_1**, then click **OK**. In the **Motion Measure** dialog, set **Characteristic** to **Torque**, which is used to measure torque of right drive shaft. Output rotational speed of left drive shaft and torque measurement of DPSD housing is the same as above.

In Adams/Processor, select **Last_Run** in **Simulation** list. Set **Source** to **Measures** and set **Measures** to the measured rotational speed and torque, then click **Add Curves**, as shown in Fig. 8.61. Output rotational speed of left drive shaft, output torque of right drive shaft and output torque of housing are shown in Fig. 8.62.

In Adams/PostProcessor, click the icon to read the average of the simulation results and compare it to the experimental results. The results are shown in Table 8.1.

8.2.3.2 Dynamic Simulation of DPSD

1. Comparison of rotational speeds of planet gears under two conditions

Table 8.2 shows the output speed and torque of each terminal when the planet gear shaft fails in the DPSD bench test. From Fig. 8.63, under normal condition, the rotational speed of planet gears is stable, and in the failure conditions, the rotational speed of planet gears is stable at first, and then violent fluctuations are produced.

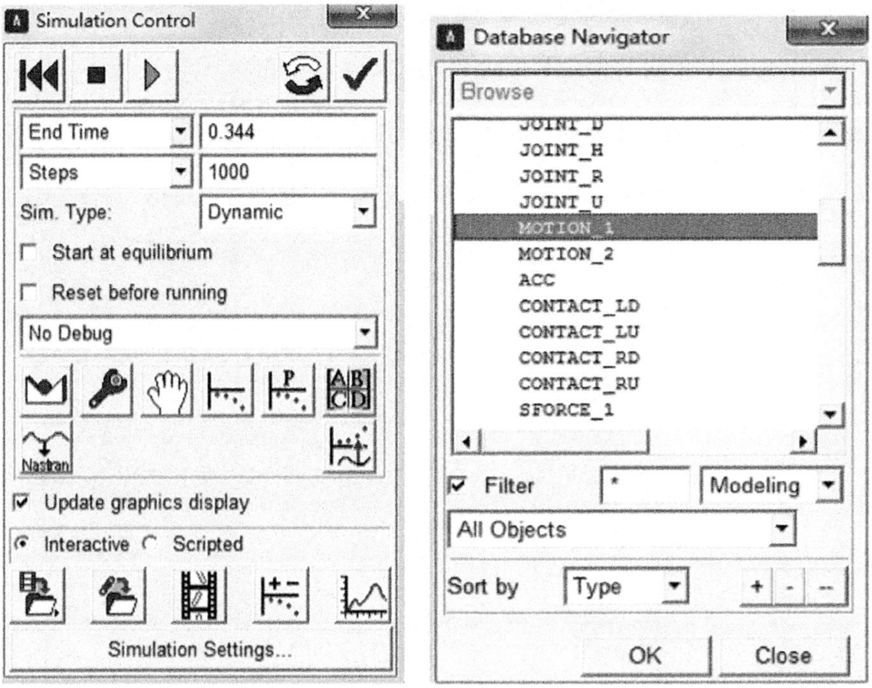

Fig. 8.59 Control tool settings

Fig. 8.60 Output rotational speed and torque measurement

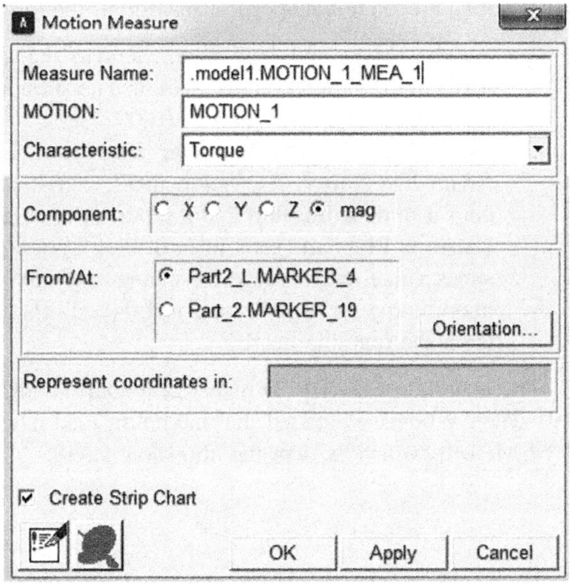

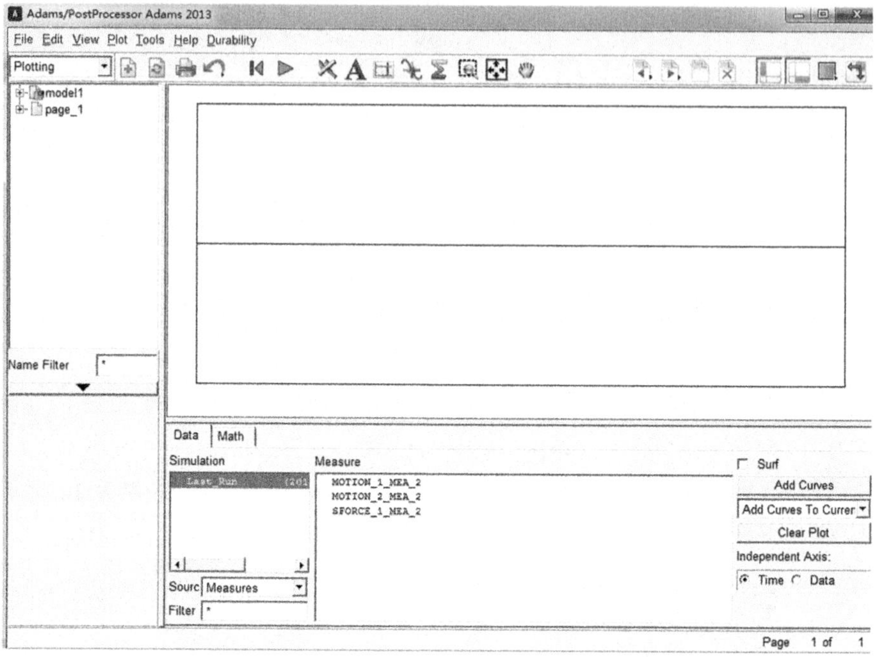

Fig. 8.61 Curve drawing

2. Stress of planet gear shaft under maximum load condition
   Torque and rotational speed at maximum load are shown in Table 8.3.
   (1) Set the driver and load parameters to the data shown in Table 8.3 and perform the simulation calculation. The simulation time is set to 0.344 s and the simulation step is set to 1/1000.
   (2) Click **Tools—Plugin Manager**, then check **Load** and **Load at StartUp** in Adams/Durability. As shown in Fig. 8.64, in Adams/PostProcessor, right click **Load Animation** and the pop-up dialog is shown in Fig. 8.65. Click **Contour Plots** and set **Contour Plot Type** to **Von Mises Stress**. Read the stress value and observe the changes of stress as shown in Fig. 8.66. The maximum Von Mises Stress is 1.04e+07 Pa and the maximum stress point is on the revolute nearby.

The material of the DPSD planet gear shaft is 20CrMnTi. Its allowable stress is 291 MPa. When it is loaded the maximum load, stress of the planet gear shaft is 10.4 MPa. It is far less than the allowable stress.

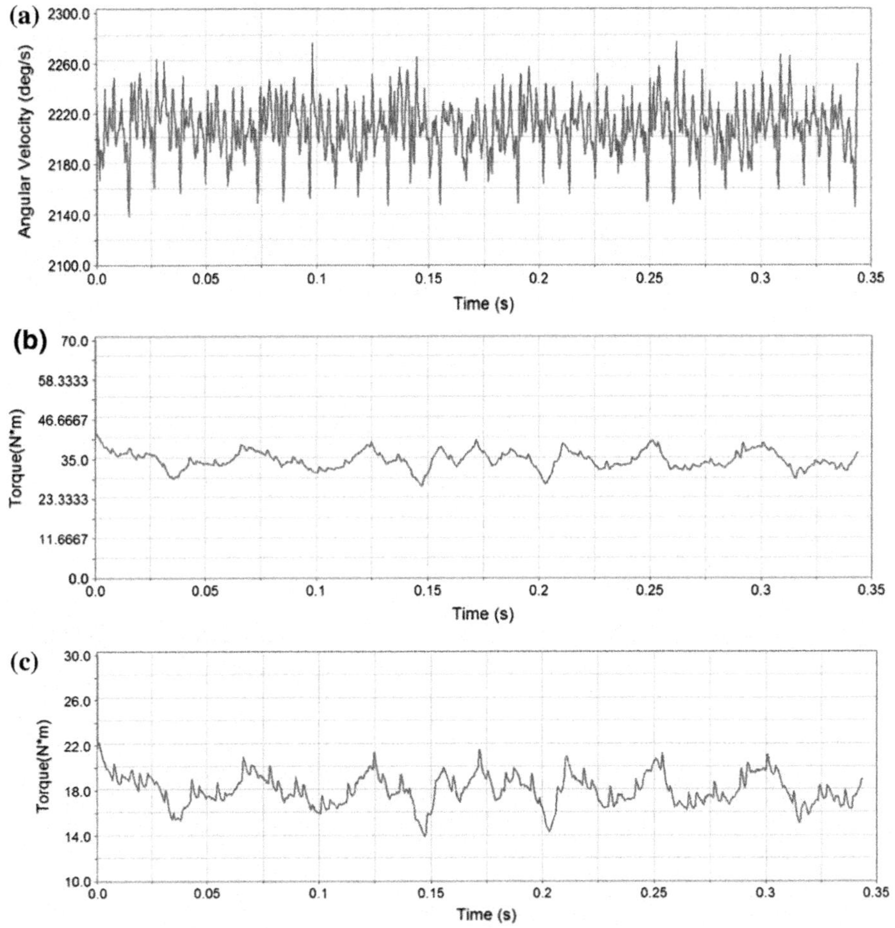

Fig. 8.62 **a** Output rotational speed of *left* drive shaft. b Output torque of *right* drive shaft. **c** Output torque of DPSD housing

Table 8.1 Comparison of simulation results and experimental results

Item	Average of the experimental results	Average of the simulation results	Relative error (%)
Output rotational speed of left driveshaft (rpm)	358	388.2	2.8
Output torque of DPSD housing (Nm)	38.4	34.8	4.4
Output torque of right drive shaft (Nm)	−20.8	−18.1	13.0

Table 8.2 Output rotational speed and torque of each terminal

Item	Left drive shaft	DPSD housing	Right drive shaft
Torque (Nm)	−41	130	−57
Rotational speed (rpm)	372	4	−358

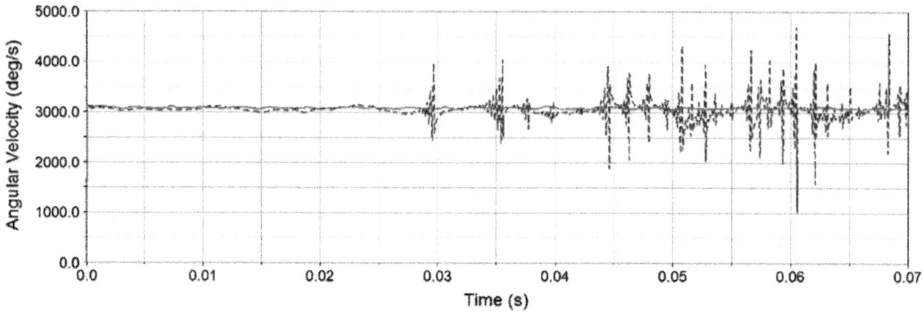

Fig. 8.63 Rotational speed of planet gear *curves* under normal and failure conditions

Table 8.3 Torque and rotational speed at maximum load

Torque of left drive shaft (Nm)	Rotational speed of DPSD housing (rpm)	Rotational speed of right drive shaft (rpm)
−30	1023	230

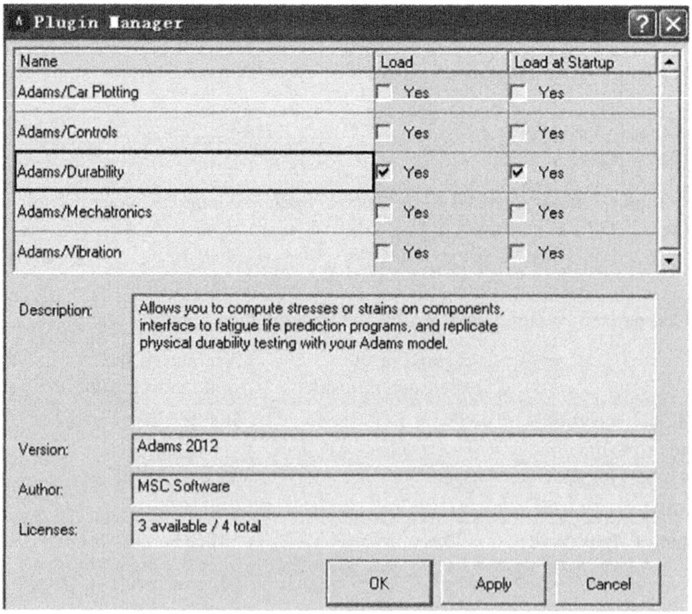

Fig. 8.64 Plugin manager

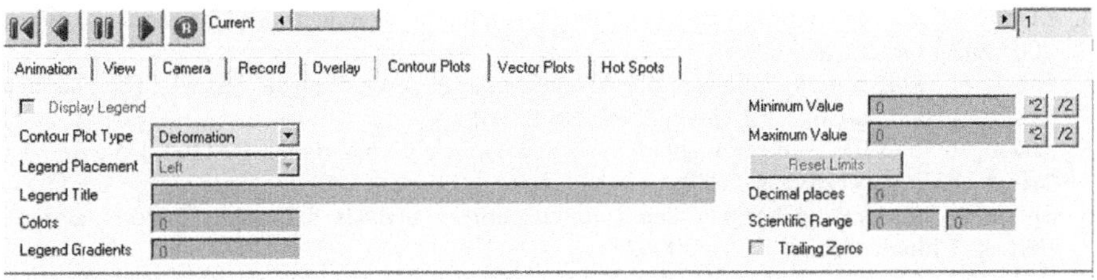

Fig. 8.65 Stress reading settings

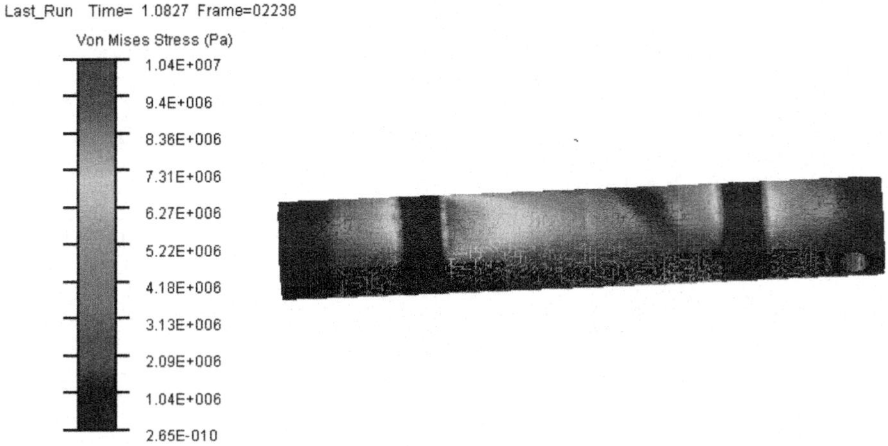

Fig. 8.66 Stress of planet gear shaft at maximum load condition

8.3 Summary

Computer aided engineering (CAE) analysis is a necessary means for the design and development of products. It provides assist for accurate and reliable design and analysis solutions, which avoids duplication in the design, manufacture and testing. In this chapter, the nonlinear contact behavior of the spur bevel gears in operation is analyzed. The dynamic behavior of the planet gear shafts is analyzed by rigid-flexible coupling model in the analysis software.

References

1. Gonzalez-Perez I, Iserte JL, Fuentes A. Implementation of Hertz theory and validation of a finite element model for stress analysis of gear drives with localized bearing contact. Mech Mach Theor. 2011;46(6):765–83.
2. Yang SH. Finite element analysis of gear contact. Chinese J Comput Mech. 2003;20(02): 189–94.

3. Mao K. Gear tooth contact analysis and its application in the reduction of fatigue wear. Wear. 2007;262(11):1281–8.
4. Hibbit H, Karlsson B, Sorensen P. Abaqus analysis user manual version. 10. Dassault Systemes Simulia Corp. Providence, RI, USA; 2011.
5. Zhuang Z, Liao JH. Finite element analysis and application based on ABAQUS [M]. Beijing: Tsinghua University Press; 2009.
6. Shi YP, Zhou YR. ABAQUS detail finite element analysis example (Chinese edition). Beijing: China Machine Press; 2006.
7. Zheng JR. ADAMS—Introduction and virtual prototyping technology to improve (Chinese edition). Beijing: China Machine Press; 2001.
8. Shi BQ. ADAMS BASIC example tutorial (Chinese edition). Beijing: China Railway Publishing House; 2007.
9. Ma XG, Yang W, Mei XM, et al. Multi-body dynamical analysis on rigid-flexible coupling for planetary gear system. Chin J Constru Mach. 2009;7(02):146–52.
10. Wu X, Meagher J, Sommer A. A differential planetary gear model with backlash and teeth damage. Rotating machinery, structural health monitoring. Shock and Vibration, Springer New York, Vol. 5, 2011.
11. Ji JF. Dynamic characteristics analysis of power split device in HEV. (Master's Degree Thesis of Jilin University 2013, in Chinese).
12. Liu Y, Ji JF, Kong I, et al. Failure analysis of planetary gear shaft of power split device based on multi-body dynamics. Appl Mech Mater. 2014;433:17–20.
13. Zheng S. Failure analysis and test validation on differential coupling device in HEV (Master's Degree Thesis of Jilin University 2011, in Chinese).

Chapter 9
Multi-objective Optimization for Differential-Based PSD Based on Surrogate Model

In Chap. 8, engineering analysis has been carried out by finite element model of DPSD. Another important aspect is to minimize DPSD's volume and weight, so the light-weight design of structure is necessary. This chapter introduces a multi-objective optimization method for housing and drive gears of DPSD based on surrogate model. This optimization method is efficient to reduce the computational cost.

9.1 Introduction

In traditional vehicles, driveline is considerably compact in structure. HEVs have an increased complexity in driveline because of the new energy storage and other auxiliaries. In this chapter, the design optimization for the housings and drive gears of DPSD [1, 2] are introduced. Design optimization is applied to downsize the volume and mess, or to improve the performance of components in engineering. With the help of computer technology, design optimization nowadays is much more convenient. Optimization methods researchers may use includes Karush-Kuhn-Tucker Method [3], Non-gradient Optimization Method [4], Genetic Algorithm [5] and Swarm Optimization [6].

The optimization method divides into traditional optimization and modern optimization. In traditional optimization many methods, such as Sequential Unconstrained Minimization Techniques (SUMT), Compound Form Method, Constrained Variable Scale Method and Random Direction Method, Generalized Reduced Gradient (GRG), Feasible Direction Method, are commonly applied to the mechanical design. However, with the increase of scale and complexity of optimization problems, the traditional optimization is prone to the unexpected local optimal solution. Thus, modern optimization based on intelligent algorithm is proposed. It has high efficiency and strong ability of global research. It is also convenient to combine with CAD/CAE technology, so it develops rapidly in recent years.

X. Zeng and J. Wang, *Analysis and Design of the Power-Split Device for Hybrid Systems*, https://doi.org/10.1007/978-981-10-4272-0_9

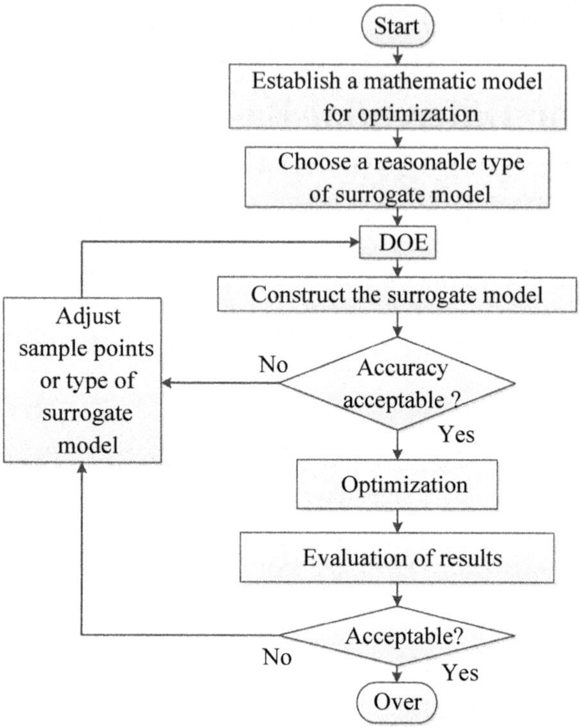

Fig. 9.1 Flowchart of optimization based on surrogate model

With maturity of CAD/CAE technology, surrogate model [7, 8] applied to the structure optimization has become more popular. Surrogate model method accelerates the optimization speed and saves the computational cost. The establishment of surrogate model is introduced in Sect. 9.2.

To optimize based on surrogate models, the first step is to analyze the problem and choose the reasonable type of surrogate model. According to the requirement of different surrogate models, the sample points are selected by experimental design method. The surrogate model is established based on those experiment results. And then the accuracy of the surrogate model is tested and if the accuracy is acceptable, the optimization will be implemented based on this surrogate model. Otherwise, it going to restart from selecting the sample points. Then to run and update the process until the design is optimal. The flowchart is shown in Fig. 9.1.

9.2 Design of Experiment (DOE)

The first step of establishing the surrogate mode is to design the experiment. The aim is to seek the orthogonal and uniform sample points. Practical implementation includes orthogonal arrays (OA), uniform arrays (UA) and latin hypercube sampling (LHS) [9, 10].

The design itself is an optimization method of discretization. The key question is how to access the optimal points. Though its simplicity, reliability and less calculation, the design of experiment is hard to realize the global optimization. The prominent approaches related to DOE is presented next, which includes full factorial experiment design, orthogonal experimental design, uniform design and Latin hypercube sampling design.

9.2.1 *Full Factorial Experiment Design*

A full factorial experiment contains two or more factors and each factor consisting of several discrete possible values or "levels". And its experimental condition takes on all possible combination of these levels across all factors. Full factorial experiment allows researchers to study the effect of each factor on each level and interaction between all factors. When a design is noted as 3^3 factorial, this identifies that there are 3 factors; each factor has 3 experimental levels; and the experimental conditions (red points) here is $3^3 = 27$, as shown in Fig. 9.2.

To determine the main effect, the number of all experimental runs is $n_1 \times n_2 \times \cdots n_i \times \cdots n_m$. "$i$" represents the number of the levels, and "m" represents the number of factors. It is obvious that the experiment needs a large amount of calculation, which is a big challenge due to high cost. Thus, full factorial experiment can be used because it needs only several runs of experiment to get the satisfied accuracy.

9.2.2 *Orthogonal Experimental Design*

Orthogonal experimental design is another method to the problems of multiple factors and multiple levels. When the experiment runs too many times, a logical

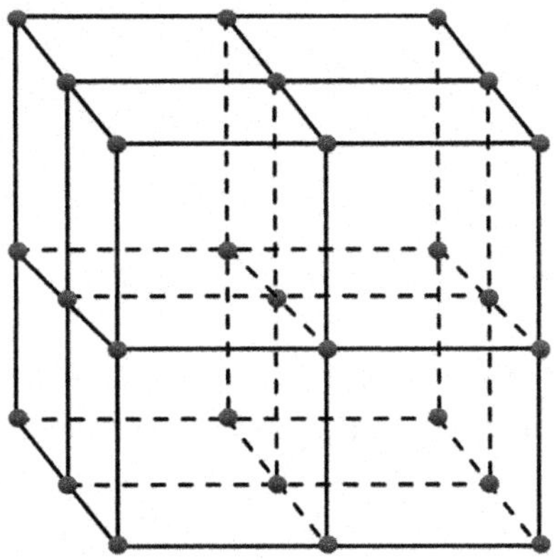

Fig. 9.2 3^3 full factorial experiment

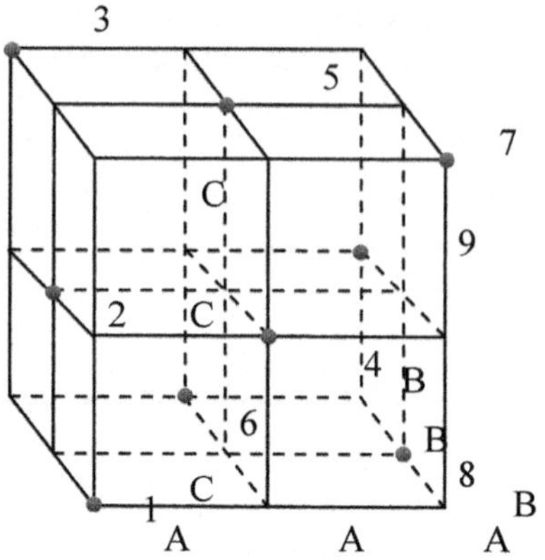

Fig. 9.3 Orthogonal experimental design

thought is to elect some representative condition to experiment. Hence here comes the orthogonal experimental design. It selects the experimental points according to OA, and the selected points are uniform, neat and comparable. So the fewer times of experiment can fully reflect the effect of the different levels of each factor. And orthogonal experimental design is an efficient, fast and economy method.

The full factorial experiment in Fig. 9.2, 3 factors and each with 3 levels, runs 27 times to find the optimal solution. However, it is simplified in orthogonal experimental design, as shown in Fig. 9.3. The first step is to presume that factor A and factor B are fixed in level A1 and level B1. Then A1, B1 is combined with three levels of factor C, namely A1B1C1, A1B1C2 and A1B1C3. The factor C is fixed in C3, if the A1B1C3 is the best combination. The second step is to presume that factor A is fixed in A1. The factor B is fixed in B2, if the best combination is A1B2C3 of A1B1C3 and A1B2C3. Finally, if the A3B2C3 is the best combination of A2B2C3 and A3B2C3, the A3B2C3 is the optimal solution of 27 experimental conditions. There are only 7 experiments.

As a summary, the steps of orthogonal experimental design are list as follows:

(1) Choose the factors, and determine a suitable number of levels for each factor.
(2) Choose a suitable orthogonal array related to the number of factors and levels.
(3) Record the response of every experiment.
(4) Find the "best" combination that maximizes/minimizes the response.

9.2.3 *Uniform Design*

The experimental points generated by orthogonal array are efficient to some degree. However, due to many levels and high experiment cost, orthogonal experimental design is not desirable enough. A more simplified design method, uniform design, has more advantages.

To reduce the experiment run further, the uniform design ignores the comparability of experimental points. The experimental points have the following features: distributing uniformly and appearing only once on each level of factors. So the experiment runs once on each level, which saves more time and cost. However, uniform design is complex in analysis of experiment results.

9.2.4 *Latin Hypercube Sampling Design*

Latin Hypercube Sampling (LHS) is sampling method which is suitable for the experiment factors that have many levels. Same as the uniform design, the LHS design is "space filling" experimental design. When sampling a function of M variables, the range of each variables is equally divided into N probable intervals, $N \geq M$. A Latin square is a square grid containing only one experimental point in M factor's N level. In experiment, experimental points are picked randomly and each point is only used once.

LHS design can be improved in uniformity, which makes the fitting of factors and response more accurate. It adds the requirement that the experimental points are uniformly contributed. In Fig. 9.4, picture (a) shows the random experimental points provided by LHS. And to improve the uniformity of the experimental points, optimization algorithm is used and result is shown in picture (b). However, because the addition of optimization process, the sampling time is increasing.

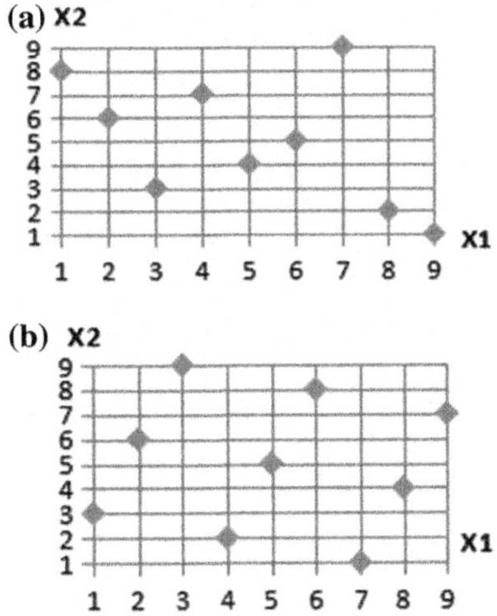

Fig. 9.4 **a** Experimental points of LHS. **b** Experimental points of the optimized LHS

9.3 Surrogate Model

A surrogate model is used in engineering when an outcome cannot be easily directly measure, so a model of outcome is used instead [11]. Most engineering design problems require experiments or simulations to evaluate design objective and constraint functions. However, a signal simulation can take many minutes, hours, or even days to complete. One way to mimic the behavior of simulation or analysis as closely as possible while be computationally cheaper is surrogate model [12, 13]. It uses the mathematical model to replace the original analysis module, and realizes the integration of the analysis module and the optimization of calculation framework. So the multi-discipline optimization design of complex system becomes feasible. The instruction of surrogate model is shown in Fig. 9.5.

There are two broad families under the surrogate model, interpolation-type surrogate model and regression-type surrogate model. For most little-error computer simulation, regression-type surrogate model is more suitable because it can filter digital noise to some degree. For little-error experiments, the interpolation-type surrogate model is better to deal with highly nonlinear problems. In sum, different types needs to fit to different situations [14, 15]. The most popular methods to establish surrogate model are polynomial response surface methodology [16, 17], Radial basis function (RBF) [18] and Kriging [19].

9.3.1 Polynomial Response Surface Methodology

Response Surface Methodology (RSM) belongs to regression-type. It expresses the relationship between design variables and responses by polynomials. Low-order polynomials is popular in engineering application, especially the second-order polynomial. Second-order polynomial is inexpensive to calculate, good for nonlinear response and easy to solve extremum.

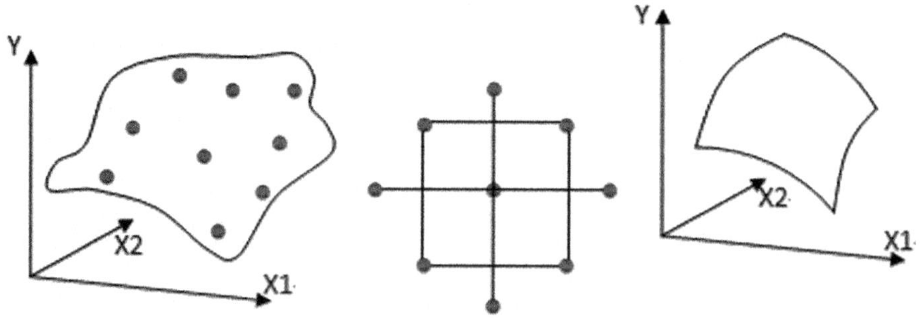

Fig. 9.5 Instruction of surrogate model

A second-order polynomial can be expressed as:

$$y = \beta_0 + \sum_{i=1}^{k} \beta_i x_i + \sum_{i=1}^{k} \beta_{ii} x_i^2 + \sum_{i=1}^{k-1} \sum_{j=i+1}^{k} \beta_{ij} x_i x_j + \varepsilon \tag{9.1}$$

where k is the number of variables, and ε is the error. β_i, β_{ii} and β_{ij} are regression coefficients.

When $k = 2$, the Eq. (9.1) turns into:

$$y = \beta_0 + \beta_1 x_1 + \beta_2 x_2 + \beta_3 x_1^2 + \beta_4 x_2^2 + \beta_5 x_1 x_2 \tag{9.2}$$

When $x_3 = x_1^2$, $x_4 = x_2^2$, $x_5 = x_1 x_2$, the Eq. (9.2) turns into:

$$y = \beta_0 + \beta_1 x_1 + \beta_2 x_2 + \beta_3 x_3 + \beta_4 x_4 + \beta_5 x_5 \tag{9.3}$$

When $k = n$, the set of equations can be expressed in matrix form as:

$$Y = X\beta + \varepsilon \tag{9.4}$$

and

$$Y = \begin{bmatrix} y_1 \\ y_2 \\ y_3 \\ \vdots \\ y_n \end{bmatrix} \quad X = \begin{bmatrix} 1 & x_{11} & x_{12} & \cdots & x_{1k} \\ 1 & x_{21} & x_{22} & \cdots & x_{2k} \\ 1 & x_{31} & x_{32} & \cdots & x_{3k} \\ \vdots & \vdots & \vdots & \ddots & \vdots \\ 1 & x_{n1} & x_{n2} & \cdots & x_{nk} \end{bmatrix} \quad \beta = \begin{bmatrix} \beta_0 \\ \beta_1 \\ \beta_2 \\ \vdots \\ \beta_k \end{bmatrix} \quad \varepsilon = \begin{bmatrix} \varepsilon_0 \\ \varepsilon_1 \\ \varepsilon_2 \\ \vdots \\ \varepsilon_k \end{bmatrix}$$

where n is the test number. When $n \geq (k + 1)(k + 2)/2$, the accuracy of the surrogate model is satisfied.

The vector of regression coefficient can be obtained by

$$\hat{\beta} = \left(X^T X\right)^{-1} X^T y \tag{9.5}$$

and finally the regressing model is the surrogate model as:

$$\hat{y} = X\hat{\beta} \tag{9.6}$$

Polynomial response surface methodology is a simple and practical choice for constructing surrogate models. It is widely used in the engineering especially when the number of variables is not big (less than 10).

9.3.2 Radial Basis Function

Radial Basis Function [20] is to approximate given functions, and this approximate process can be interpreted as a kind of simple neural network. It has been developed for the interpolation of scattered multivariate data. Considering the accuracy and robustness, RBF is better than other surrogate models [21, 22]. When increasing the design variables, PSM needs to execute [$(n + 1)(n + 2)/2$, n is the number of variables] times to guarantee the accuracy. However, RBF only needs $(2n + 1)$ times to obtain the satisfied accuracy.

Here are some description of RBF.

Assume a random kernel function as:

$$\psi_j(\boldsymbol{x}) = \psi(\|\boldsymbol{x} - \boldsymbol{x}_j\|),\ j = 1, 2, \cdots, n \tag{9.7}$$

where $\|\boldsymbol{x} - \boldsymbol{x}_j\|$ is Euclidean distance. There are the responses values $(y_1, y_2, \cdots, y_n)$ of some given points $(\boldsymbol{x}_1, \boldsymbol{x}_2, \cdots, \boldsymbol{x}_n)$ to approximate response function as:

$$f(\boldsymbol{x}) = \sum_{j=1}^{n} [\alpha_j \psi_j(\boldsymbol{x})] + \alpha_{n+1} \tag{9.8}$$

Substitute the given values into (9.8), the $(n + 1)$ linear equations are:

$$\begin{cases} \sum_{j=1}^{n} [\alpha_j \psi_j(\boldsymbol{x}_i)] + \alpha_{n+1} = y_i,\ i = 1, \cdots, n \\ \sum_{j=1}^{n} \alpha_j = 0 \end{cases} \tag{9.9}$$

$(n + 1)$ equations for $(n + 1)$ unknown coefficients guarantees the unique solution.

Vectors and matrix are introduced as:

$$\mathsf{p} = [1 \cdots 1]^T \in \mathsf{R}^n \tag{9.10}$$

$$\phi = \begin{bmatrix} \psi_1(x_1) & \cdots & \psi_n(x_1) \\ \vdots & \vdots & \vdots \\ \psi_1(x_1) & \cdots & \psi_n(x_1) \end{bmatrix} \in \mathsf{R}^{n \times n} \tag{9.11}$$

$$\mathsf{H} = \begin{bmatrix} \phi & \mathsf{p} \\ \mathsf{p}^T & 0 \end{bmatrix} \in \mathsf{R}^{(n+1) \times (n+1)} \tag{9.12}$$

$$\boldsymbol{\alpha} = [\alpha_1, \cdots, \alpha_{n+1}]^T \tag{9.13}$$

$$\boldsymbol{y} = [y_1, \cdots, y_n]^{\mathrm{T}} \tag{9.14}$$

and the Eq. (9.9) turns into:

$$\boldsymbol{H\alpha} = \boldsymbol{y} \tag{9.15}$$

So the coefficient matrix of interpolation function is solved as:

$$\boldsymbol{\alpha} = \boldsymbol{H}^{-1}\boldsymbol{y} \tag{9.16}$$

From the above process, we learn that the kernel function determines the properties of the surrogate model.

There are some different kernel functions commonly used.

(1) Linearsplines

$$\psi_j(\boldsymbol{x}) = \left\|\boldsymbol{x} - \boldsymbol{x}_j\right\| \tag{9.17}$$

(2) Cubicsplines

$$\psi_j(\boldsymbol{x}) = \left\|\boldsymbol{x} - \boldsymbol{x}_j\right\|^3 \tag{9.18}$$

(3) Multiquadrics

$$\psi_j(\boldsymbol{x}) = \sqrt{1 + \frac{\left\|\boldsymbol{x} - \boldsymbol{x}_j\right\|^2}{c_j^2}} \tag{9.19}$$

(4) Gaussian

$$\psi_j(\boldsymbol{x}) = \exp\left(-\frac{\left\|\boldsymbol{x} - \boldsymbol{x}_j\right\|}{c_j^2}\right) \tag{9.20}$$

9.3.3 *Kriging*

In statistics, originally in geostatistics, Kriging [23] is a method of interpolation and gives the best non-linear unbiase prediction of intermediate values. The theoretical basis for the method was developed based on the thesis of DanieKiging, the pioneering plotter of distance-weight average grades in South Africa.

The model assumes that the relationship between the response and variables as:

$$f(x) = g(x) + z(x) \tag{9.21}$$

where *g(x)* is deterministic function; *z(x)* is a systematic departure, and its Statistic properties can be formulated as:

$$\begin{cases} E[z(x)] = 0 \\ Var[z(x)] = \sigma^2 \\ E[z(x^i), z(x)] = \sigma^2 R(c, x, x^i) \end{cases} \tag{9.22}$$

where $E[z(x)]$ is the mean; $Var[z(x)]$ identifies the deviation; $E[z(xi), z(x)]$ is the covariance. *R(c, x, xi)* is a correlation function with a parameter *c*. Popular correlation function includes Gaussian function and exponential function. Through the interpolation of linear weighting superposition of y^i, the function to calculate the response of x is:

$$\hat{f}(x) = w(x)^T Y \tag{9.23}$$

where $w(x) = (w_1, w_2, \ldots, w_n)^T$ is the unknown vector of weight coefficient.

The vector of weight coefficient $Y = (y^1, y^2, \ldots, y^n)^T$ is:

$$w(x) = R^{-1}(r(x) + G(G^T R^{-1} G)^{-1})(G^T R^{-1} r(x) - g(x)) \tag{9.24}$$

and substitute (9.24) into (9.23), the Eq. (9.23) turns into:

$$\hat{f}(x) = g(x)\beta^* + r(x)^T \gamma^* \tag{9.25}$$

where

$$\begin{cases} \beta^* = (G^T R^{-1} G)^{-1} G^T R^{-1} Y \\ \gamma^* = R^{-1}(Y - G\beta^*) \\ G^T = (g(x^1), g(x^2), \ldots, g(x^k)) \\ R = [R_{ij}] = [R(c, x^i, x^j)] \end{cases} \tag{9.26}$$

and the one-dimensional Kriging model is:

$$r(x) = (R(c, x, x^1),\ R(c, x, x^2), \ldots, R(c, x, x^n))^T \tag{9.27}$$

Under the correlation function, the Kriging method has the characteristics of local estimation, which makes it easy to obtain the ideal fitting result when solving the problem of higher degree of nonlinearity. In addition, because the direction of the input vector and the parameter c_j of the function can take different values, the Kriging method can be used to solve the problem of isotropy and anisotropy.

9.3.4 Accuracy Appraisal of the Model

In engineering application, the accuracy of the surrogate model is represented by multiple correlation coefficient R^2. n is the number of sampling points. S_T represents the quadratic sum of total departure. S_R represents the quadratic sum of regression. S_E represents the quadratic sum of residual.

$$R^2 = 1 - \frac{S_E}{S_T} = \frac{S_R}{S_T} \tag{9.28}$$

and

$$S_R = \sum_{i=1}^{n} (\hat{y}_i - \bar{y})^2 \tag{9.29}$$

$$S_E = \sum_{i=1}^{n} (y_i - \hat{y}_i)^2 \tag{9.30}$$

$$S_T = S_R + S_E = \sum_{i=1}^{n} (y_i - \bar{y})^2 \tag{9.31}$$

The value of R^2 is from 0 to 1. And the higher it is, the more accurate the surrogate model is. In engineering application, the acceptable R^2 needs to be over 0.9.

9.4 Optimization Algorithm

Optimization algorithms are used as the search method to figure out the best solution, namely to have a given variable optimized under a set of constraints. Although the concept is simple, it contains a lot of complex problems to process. For example, the solution may be consisted by different types of data; Nonlinear constraints may limit the search area; There are likely many possible candidate solution search area; The characteristics root of the problem is to change with time; Or the variable may have many conflicting objective functions.

The traditional optimization algorithm is mainly used to the problems with explicit mathematical expressions. The first step is to determine the necessary conditions of the existence of extreme points by means of derivation or variational methods. And the necessary condition can be expressed by a group of equations or inequalities. Then, solve the equation or inequality. However, the traditional optimization algorithm must be based on the existence of the derivative of the objective function. But in many practical problems, their objective functions do not have derivative. In recent years, based on search method designed for simulation of the changing process of material and being, many algorithms are proposed. This kind of algorithms are called intelligent algorithm, also known as bionic algorithm, or evolutionary algorithm. The biggest advantage of this kind of algorithms is that it does not need to have the derivable the objective function, and even does not need the explicit expression form of the objective function. They only needs the input and output. Intelligent algorithms cater to the requirements of the development of science and technology. And with its success in the field of optimization, they have attracted increasing attention. This section focuses on several commonly used intelligent algorithms.

9.4.1 Multi-island Genetic Algorithm

A genetic algorithm is a metaheuristic inspired by the process of natural selection that belongs to the larger class of evolutionary algorithm (EA). Genetic algorithms are commonly used to generate high-quality solution to optimization and search problem by relying on bio-inspired operators such as mutation, crossover and selection. Genetic algorithm became popular through the work of John Holland in the early 1970s, particularly after the publishment of his book *Adaptation in Nature and Artificial systems* (*1975*). His work originated with studies of cellular automata, conducted by Holland and his students at University of Michigan.

The evolution usually starts from a population of randomly generated individuals, and is an iterative process, with the population in each iteration called a generation. In each generation, the fitness of every individual in the population is evaluated; the fitness is usually the value of objective function in the optimization problem be solved. The fitter individuals are selected from the current population, and each individual's genome is modified to form a new generation. The new generation of candidate solution is then used in the next iteration of the algorithm. Commonly, the algorithm terminates when either a maximum number of generations has been produced, or a satisfactory fitness level has been reached for the population. The flowchart of GA is shown in Fig. 9.6.

In a generation algorithm, a population of candidate solution (called individuals) to an optimization problem is evolved toward better solution. Each candidate solution has a set of properties which can be mutated and altered. To generate a second generate population of solutions from those selected is to combine genetic operators: crossover and mutation.

Three main parameters in GA are mutation probability, crossover probability and population size. The probabilities of crossover and mutation greatly determine the degree of solution accuracy and the convergence speed that genetic algorithm can obtain. A very small mutation rate may lead to genetic drift (which is mon-ergodic in nature). However, a very high may lead to loss of good solutions. A crossover rate that is too high may lead to premature convergence of the genetic algorithm. The population size depends on the nature of the problem, but typically contains

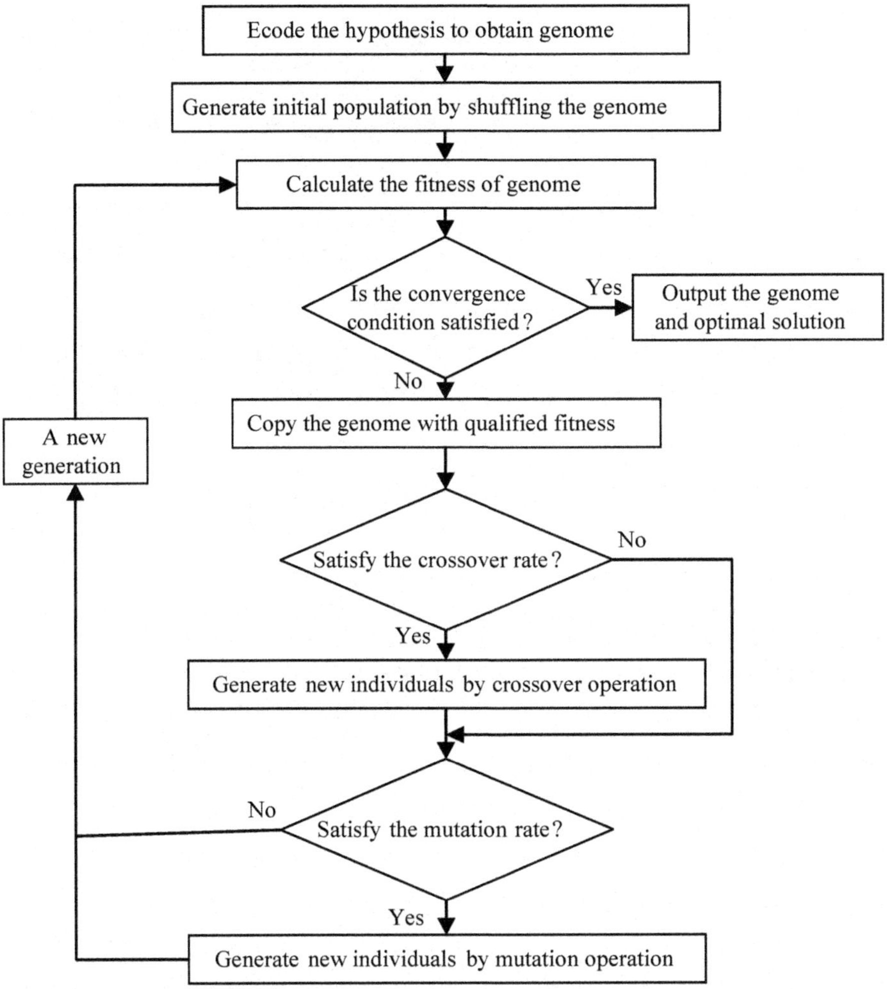

Fig. 9.6 Flowchart of NSGA-II

several hundreds of possible solution, and the solutions may be "seeded" in areas where optimal solutions are likely to be found.

Multi-Island genetic algorithm (MIGA) is a parallel GA based on the traditional GA. MIGA mainly involves the division of the initial population into several independent sub-population by using the "island" concept, which is shown in Fig. 9.7. GA in the sub-population is performed. To ensure the diversity of the population, the individuals of the sub-population are allowed to among sub-populations on the basis of mobility [24]. MIGA can easily overcome the shortcomings that GA traps into local optimum and solve the problem of premature convergence to some degree.

Fig. 9.7 Diagram of MIGA

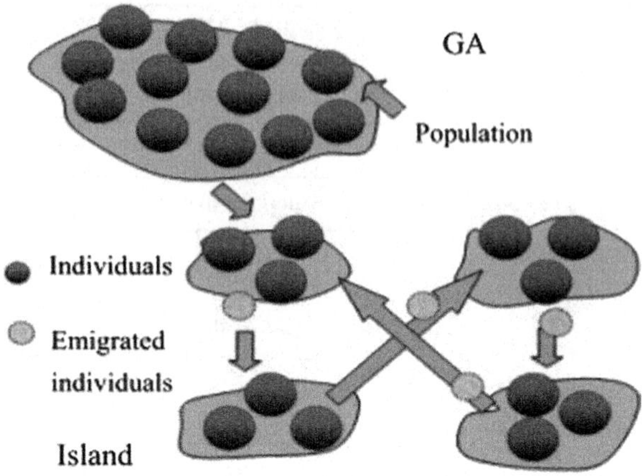

9.4.2 Non-dominated Sorting Genetic Algorithm-II

Non-dominated Sorting Genetic Algorithm (NSGA) is first by Srinivas and Deb [25] in 1995. NSGA-II, the improved algorithm of NSGA, adopts an elitist strategy that selects the best individuals from parent and offspring population to form a new parent population. Therefore, NSGA-II [26] can maintain the diversity of solution set and overcome the shortcomings of NSGA. However, NSGA-II requires considerable computational effort because it combines the offspring population with parent population and uses non-dominated sorting to classify the entire population [27]. The flowchart of NSGA-II is shown in Fig. 9.8.

9.4.3 Multi-objective Particle Swarm Optimization

Particle swarm optimization (PSO) [28, 29] is a computation method that optimizes a problem by iteratively trying to improve a candidate solution with regard to given measure of quality. PSO is originally attributed to Kennedy, and was first intended for simulating social behavior, as a stylized representation of the movement of organisms in a bird flock or fish school. A basic variant of the PSO algorithm works by having a population (called a swarm) of candidate solutions. These particles are moved around in the search-space according to a few simple formula. The movements of the particles are guided by their own best known position in the search-space as well as the entire swarm's best known position. When improved positions are being discovered these will then come to guide the movements of the swarm. The process is repeated and by do it that a satisfactory solution will eventually be discovered. The basic version of the algorithm uses the global

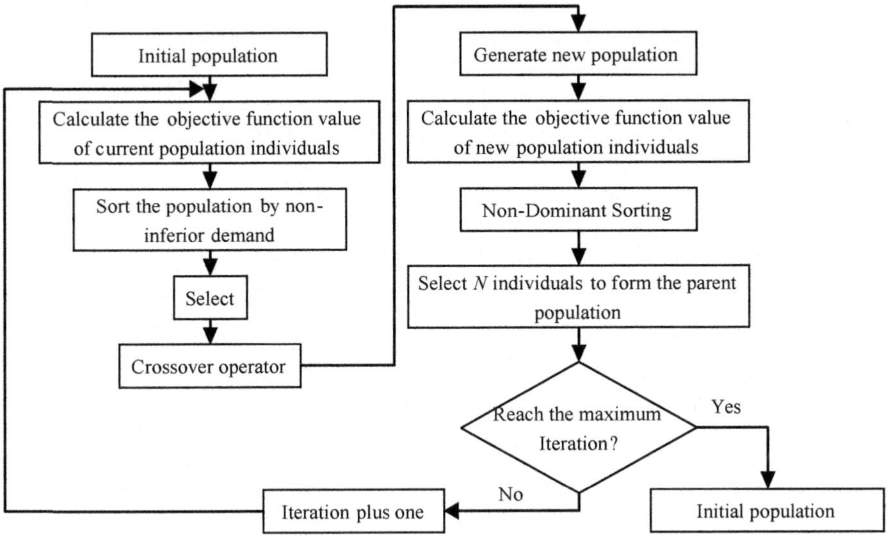

Fig. 9.8 Flowchart of NSGA-II

topology as the swarm communication structure. This topology allows all particles to communicate with all the other particles, thus the whole swarm share the same best position from a single particle (Fig. 9.9).

PSO has also been applied to multi-objective problem, in which the objective function comparison takes Pareto dominance into account when moving the PSO particles and non-dominated are stored so as to approximate the Pareto front.

The optimization comprises three major steps as follows:

1. Describe the physical model by mathematical formulations.
2. Choose design variables.
3. Determine the objective function. The objective function indicates how much each variable contributes to the value to be optimized in the problem.
4. Choose the most appropriate algorithm to solve objective function under the given constraints. The chosen algorithm is applicable to the design problem and is efficient to obtain the optimal value.

To show the optimization process clearly, two examples, design and optimization of PSD housing and multi-objective optimization of drive gears for PSD, are presented. The content includes the creation and sensitivity analysis of surrogate model, the determination of constraints and objective function, and the whole calculation.

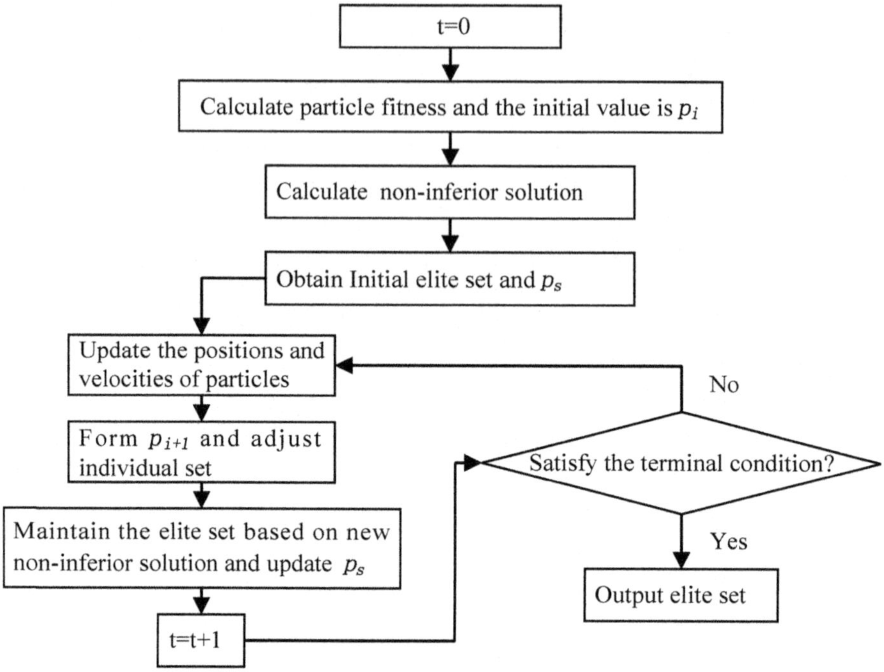

Fig. 9.9 Flowchart of PSO

9.5 Design and Optimization of PSD Housing

PSD is a key component in the energy coupling and decoupling of parallel-series HEV. PSD is mainly composed of a planetary gear train to meet the torque and control requirements of rotation speed decoupling. Good fuel economy may be achieved by the optional control of engine operating mode when an HEV with a PSD continuously operates under different load conditions. And lightweight design is always the emphasis in vehicle design due to its great contribution to the size and mass of vehicles, the cost and energy-saving. Section 9.4.1 takes PSD housing in HEV as an example to give more details about the signal-objective optimization process [30].

The first step is to successively determine the design variables, constraints and objective function. Then create the surrogate model of PSD housing. Then carry out the signal-objective optimization of PSD housing in Isight where the optimization goal is to minimize the mass in different stiffness. The schematic diagram of operation steps in Isight is appended for readers' convenience.

To create the surrogate model for optimization, the first step is to select the way of creation which is determined through the problem analysis. And according to the requirement of different surrogates, select the test simple points by experimental design. After the experiment, use the test results to construct the surrogate model.

Then test accuracy of surrogate model. After all that, the optimization is operated based on the surrogate model.

9.5.1 Mathematical Model of Optimization

9.5.1.1 Design Variables

Those independent parameters chosen and determined in the design are called design variables. In mechanical design, common design variables are assembly dimensions, geometry dimensions of parts and physical and mechanical characteristics etc.

In this design, the structure size of the inner ring, the outer ring and the radial plate are selected as design variables, as shown in Fig. 9.10. On the premise of noninterference, the chosen parameters and their ranges are shown in Table 9.1. What the design variables $x_1 \sim x_8$ represent for are shown in Table 9.2.

The interface when setting the parameters and their ranges in Isight is shown in Fig. 9.11.

9.5.1.2 Objective Function

The objective function indicates how much each variable contributes to the value to be optimized in the problem. To formulate the optimization goal by design variables is called the creation of objective function.

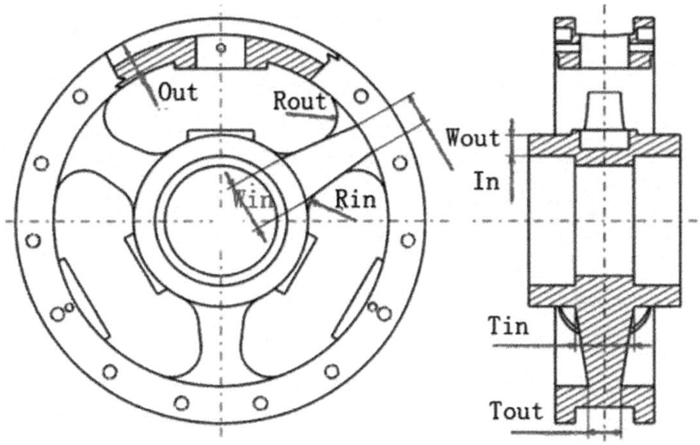

Fig. 9.10 Design variables

Table 9.1 Chosen parameters

Parameters	Ranges(mm)
Out	3–8
In	3–8
Wout	5–20
Win	5–17
Tout	5–20
Tin	5–20
Rout	5–12
Rin	5–15

Table 9.2 Corresponding parameters to $x_1 \sim x_2$

x_1	x_2	x_3	x_4	x_5	x_6	x_7	x_8
Tin	Tout	Rin	Rout	Win	Wout	In	Out

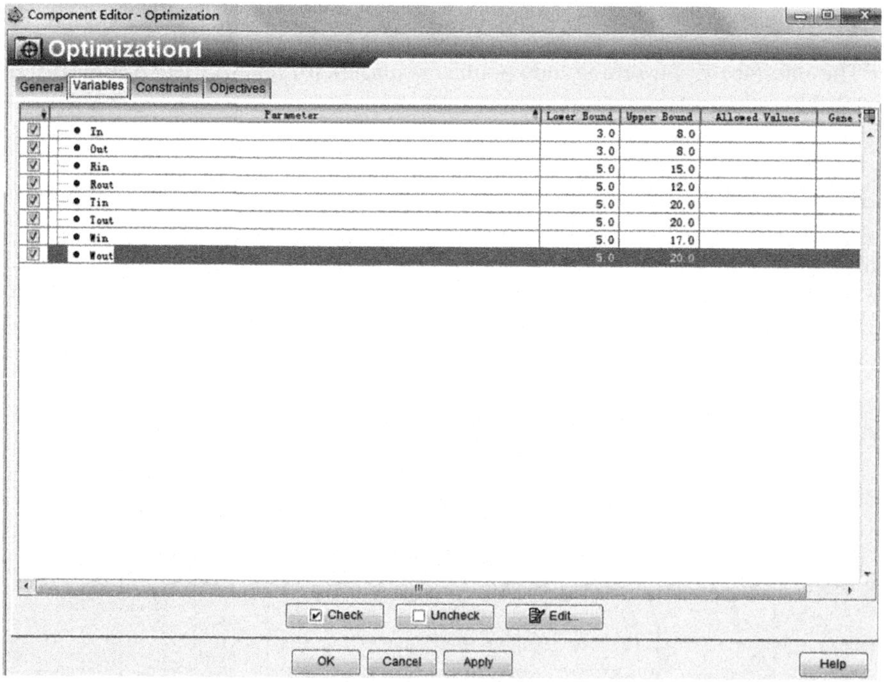

Fig. 9.11 Set of the parameters and their ranges

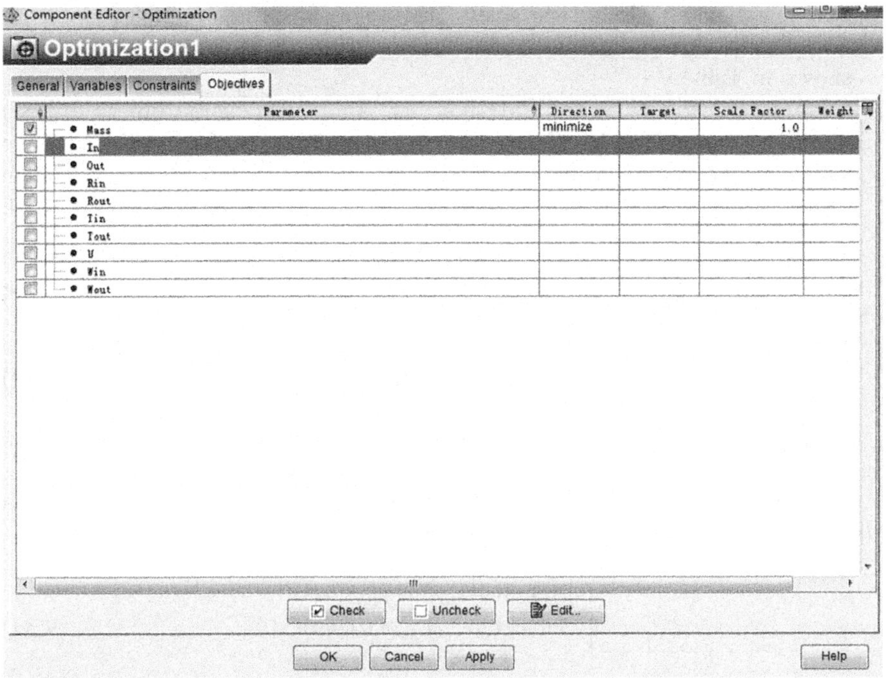

Fig. 9.12 Set of the objective function

To meet the lightweight design requirement, minimizing the weight of the PSD housing is considered the objective function, as shown in (9.32).

$$\text{Min } M = f(X) \tag{9.32}$$

where W is the weight of the PSD housing and $\boldsymbol{X}$ is the design variable matrix related to W.

The interface when setting objective function in Isight is shown in Fig. 9.12.

9.5.1.3 Constraint Conditions

In optimization, the design requirement added to design variables is called constraints which the solution must be satisfied. Constraints can be represented by equations or inequations. There are two constraints in this design. One is self-constraints of design variables. The other is stiffness constraints of the PSD housing.

(1) Self-constraints of design variables: The ranges of design variables are determined under the geometric noninterference of the PSD housing structure, as shown in Table 9.1.
(2) Stiffness constraints of the PSD housing: The insufficient stiffness of PSD housing will make the bevel gears deflect along the axle, which will lead to the stress concentration. With the variation of the skew angle β, the maximum contact stress of the modified and unmodified tooth profile is shown in Fig. 9.13.

In the PSD housing optimization, the skew angle of bevel gear is restricted first to less than 0.02°, namely the deflection of axle endpoint is 0.01 mm. Thus, the constrains can be defined in the following form.

$$U \leq 0.01 \text{ mm} \tag{9.33}$$

Maximum contact stress of modified and unmodified tooth profile is shown in Fig. 9.13.

From the above, the optimization problem PSD housing can be described as follow:

$$\begin{cases} \text{Minimize} : M = f(x) \\ \text{Subject to} : U \leq 0.01 \text{ mm} \\ X_L \leq X \leq X_U \end{cases} \tag{9.34}$$

where **X** represents the design variable matrix and $\mathbf{X_L}$ and $\mathbf{X_U}$ is the upper and lower bound of design variables.

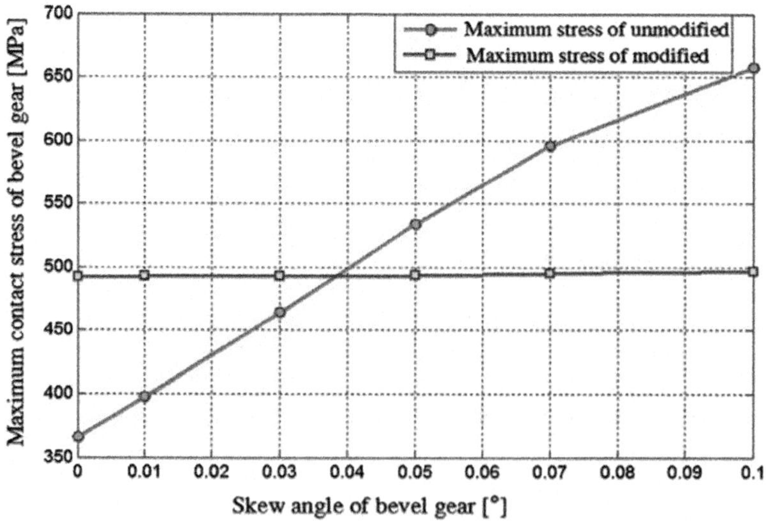

Fig. 9.13 Maximum contact stress of modified and unmodified tooth profile

9.5.2 Establishment of the Surrogate Model of PSD Housing

9.5.2.1 Design of Experiment

After determining of the factors and levels of the experiment, the next steps are to design the experiment scheme, and then operate the experiment. The experiment design is completed in the Isight, the multi-objective optimization software. The sample points are generated using the Latin hypercube method. Before the operation, it is necessary to build the connection between the Isight and ABAQUS, and the interface is shown in Fig. 9.14. In the experimental process, the data of experiment points transfer from Isight to the computing module in ABAQUS. The module calls ABAQUS/Standard solver in the background. The circulation of data transfer and modification of FE model parameters are operated without manual intervention.

This section uses Latin hypercube method to elect the sample points, and Quadratic Polynomial Response Surface Method to establish the surrogate model. Quadratic Polynomial Response Surface Model formulates as follow:

$$\begin{cases} y = \alpha_0 + \sum\limits_{i=1}^{n} \alpha_i x_i + \sum\limits_{i=1}^{n} \alpha_i x_i^2 + \sum\limits_{j=2}^{n} \sum\limits_{i=1}^{j-1} \alpha_{ij} x_i x_j \\ \boldsymbol{\alpha} = \left[\alpha_0, \alpha_1, \ldots, \alpha_n, \alpha_{11}, \alpha_{22}, \ldots, \alpha_{nn}, \alpha_{12}, \alpha_{13}, \ldots, \alpha_{(n-1)n}\right] \end{cases} \tag{9.35}$$

where y is the output variable, and x is the design variable. n is the number of design variable, α is the undetermined coefficient vector, which obtained by least square fitting.

The operation of election of sample points by Latin hypercube method is shown in Fig. 9.15. And the data reading is shown in Fig. 9.16. These two experiments results are mutual independence data. One set of experiment data of surrogate model can be used to test the accuracy of the other surrogate model.

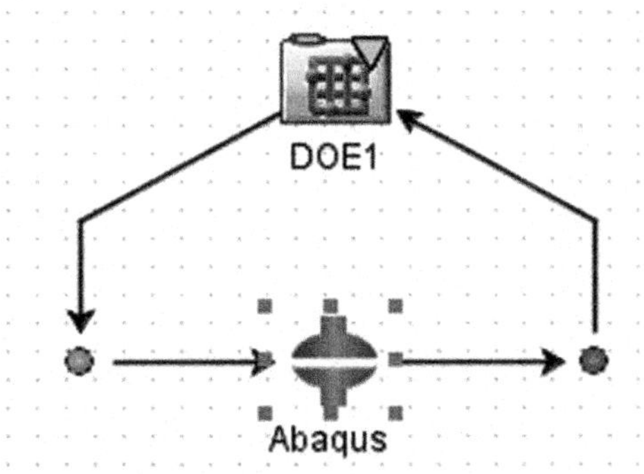

Fig. 9.14 Isignt model of DOE

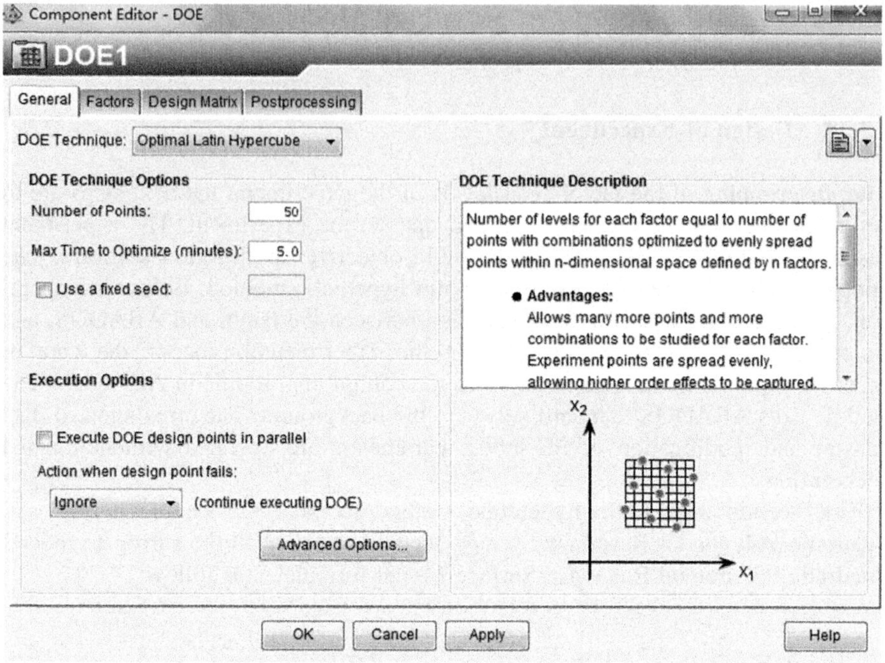

Fig. 9.15 Selection of LHS points

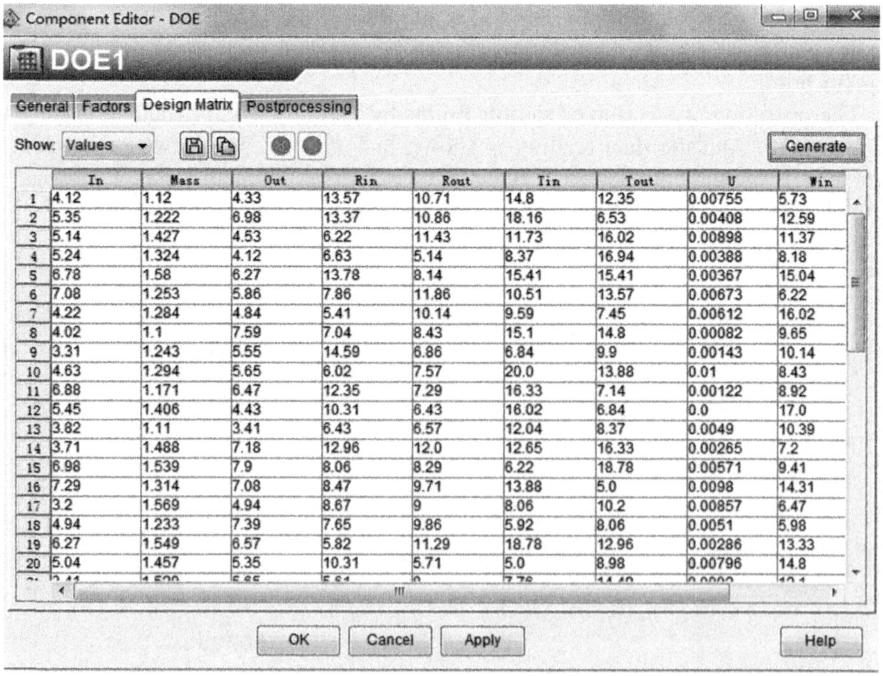

	In	Mass	Out	Rin	Rout	Tin	Tout	U	Win
1	4.12	1.12	4.33	13.57	10.71	14.8	12.35	0.00755	5.73
2	5.35	1.222	6.98	13.37	10.86	18.16	6.53	0.00408	12.59
3	5.14	1.427	4.53	6.22	11.43	11.73	16.02	0.00898	11.37
4	5.24	1.324	4.12	6.63	5.14	8.37	16.94	0.00388	8.18
5	6.78	1.58	6.27	13.78	8.14	15.41	15.41	0.00367	15.04
6	7.08	1.253	5.86	7.86	11.86	10.51	13.57	0.00673	6.22
7	4.22	1.284	4.84	5.41	10.14	9.59	7.45	0.00612	16.02
8	4.02	1.1	7.59	7.04	8.43	15.1	14.8	0.00082	9.65
9	3.31	1.243	5.55	14.59	6.86	6.84	9.9	0.00143	10.14
10	4.63	1.294	5.65	6.02	7.57	20.0	13.88	0.01	8.43
11	6.88	1.171	6.47	12.35	7.29	16.33	7.14	0.00122	8.92
12	5.45	1.406	4.43	10.31	6.43	16.02	6.84	0.0	17.0
13	3.82	1.11	3.41	6.43	6.57	12.04	8.37	0.0049	10.39
14	3.71	1.488	7.18	12.96	12.0	12.65	16.33	0.00265	7.2
15	6.98	1.539	7.9	8.06	8.29	6.22	18.78	0.00571	9.41
16	7.29	1.314	7.08	8.47	9.71	13.88	5.0	0.0098	14.31
17	3.2	1.569	4.94	8.67	9	8.06	10.2	0.00857	6.47
18	4.94	1.233	7.39	7.65	9.86	5.92	8.06	0.0051	5.98
19	6.27	1.549	6.57	5.82	11.29	18.78	12.96	0.00286	13.33
20	5.04	1.457	5.35	10.31	5.71	5.0	8.98	0.00796	14.8

Fig. 9.16 Data reading

9.5.2.2 Establishment of the Surrogate Model

The surrogate model uses a simple mathematical model to replace a complex relation between the design variables and corresponding response during optimization process. Based on 50 sets of the sample points in Table 9.3, the surrogate models are established using the Quadratic Polynomial Response Surface Model. The surface of importing 50 sets of the sample points to the Approximation module in Isight is shown in Fig. 9.17.

Table 9.3 Parameters of MIGA

Size of operartion	Number of islands	Number of generation	Rate of crossover	Rate of mutation	Interval of migration	Rate of migration	Iteration
10	5	40	0.7	0.02	2	0.3	2000

Fig. 9.17 **a** First step. **b** Second step. **c** Third step. **d** Forth step

Table 9.4 Optimization results

Parameters	Original value	Optimal results	Modified optimal results
Out/mm	5	3.000076	3
In/mm	6	4.000366	4
Wout/mm	15	11.149462	11
Win/mm	15	9.250492	9.5
Tout/mm	15	5.272602	5.5
Tin/mm	15	12.991531	13
Rout/mm	7	7.846784	8
Rin/mm	7	13.166476	13
U/mm	\	0.0099882	\
Mass/kg	\	1.22807	\
U/mm	0.00374	\	0.00956
Mass/kg	1.592	\	1.2348

9.5.2.3 Accuracy Test of Surrogate

As the surrogate model is an approximate equation between design variables and response function, conducting an error analysis of a surrogate model is necessary. The accuracy of surrogate model is measured via the multiple correlation coefficient R^2 in engineering application. 20 sets of sample points in Table 9.4 is used to test the accuracy of the surrogate model by the Approximation module in Fig. 9.18. The error analysis of a surrogate model using the Quadratic Polynomial Response Surface Model is shown in Fig. 9.19. Where, (a) shows error of the model of housing weight with $R^2 = 1$ and (b) shows the error of the model of maximum deflection with $R^2 = 0.96$.

In the engineering application, the acceptable R^2 needs to be over 0.9, and the higher it is, the more accurate the surrogate model is. And the above results indicate that the surrogate models can be used to predict performance.

9.5.3 Optimization and Results

A multi-island genetic algorithm (MIGA) is used to optimization of the housing weight. And minimizing the weight of the PSD housing is considered the objective function. According to the above analysis, the deflection of the PSD hosing is restricted to less than 0.01 mm. The optimization process in Isight needs to build the data connection between the surrogate model and integrated toolkit, as shown in Fig. 9.20. Some parameters are set as in the Table 9.3, and the other parameters are set as defaults.

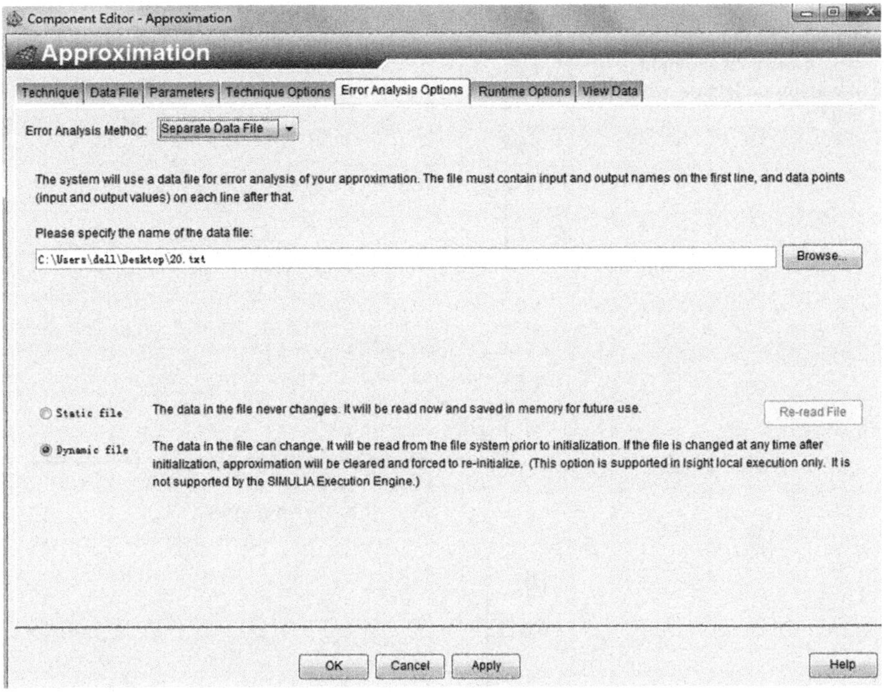

Fig. 9.18 Approximation module

After setting all the parameters in Optimization and Approximation to meet the above requirement, the optimal calculation is operated, and the step to see optimization results are shown in the Fig. 9.21.

The MIGA terminates after meeting the iteration steps, and the output is the optimum calculation. The total iteration of the MIGA is 2000 steps, which is shown in the Fig. 9.22. The return value of housing weight is shown in Fig. 9.23 and the objective function begins to converge when iterating at about 40 steps.

From Fig. 9.23 the housing weight is decreasing with the increasing of iteration steps, and the value converges to 1.23 kg. The optimal results after modification are shown in Table 9.4. The FEA result is 1.2348 kg. Compare to the initial design, the PSD housing weight is reduced by 18% and the maximum reflection is 0.00956 mm which is less than constraint condition 0.1 mm. The above analysis shows that the optimization of PSD is reasonable.

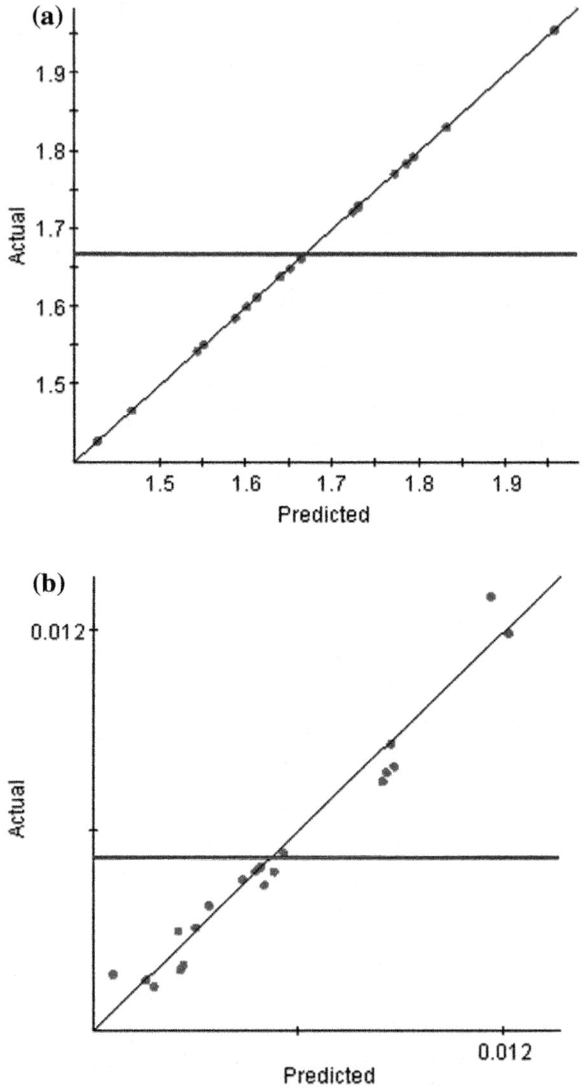

Fig. 9.19 **a** Error of the model of housing weight with $R^2 = 1$. **b** Error of the model of maximum deflection with $R^2 = 0.96$

9.6 Multi-objective Optimization of Drive Gear for PSD

Multi-objective optimization problem is common in daily life. And as there are multiple solutions for multi-objective optimization problem, that means to solve such a problem is not as straightforward as it is for a conventional signal-objective optimization problem. In engineering application, there does not exist a signal

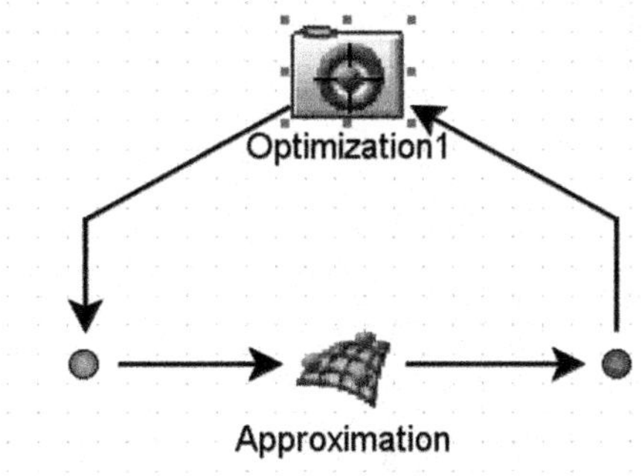

Fig. 9.20 Isight model based on surrogate model

solution that simultaneously optimizes each objective. In that case, the objective function is conflicting.

9.6.1 Establishment of the Surrogate Model of Drive Gears

According to the different performance requirements, the design of drive gears for PSD involves a complex problem with multi-objective and constraint conditions. Multiple power sources (ICE, battery, motor, etc.) leads to a highly non-linear relation between design variables and objective function in optimization design of drive gears for PSD. In addition, the relation is difficult to describe by explicit equation. One way to alleviating the burden of complexity and large calculation is constructing approximate model, known as the surrogate model. The model is constructed based on modeling the response to a limited number of intelligently chosen sample points. The approach also helps to reduce the cost and shorten the design period, and is useful to obtain the mathematical model. Those properties make it widely used in complex engineering optimization. Thus, multi-objective optimization of drive gears is performed based on the surrogate model.

9.6.1.1 Design Variables

The design variables of gear optimization are generally composed of basic geometric parameter and performance parameters, such as gears transmission ratio (μ), modulus (m), tooth number (z), face width (b), and pressure angle. To diminish the complexity and consuming time of optimization, considering the cost and the requirement of sample points comprehensively, m, z and μ are chosen as the design variables. In practical application, less rigidity and manufacturing error of housing, assembly error and needle bearing clearance cause the deviation of gear axis, which causes load concentration. Influence of axial force, bearing stiffness, different preload, housing stiffness and etc. makes the planetary gear of DPSD displaces

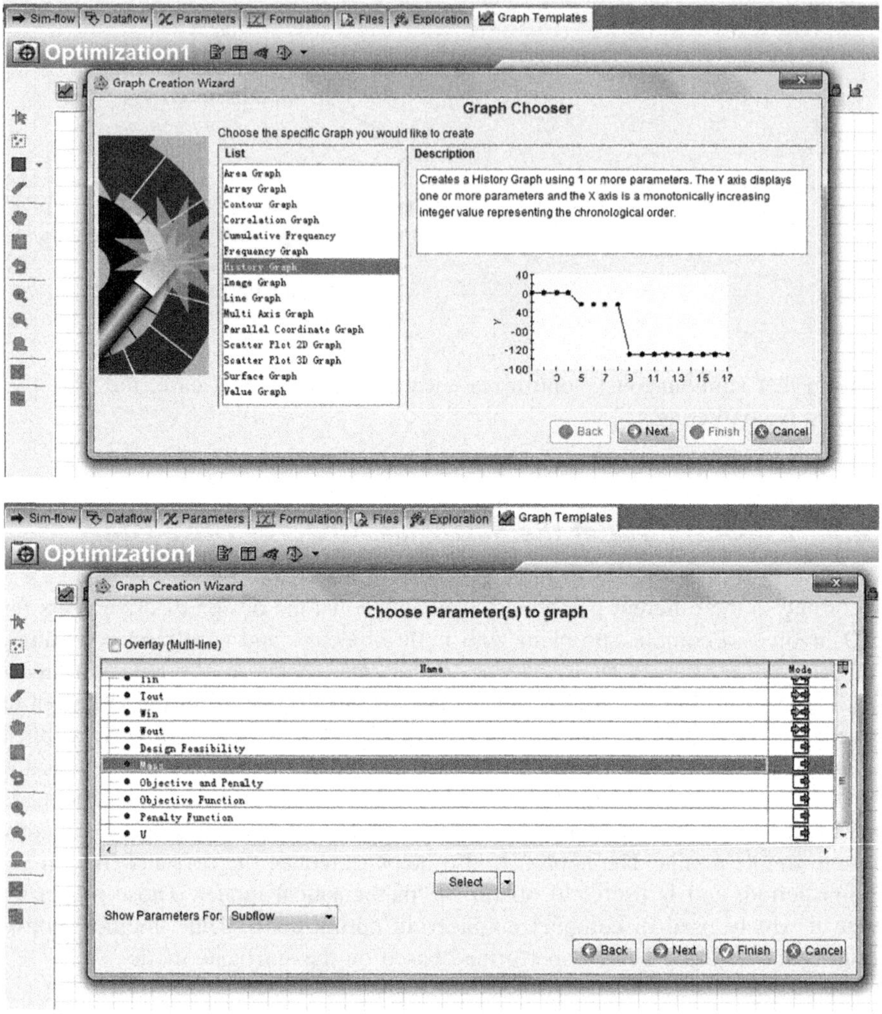

Fig. 9.21 View the optimization results

axially in meshing period, namely the planetary gear deviates back from the original meshing center point along the axial direction of a certain distance. So the deflection of axis and axial displacement are chosen as design variables.

The vector of design variables is

$$\mathbf{x} = [x_1, x_2, x_3, x_4, x_5]^T = [m, \mu, z_1, a, d]^T \tag{9.36}$$

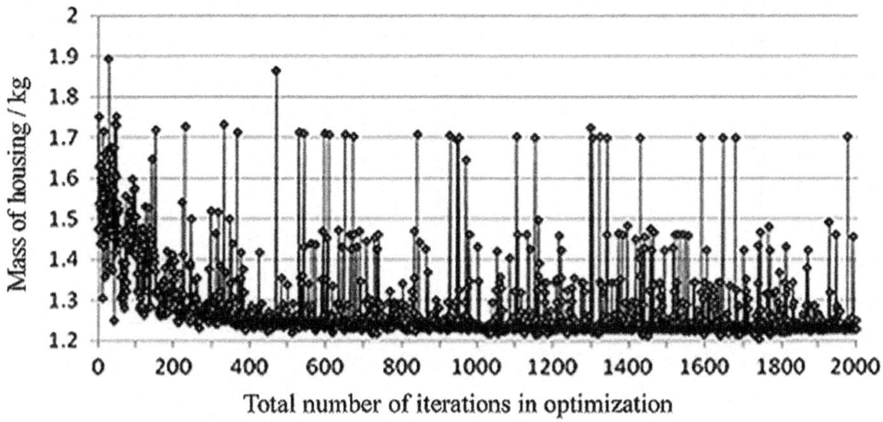

Fig. 9.22 Iteration history of the MIGA

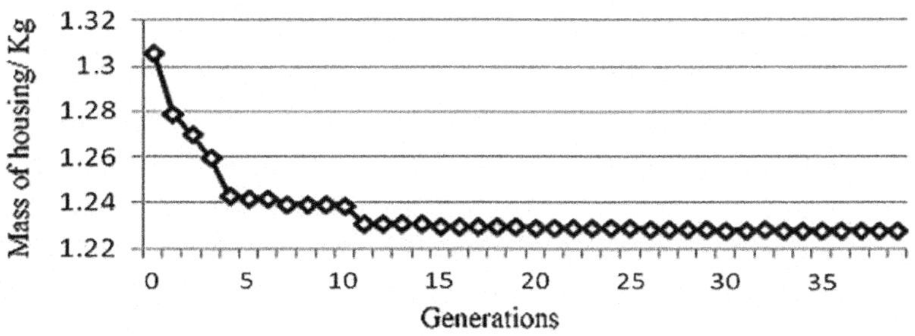

Fig. 9.23 Return values of housing weight

The upper and lower bounds on the design variables are shown as follows:

$$\begin{cases} 1 \leq m \leq 6 \\ 1 \leq \mu \leq 2 \\ 14 \leq z_1 \leq 25 \\ 0 < a < 0.05 \\ 0 < d < 0.5 \end{cases} \quad (9.37)$$

where m is the modulus, $u = z_2/z_1$ is the transmission ratio, and z_1 and z_2 are the tooth numbers of the planetary and half-axle gears respectively. d is axial displacement, and α is the deflection of axis.

30 sets of experiment data (Table 9.5) obtained by Latin hypercube method are used to build the surrogate model.

The other 10 sets (Table 9.6) of experiment data are used to analysis the predicting performance of the surrogate model.

Table 9.5 Original sample points

Variables NO.	Modulus (m)	Tooth number of planetary (z_1)	Transmission ratio (u)	Axial displacement d(0–0.5)	Defection of axis a(0–0.05)
1	2	24	1.166667	0.3795	0.0207
2	4	16	1.125	0.0345	0.0379
3	5	23	1.043478	0.155	0.0276
4	1.375	20	1.6	0.431	0.0414
5	4.5	15	1.733333	0.362	0.0017
6	4	19	1.052632	0.414	0.0431
7	3.75	24	1.458333	0.1895	0
8	5	17	1.764706	0.483	0.0345
9	3.25	15	1.533333	0.3105	0.05
10	6	20	1.5	0.207	0.0466
11	3	19	1.947368	0.086	0.0069
12	6	22	1.909091	0.2585	0.0172
13	5	15	1.8	0.1205	0.031
14	4.5	25	1.6	0.3965	0.0397
15	1.75	17	1.705882	0.0515	0.0362
16	1.5	25	1.76	0.1725	0.0224
17	3.5	18	1	0.276	0.0052
18	5.5	14	1.214286	0.3275	0.0259
19	3	22	1.863636	0.4655	0.0121
20	2.5	23	1.304348	0.138	0.0483
21	2.75	14	1.428571	0.1035	0.0103
22	4.5	23	1.652174	0	0.0293
23	1	18	1.555556	0.293	0.0034
24	5.5	21	1.285714	0.4485	0.0138
25	2.25	16	1.9375	0.345	0.0241
26	1.125	16	1.125	0.2415	0.0328
27	2.5	15	1.333333	0.5	0.019
28	5.5	18	1.388889	0.069	0.0086
29	1.75	21	1.190476	0.017	0.0155
30	3.5	20	2	0.224	0.0448

9.6.1.2 Error Analysis of Surrogate

This design chooses three kinds of commonly used method, Kriging, RBF and PSM, to set up the surrogate for the gear optimization problem. Error analysis of the surrogate model is established by the root mean square method. Then the accuracy of each model is compared, and the optimal design of high accuracy is chosen for further optimization design.

Table 9.6 Test points

Variables NO.	Modulus (m)	Tooth number of planetary (z_1)	Transmission ratio (u)	Axial displacement d(0–0.5)	Deflection axis a(0–0.05)
1	5.5	16	1.3125	0.11	0.0444
2	2.75	15	1.8	0.055	0.0111
3	5	20	1.1	0.165	0
4	4.5	24	1.208333	0.445	0.0389
5	3.25	25	1.56	0	0.0333
6	3.75	14	1.428571	0.5	0.0167
7	2	18	1.888889	0.335	0.05
8	1	17	1	0.22	0.0278
9	6	21	2	0.28	0.0222
10	1.5	22	1.681818	0.39	0.0056

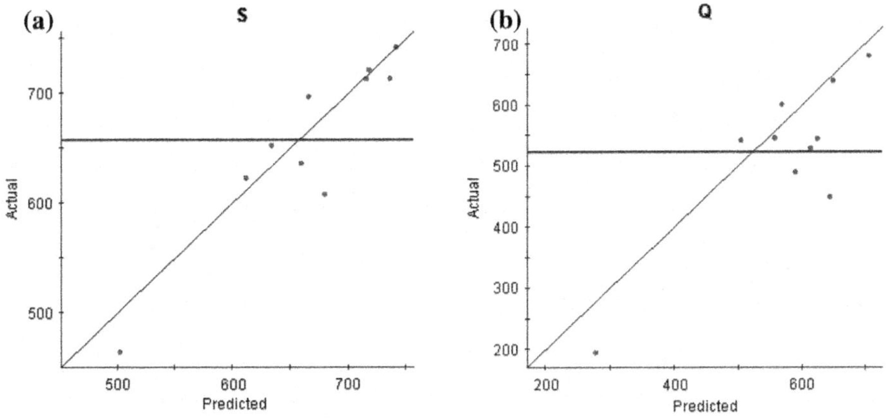

Fig. 9.24 **a** Error of stress based on Kriging. **b** Error of friction loss based on Kriging

In the surrogate model based on Kriging, the error of stress is 0.10787 and error of the friction loss is 0.17188, as shown in the Fig. 9.24.

In the surrogate model based on RBF, the error of stress is 0.09255, and the error of friction loss is 0.14662, as shown in the Fig. 9.25.

In the surrogate model based on PSM, the error of stress is 0.14085, and the error friction loss is 0.17625, as shown in the Fig. 9.26.

After the comparison, the surrogate model based on Kriging is chosen.

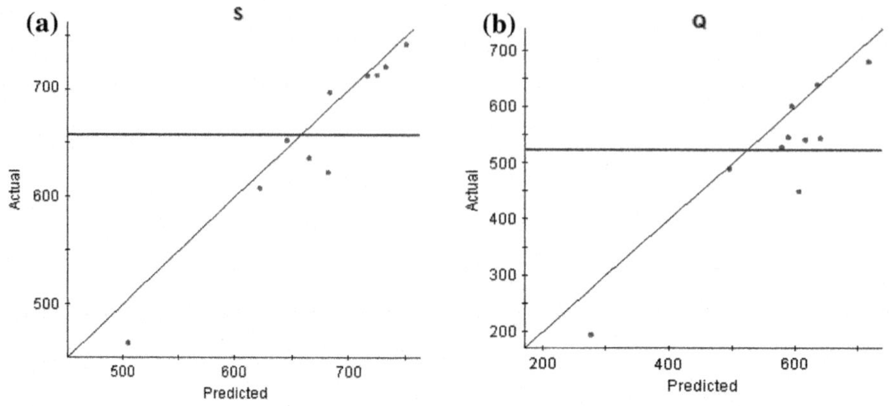

Fig. 9.25 **a** Error of stress based on RBF. **b** Error of friction loss based on RBF

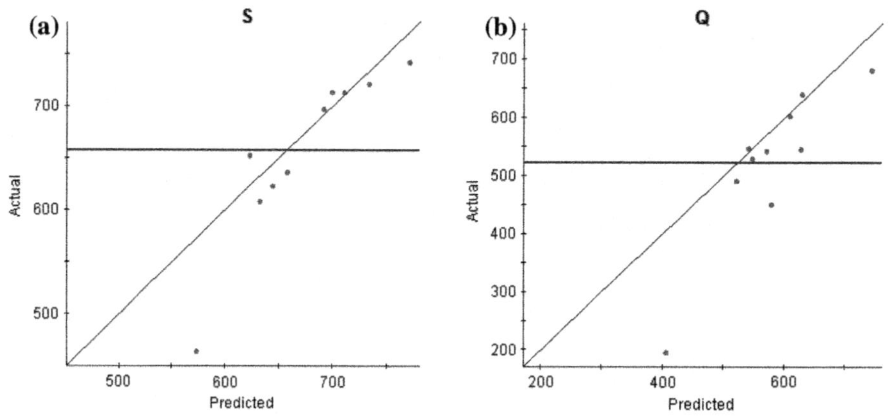

Fig. 9.26 **a** Error of stress based on PSM. **b** Error of friction based on PSM

9.6.2 Multi-objective Optimization Design [2, 31]

9.6.2.1 Model Optimization

The influence of meshing includes axial deviation, displacement of planetary gear, heat and stress of PSD in operation. By optimization design, the PSD is still safe and efficient under all above factors. Mission requirements are as follows:

(1) Lightweight design of PSD, namely the smallest total mass of gear;
(2) Efficient meshing of gear in the operation;
(3) Meshing stress of gear below the safety value in the operation;

According to the above requirements, it is determined that the PSD has the following functions:

(1) The weight lightest of the PSD;
(2) Satisfaction of the bending and contact stress of the tooth as well as the torsional deformation and stress of the shaft;
(3) The highest transmission efficiency.

For certain gear material, the volume is the smallest, when the weight is the lightest. So the volume is one of the objective functions. For certain gear parameters, the tooth root bending stress at the specified location is related to the tooth surface contact stress. For the certain location and parameters, the bigger contact stress is, the larger bending stress is, which is similar to the mechanical principle of the root and end of cantilever beam. Therefore, the contact stress also is the objective function. And when it is the minimum value, the bending stress at this time is the minimum. The transmission power loss mainly includes: the friction energy dissipation in the meshing, the oil resistance loss, the friction loss in the bearing. Due to the friction energy dissipation of oil resistance can be neglected, the efficiency objective function can be converted to the minimum friction energy consumption. In sum, the optimization problem consists of three objective functions: gear volume, contact stress and friction energy consumption.

(1) Objective function 1: Volume (V)

Reducing the volume of DPSD drive gears benefits not only the compact structure to reduce the technical requirement of drivetrain assembly within a limited space, but also the light-weight design of DPSD. Therefore, the miniaturization of drive gears is a critical requirement of DPSD design.

The volume of DPSD drive gears is formulated as follows:

$$\begin{aligned} V_T &= V(m, z_1, \mu) \\ &= \frac{n_1 \pi b \cos\delta_1 (3m^2 z_1^2 - 6mz_1 b \sin\delta_1 + 4b^2 \sin^2\delta_1)}{12} \\ &\quad + \frac{n_2 \pi b \cos\delta_2 (3m^2 z_2^2 - 6mz_2 b \sin\delta_2 + 4b^2 \sin^2\delta_2)}{12} \end{aligned} \tag{9.38}$$

where n_1 is the number of the planetary gear, n_2 is the number of half-axle gear, and b is the face width. δ_1 and δ_2 can be calculated as:

$$\begin{cases} \delta_1 = \tan^{-1}\left(\frac{z_1}{z_2}\right) \\ \delta_2 = \frac{\pi}{2} - \delta_1 \end{cases} \tag{9.39}$$

b can be computed as:

$$b = \gamma \cdot \varphi_R \tag{9.40}$$

where φR is the coefficient of the face width, which is usually equal to 1/3; and γ is the cone distance of the planetary gear, which can be obtained as:

$$\gamma = \frac{(mz_1)}{2 \sin \delta} = \frac{m\sqrt{z_1^2 + z_2^2}}{2} = \frac{mz_1\sqrt{1+\mu^2}}{2} \tag{9.41}$$

Thus, the face width is expressed as:

$$b = \frac{mz_1\sqrt{1+\mu^2}}{6} \tag{9.42}$$

(2) Objective function 2: Maximum contact stress (S)

Under high-speed condition, the wear of transmission components in DPSD is the main failure form, especially tooth face wear. Cyclic contact stress will emerge on the tooth surface during meshing. Tooth surface damage may occur when the maximum contact stress exceeds the contact fatigue limit. Therefore, reducing the contact stress is selected as the second objective function.

(3) Objective function 3: Frictional energy dissipation (Q)

The working condition analysis of DPSD shows that the drive gears in the DPSD operate at high-speed difference conditions. The large difference between the left and right half-axle causes high sliding friction, which generates a large amount of heat. With the increase of temperature, the performance of lubricant will worsen, further increasing the friction energy dissipation gearing and resulting in drive gears failure. To achieve high transmission efficiency and low power consumption, the frictional energy dissipation must be reduced during DPSD operation. Therefore, frictional energy dissipation is selected as the third objective function.

The response surface functions (V, S and Q) according to each two design variables are shown in Fig. 9.27. The variable responses *V, S* and *Q* with respect to design variables all exhibit a high nonlinear relationship.

9.6.2.2 Optimization Process and Results

MIGA, NSGA-II, and MOPOS are adopted to control the optimization process in Isight. First, the design variables are random read, and then are substituted to the maximum contact stress and frictional energy dissipation surrogate model to calculate. And then comparing the results to the setting maximum, if result is smaller than the bound, the design variable will be kept, otherwise the design variable will be abandoned. Then the left design variables are substituted to volume surrogate

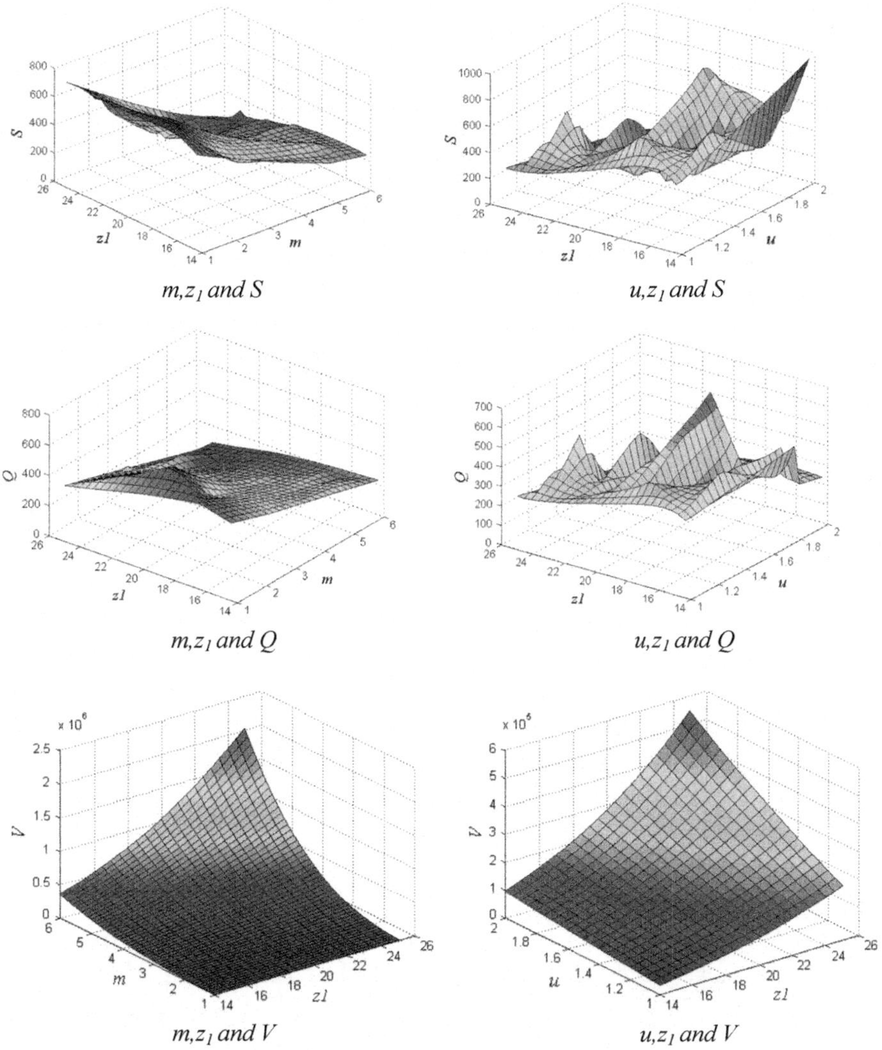

Fig. 9.27 Relationship between objective function values and design variables

model, and in the same way the qualified design variables will be kept. The optimum Pareto solution will be found by iterating the above steps until out of all the design variables [32]. The optimization process is shown in the Fig. 9.28.

The optimization is operated using MIGA, NSGA-II and MOPSO as follows. The iteration processes and Pareto frontier of three optimization algorithms are diagrammed. Tables 9.7, 9.8 and 9.9 list the detailed final settings of MIGA, NSGA-II and MOPSO used.

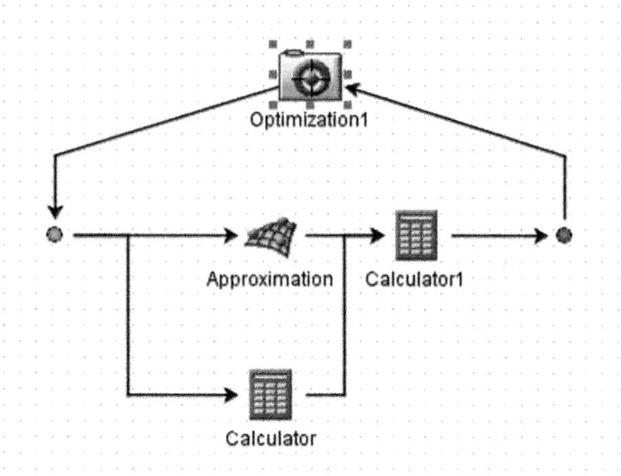

Fig. 9.28 Optimization process

Table 9.7 MIGA parameters

Size of sub-population	The number of island	The number of generation	Rate of mutation	Rate of migration	Interval of migration	Rate of crossover	Iterations
10	10	10	0.6	0.02	5	0.1	2000

Table 9.8 NSGA-II parameters

Size of population	Number of generation	Rate of migration	Cycles of migration	Cycles of crossover	Iterations
20	20	0.6	20	10	400

Table 9.9 MOPSO parameters

Maximum generation	Particle population	Inertia weight	Global increment	Particle increment	Maximum velocity
50	10	0.9	0.9	0.9	50

The number of generations reaches 1000, the iteration history is shown in Fig. 9.29.

The number of generations reaches 400, the iteration history is shown in Fig. 9.30.

The number of generations reaches 500, the iteration history is shown in Fig. 9.31.

The design variables are rounded to the nearest discrete values, as shown in Table 9.10.

Comparing all the optimization results, it is found that the optimal solution of three optimization algorithms are not very different from each other. All of them can accurately find the range of solutions of Pareto, which means the Pareto solution is

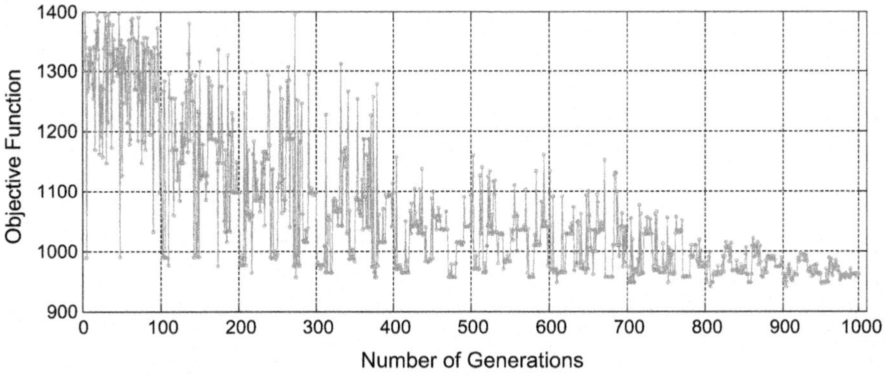

Fig. 9.29 Iteration history of MIGA

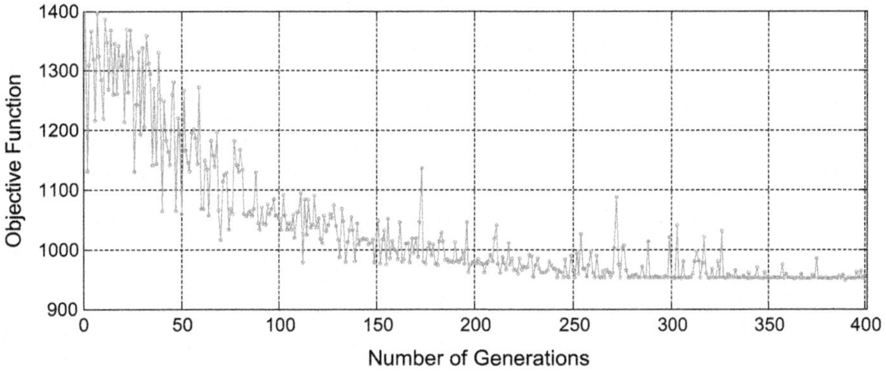

Fig. 9.30 Iteration history of NSGA-II

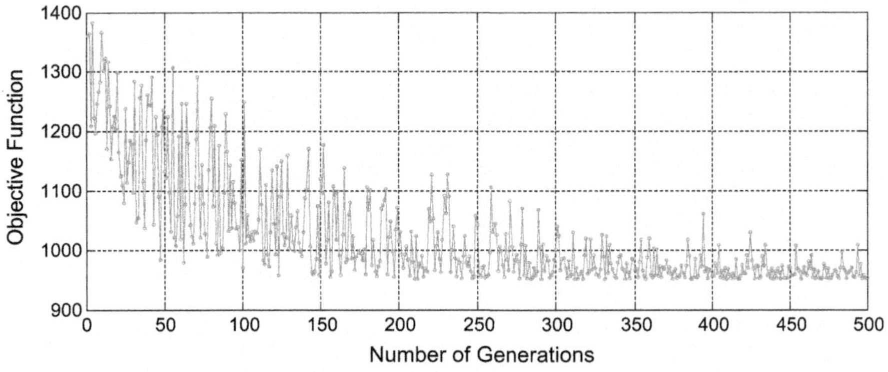

Fig. 9.31 Iteration history of MOPSO

Table 9.10 Results and rounded values

	Results of three optimization algorithms			Rounded values of optimization results		
	m	μ	z_1	m	z_2 $(z_2 = \mu z_1)$	z_1
MIGA	3.302	1.206	19.386	3	24	20
NSGA-II	2.876	1.253	18.589	3	24	19
MOPSO	3.611	1.298	19.238	3.5	25	19

accurate. The trend of NSGA-II in iteration process is the best. Therefore, the optimization parameters are selected as $m = 3$, $z_1 = 19$, $z_2 = 24$.

9.7 Summary

This chapter firstly introduces the different types of design of experiment and their applicability respectively. And then surrogate model and their features are introduced. Finally two examples, the optimization of housing and drive gear of DPSD, are given to introduce more details about surrogate model step by step to minimize the volume and weight.

References

1. Yao MY. Gear optimization of power split device for hybrid electric vehicle. Changchun: Jilin University; 2014.
2. Shen WH. Surrogate model based optimization design method research for gear transmission components of HEV. Changchun: Jilin University; 2014.
3. Jafari S, Hojjati MH, Fathi A. Classical and modern optimization methods in minimum weight design of elastic rotating disk with variable thickness and density. Int J Press Vessels Pip. 2012;92:41–7.
4. Hare W, Nutini J, Tesfamariam S. A survey of non-gradient optimization methods in structural engineering. Adv Eng Softw. 2013;59:19–28.
5. Montazeri-Gh M, Poursamad A, Ghlichi B. Application of genetic algorithm for optimization of control strategy in parallel hybrid electric vehicle. J Franklin Inst. 2006;343(4):420–35.
6. Navalertporn T, Afzulpurkar NV. Optimization of tile manufacturing process using particle swarm optimization. Swarm Evol Comput. 2011;1(2):97–109.
7. Wang GG, Shan S. Review of metamodelling techniques in support of engineering design optimization. J Mech Des. 2007;129(4):370–80.
8. Park HS, Dang XP. Structural optimization based on CAD-CAE integration and metamodeling techniques. Comput Aided Des. 2010;42(10):889–902.
9. Montgomery DC, Montgomery DC. Design and analysis of experiments [M]. New York: Wiley; 1997.
10. Bement TR. Taguchi techniques for quality engineering. Technometrics. 1989;31(2):253–5.
11. Ursache NM, Bressloff NW, Keane AJ. Aircraft roll enhancement via mutli-objective optimization using surrogate modeling. AIAA J. 2011;49(7):1525–41.

12. Li M, Li G, Azarm S. A Kriging metamodel assisted multi-objective genetic algorithm for design optimization. J Mech Des. 2008;130(3):030401.
13. Chen Y, Zou X, Xie W. How can surrogates influence the convergence of evolutionary algorithm. Swarm Evol Comput. 2013;12:18–23.
14. Forrester AIJ, Keane AJ. Recent advances in surrogate-based optimization. Prog Aerosp Sci. 2009;45(1):50–79.
15. Krisch U, Papalambros PY. Structural reanalysis for topological modifications–a unified approach. Structural and multidisciplinary Optimization. 2001;21(5):333–44.
16. Myers RH, Montgomery DC. Response surface methodology: process and product optimization using designed experiments. New York: Wiley; 1995.
17. Broomhead DS, Loewe D. Multivariable function interpolation and adaptive networks. Complex Syst. 1988;2:321–55.
18. Forrester A, Sobester A, Keane A. Engineering design via surrogate modeling: a practical guide. Chichester: Wiley; 2008.
19. Makadia AJ, Nanavati JI. Optimization of machining parameters for turning operations based on response surface methodology. Measurement. 2013;46(4):1521–9.
20. Dai HZ, Zhao W, Wang W, Gao ZG. An improved radial basis function network for structure reliability analysis. J Mech Sci Technol. 2011;25(9):2151–9.
21. Jin R. Chen.W. Simpson T W. Comparative studies of metamodelling techniques under multiple modeling criteria. Struct Multidiscip Optim. 2001;23(1):1–13.
22. Lophaven SN, Nielsen HB, Soudergaard J. Aspects of the matlab toolbox DACE. Report IMM-REP2002-12, Informatics and Mathematical Modeling [DB/OL]. Denmark: Technical University of Denmark; 2002.
23. Sakata S, Ashida F, Tanaka H. Stabilization of parameter estimation for Kriging-based approximate with empirical semivariogram. Comput Methods Appl Mech Eng. 2010;199 (25):1710–21.
24. Liu CL. Dynamic analysis and braking performance optimization of drum brake. Changchun: Jilin University; 2011.
25. Srinivas N, Deb K. Multiobjective optimization using nondominated sorting in genetic algorithms. Evol Comput. 1994.
26. Li HY, Ma MX, Jing YW. A new method based on LPP and NSGA-II for multiobjective robust collaborative optimization. J Mech Sci Technol. 2011;25(5):1071–9.
27. Wang JH, Yin ZY. Multi-objective optimization algorithm based on the combination of NSGA-II and MOPSO. J Comput Appl. 2007;27(11): 2817–21.
28. Savani V, Rao RV, Vakharia DP. Optimal weight design of a gear train using particle swarm optimization and simulation and simulated annealing algorithms. Mech Mach Theory. 2010;45(3):531–41.
29. Reyes-Sierra M, Coello CAC. Multi-objective particle swarm optimizers: a survey of the sate-of-the-art. Int J Comput Intell Res. 2006;2(3):287–308.
30. Li F, Qin YM, Pang Z, et al. Design and optimization of PSD housing using a MIGA-NLPQL hybrid strategy based on a surrogate model. Strojniski vestink-J Mech Eng. 2014; 60(7–8):525–35.
31. Wang JX, Shen WH, Wang Z, et al. Multi-objective optimization of drive gears for power split device using surrogate models. J Mech Sci Technol. 2014;28(6):2205–14.
32. Chen B. Modeling, programming and solving strategies in multidisciplinary optimization design. Wuhan: Huazhong University of Science and Technology; 2000.

Printed by Printforce, the Netherlands